ELEMENTARY SURVEY SAMPLING

FIFTH EDITION

Richard L. Scheaffer
University of Florida

William Mendenhall III
University of Florida, Emeritus

Lyman Ott
Marion Merrell Dow, Inc.

An Alexander Kugushev Book

Duxbury Press
An Imprint of Wadsworth Publishing Company
An International Thomson Publishing Company

I(T)P™

Belmont • Albany • Bonn • Boston • Cincinnati • Detroit • London • Madrid
Melbourne • Mexico City • New York • Paris • San Francisco • Singapore
Tokyo • Toronto • Washington

Editorial Assistant: Cynthia Mozow
Production Editor: The Book Company
Print Buyer: Karen Hunt
Permissions Editor: Peggy Meehan
Copy Editor: Betty Duncan
Technical Illustrator: Lotus Art
Cover: Craig Hansen
Compositor: Techset Composition, Salisbury, Wiltshire
Printer: Quebecor Printing Book Group, Fairfield

This text is printed on acid-free recycled paper.

Printed in the United States of America
6 7 8 9 10—02 01 00 99 98

For more information, contact Wadsworth Publishing Company:

Wadsworth Publishing Company
10 Davis Drive
Belmont, California 94002, USA

International Thomson Publishing Europe
Berkshire House 168–173
High Holborn
London, WC1V 7AA, England

Thomas Nelson Australia
102 Dodds Street
South Melbourne 3205
Victoria, Australia

Nelson Canada
1120 Birchmount Road
Scarborough, Ontario
Canada M1K 5G4

International Thomson Editores
Campos Eliseos 385, Piso 7
Col. Polanco
11560 México D.F. México

International Thomson Publishing GmbH
Königswinterer Strasse 418
53227 Bonn, Germany

International Thomson Publishing Asia
221 Henderson Road
#05-10 Henderson Building
Singapore 0315

International Thomson Publishing Japan
Hirakawacho Kyowa Building, 3F
2-2-1 Hirakawacho
Chiyoda-ku, Tokyo 102, Japan

Library of Congress Cataloging-in-Publication Data

Scheaffer, Richard L.

Elementary survey sampling / Richard L. Scheaffer, William Mendelhall, Lyman Ott.—5th ed.

p. cm.
Includes bibliographical references (p.) and index.
ISBN 0-534-24342-8 (alk. paper)
1. Sampling (Statistics). I. Mendelhall, William. II. Ott, Lyman. III. Title.
QA276.6.S385 1995
519.5′2—dc20 95-23646
CIP

PREFACE

Purpose

Elementary Survey Sampling, 5th edition, is an introductory text on the design and analysis of sample surveys intended for students of business, the social sciences, or natural resource management. Because it is written to appeal to students of limited mathematical background, the only prerequisite is an elementary course in statistics.

Approach

This text emphasizes the practical aspects of survey problems. It begins with brief chapters on the role of sample surveys in the modern world. Thereafter, each major chapter introduces a sample survey design or estimation procedure by describing the pertinent practical problem. These chapters are structured as follows:

- The methodology proposed for solving the problem is described, followed by details of the estimation procedure, including a compact presentation of the formulas needed to complete the analysis.
- In each instance, a practical example is worked out in complete detail. Wherever possible, the computations are set up in a format that makes them readily acceptable by standard software.
- Many exercises are provided at the end of each chapter to give students ample opportunity to practice the techniques and to stretch their grasp of ideas.

We have endeavored to avoid making this text a cookbook. To justify many of the formulas and to support the choice of particular sample survey designs, we have supplied explanations that appeal to students' intuition.

We have selected examples and exercises from many fields of application. Answers given for selected exercises may be subject to small rounding errors because of the complexity of some formulas.

Experiences with Real Data

The 'Experiences with Real Data' sections found at the end of most chapters include suggestions on how the student can become involved with real sampling problems. These problems may be large or small projects, with some requiring computations to be handled by a computer, but we have found such projects to be valuable learning experiences for students taking a sampling course. Working on a real project forces students to think about every aspect of the survey and causes them to realize that some ideas that sound simple in the textbook are not so easily carried out in practice.

Organization

The text includes a review of elementary concepts (Chapters 1 and 2) and a description of terms pertinent to survey sampling, along with a discussion of the design of questionnaires and methods of data collection (Chapter 3). Chapters 4, 5, 7, and 8 present the four most common sample survey designs—namely, simple random sampling, stratified random sampling, systematic sampling, and cluster sampling, respectively. Chapter 6 discusses ratio, regression, and difference estimation. The remaining chapters deal with two-stage cluster sampling and other specialized problems that occur in survey sampling.

Practical aspects of conducting sample surveys are emphasized, with sections on sources of errors in surveys, methods of data collection, designing questionnaires, and guidelines for planning surveys. Sampling with probabilities proportional to size is introduced in Chapter 4 and applied to cluster sampling in Chapters 8 and 9. All chapters contain many examples of how the sampling concepts are used in practice.

Chapter 12 contains a review of the main sampling designs and a set of exercises that may require some thought in the selection of appropriate analyses.

The appendix also includes the mathematical derivations of many of the main results in the text. The understanding of many of these derivations requires a working knowledge of elementary probability theory.

Topical Emphasis

Topics receiving considerable attention include poststratification, double sampling, sample-size determination in two-stage cluster sampling, and estimation of population density. A major emphasis is placed on calculating relative efficiencies for sampling designs so that proper choices for designs can be made in various practical contexts.

This Revision

A main goal of this edition is an increased emphasis on the use of modern computer software for computational purposes. Throughout the text, calculations are set up in a spreadsheet format that allows easy completion of the data analysis with any statistical software system or graphing calculator.

For ease of use on a computer, a data disk is provided with the text. It contains all large data sets from the examples and exercises, as well as the eight real data sets explained in the appendix and referred to in many of the Experiences with Real Data sections.

MINITAB is used to illustrate the calculations in the text. MINITAB macros are used in Chapters 4 and 5 and are listed in the appendix, as well as on the data disk. SAS macros are also given in the appendix and on the data disk.

Other enhancements include (1) a greater emphasis on graphical descriptions of data as illustrated in Chapter 2, (2) an expanded section of sources of errors in surveys, (3) a section on comparing estimates by looking at differences, and (4) more details on using plots to detect influential observations and to study their effects on estimates, particularly ratio and regression estimates.

Acknowledgments

We wish to express our sincere appreciation to the many people who have helped in the preparation of this revision. Particular thanks are due to the reviewers for their helpful comments during the preparation of this manuscript: David Bowden, Colorado State University; Dale O. Everson, University of Idaho; Stephen Haslett, Victoria University of Wellington; Dennis M. O'Brien, University of Wisconsin, La Crosse; Larry J. Ringer, Texas A&M University; Richard J. Rossi, California Polytechnic State University, San Luis Obispo.

Survey sampling plays in increasingly important role in the information society of today. We hope this book helps students design better surveys and understand the subtleties of survey results presented to them. In short, our goal is to improve the flow of knowledge from data.

Richard L. Scheaffer
William Mendenhall III
Lyman Ott

CONTENTS

CHAPTER 6

RATIO, REGRESSION, AND DIFFERENCE ESTIMATION 197

CHAPTER 7

SYSTEMATIC SAMPLING 251

CHAPTER 8

CLUSTER SAMPLING 289

CHAPTER 9

TWO-STAGE CLUSTER SAMPLING 335

CHAPTER 10

ESTIMATING THE POPULATION SIZE 363

CHAPTER 11

SUPPLEMENTAL TOPICS 387

CHAPTER 12

SUMMARY 405

APPENDIX 1 423

APPENDIX 2 435

APPENDIX 3 447

APPENDIX 4 473

SELECTED ANSWERS 493

INDEX 497

CHAPTER 1

INTRODUCTION

"Twenty-two percent of Americans doubt that the Holocaust ever occurred" blasted a news statement in 1993. How could this be? How could so many doubt the occurrence of one of the most significant events of the twentieth century—and, perhaps, of all time? The answer lies in the poll, or survey, itself and in the difficulty of communicating with language, which always produces some degree of imprecision. The question actually asked by the Roper organization was as follows:

> Does it seem possible or does it seem impossible to you that the Nazi extermination of the Jews never happened?

Among the respondents, 22% said "it seemed possible," and another 12% said they did not know. Only 65% said it was "impossible it never happened." The double negative in the question, it seems, confused the respondents to the point that they were not sure how to phrase their answers. Picking up on this point, the Gallup organization conducted a follow-up poll that asked the question in more detailed form:

> The term *Holocaust* usually refers to the killing of millions of Jews in Nazi death camps during World War II. In your opinion, did the Holocaust: definitely happen, probably happen, probably not happen, or definitely not happen?

Among the respondents to this poll, 83% said the Holocaust definitely happened, and another 13% said it probably happened. Only 1% said it definitely did not happen. Quite a difference! In a separate poll, Gallup asked the Roper question again and found that 37% of the respondents said it seemed possible that it never happened.

Polls and surveys. They increasingly guide political, research, and business decisions, but they can be highly volatile and are little understood. An excellent study of current trends in collecting and using data is the book *Tainted Truth* by Cynthia Crossen, which points out that

> We are skeptical about statistical and factual information, but not as skeptical as we think.
>
> . . . we respect numbers and we cannot help believing them. Yet, more and more of the information we use to buy, elect, advise, acquit and heal has been created not to expand our knowledge but to sell a product or advance a cause.
>
> That's what surveys do, they basically manufacture news.

We depend on data to make intelligent decisions, yet the data we see are often tainted. An old saying on the use and misuse of computers was "garbage in–garbage out" but this has become "garbage in–gospel out" as more and more people get into the numbers game. So, what can we do? Part of the answer lies in education. Consumers and producers of data with serious, unbiased objectives to get at the "truth" must be educated in how surveys work, how good surveys can be designed, and how survey data can be properly analyzed. That education is the purpose of this book.

Introductory courses stress that modern statistics is a theory of information with inference as its objective. The target of our curiosity is a set of measurements, a *population*, that exists in fact or may be generated by repeated experimentation. The medium of inference is the *sample*, which is a subset of measurements selected from the population. We wish to make an inference about the population on the basis of characteristics of the sample—or, equivalently, the information contained in the sample.

For example, suppose a chain of department stores maintains customer charge accounts. The amount of money owed the company will vary from day to day as new charges are made and some accounts are paid. Indeed, the set of amounts due the company on a given day represents a population of measurements of considerable interest to the management. The population characteristic of interest is the total of all measurements in the population or, equivalently, the daily total credit load.

Keeping track of the daily total credit associated with charge accounts may seem to be a simple task for an electronic computer. However, the data must be updated daily, and updating takes time. A simpler method for determining the total credit load associated with the charge accounts is to randomly sample the population of accounts on a given day, estimate the average amount owed per account, and multiply by the number of accounts. In other words, we employ a statistical estimator to make an inference about the population total. Elementary statistics tells us that this estimate can be made as accurate as we wish simply by increasing the sample size. The resulting estimate either is accompanied by a bound on the error of estimation (Mendenhall and

Beaver, 1994, Chapter 8) or is expressed as a confidence interval. Thus, information in the sample is used to make an inference about the population.

Information from sample surveys affects almost every facet of our daily lives. Such information determines government policies on, for example, the control of the economy and the promotion of social programs. Opinion polls are the basis of much of the news reported by the various news media. Ratings of television shows determine which shows are to be available for viewing in the future.

One usually thinks of the U.S. Census Bureau as contacting every household in the country. Actually, in the 1990 census only 14 questions were asked of all households. Information on an additional 42 questions was obtained from only a sample of households. The resulting information is used by the federal government in determining allocations of funds to states and cities. It is used by business to forecast sales, to manage personnel, and to establish future site locations. It is used by urban and regional planners to plan land use, transportation networks, and energy consumption. It is used by social scientists to study economic conditions, racial balance, and other aspects of the quality of life.

The U.S. Bureau of Labor Statistics routinely conducts over 20 surveys. Some of the best known and most widely used are the surveys that establish the consumer price index (CPI). The CPI is a measure of price change for a fixed-market basket of goods and services over time. It is used as a measure of inflation and serves as an economic indicator for government policies. Businesses have wage rates and pension plans tied to the CPI. Federal health and welfare programs, as well as many state and local programs, tie their bases of eligibility to the CPI. Escalator clauses in rents and mortgages are based on the CPI. So we can see that this one index, determined on the basis of sample surveys, plays a fundamental role in our society.

Many other surveys at the Bureau of Labor Statistics (BLS) are crucial to society. The monthly Current Population Survey establishes basic information on the labor force, employment, and unemployment. The consumer expenditure surveys collect data on family expenditures for goods and services used in day-to-day living. The Establishment Survey collects information on employment hours and earnings for nonagricultural business establishments. The survey on occupational outlook provides information on future employment opportunities for a variety of occupations, projecting to approximately ten years ahead. Other activities of the BLS are addressed in the *BLS Handbook of Methods* (1982).

Opinion polls are constantly in the news, and the names of Gallup and Harris have become well known to everyone. These polls, or sample surveys, reflect the attitudes and opinions of citizens on everything from politics and religion to sports and entertainment. The Nielsen ratings determine the success or failure of TV shows.

Businesses conduct sample surveys for their internal operations, in addition to using government surveys for crucial management decisions. Auditors estimate account balances and check on compliance with operating rules by sampling accounts. Quality control of manufacturing processes relies heavily on sampling techniques.

One particular area of business activity that depends on detailed sampling activities is marketing. Decisions on which products to market, where to market them, and how to advertise them are often made on the basis of sample survey data. The data may come from surveys conducted by the firm that manufactures the product or may be purchased from survey firms that specialize in marketing data. The activities of three such firms are outlined next.

The Nielsen retail index is less famous than the Nielsen television ratings, but it is very important to firms marketing products for retail sale. This index furnishes continuous sales data on foods, cosmetics, pharmaceuticals, beverages, and many other classes of products. It can provide estimates of total sales for a product class, sales for a client's particular brand, sales for a competing brand, retail and wholesale price data, and the percentage of stores stocking a particular item. The data comes from auditing inventories and sales in 1600 stores across the United States every 60 days.

Selling Areas—Marketing, Inc. (SAMI), collects information on the movement of products from warehouses and wholesalers. Data are obtained in 36 major television market areas, containing 74% of national food sales, and covers 425 product categories.

The Market Research Corporation of America provides many types of marketing data through the use of surveys, but some of the more interesting results come from its National Menu Census. This survey samples families and observes their eating patterns for two weeks. As many as 4000 families may participate during a year. Data are obtained on the number of times a particular food item is served, how it is served, how many persons eat the item, and many other details, including what happens to the leftovers. Such details are important for product development and advertising.

Many interesting examples of the practical uses of statistics in general and sampling in particular can be found in *Statistics: A Guide to the Unknown* (Tanur, 1989). In this book, you might want to look at some of the methods and uses of opinion polling discussed in the articles "Opinion Polling in a Democracy" by George Gallup and "Election Night on Television" by R. F. Link. Those interested in wildlife ecology should read "The Plight of the Whales" by D. G. Chapman. Find out how interrailroad and interairline billing is handled economically through sampling by reading "How Accountants Save Money by Sampling" by John Neter.

Since the objective of modern statistics is inference, you may question what particular aspect of statistics will be covered in a course on sample survey design. The answer to this question is twofold. First, we will focus on the economics of purchasing a specific quantity of information. More specifically, how can we design sampling procedures that reduce the cost of a fixed quantity of information? Although introductory courses in statistics acknowledge the importance of this subject, they place major emphasis on basic concepts and on how to make inferences in specific situations *after* the data have been collected. The second distinguishing feature of our topic is that it is aimed at the particular types of sampling situations and inferential problems most frequently encountered in business, the social sciences, and natural resource management (timber, wildlife, and recreation) rather than in the physical sciences.

Even the terminology of the social scientist differs from that of the physical scientist. Social scientists conduct *surveys* to collect a sample, whereas physical scientists perform *experiments*. Thus, we acknowledge that differences exist from one field of science to another in the nature of the populations and the manner in which a sample can be drawn. For example, populations of voters, financial accounts, or animals of a particular species may contain only a small number of elements. In contrast, the conceptual population of responses generated by measuring the yield of a chemical process is very large indeed. (You may recall that the properties of estimators and test statistics covered in most introductory courses assume that the population of interest is large relative to the sample.) Limitations placed on the sampling procedure also vary from one area of science to another. Sampling in the biological and physical sciences can frequently be performed under controlled experimental conditions. Such control is frequently impossible in the social sciences, business, and natural resource management. For example, a medical researcher might compare the growth of rats subjected to two different drugs. For this experiment, the initial weights of the rats and the daily intake of food could be controlled to reduce unwanted variation in the experiment. In contrast, very few variables can be controlled in comparing the effect of two different television advertisements on sales for a given product; no control is possible when studying the effect of environmental conditions on the number of seals in the North Pacific Ocean.

In summary, this text is concerned with the peculiarities of sampling and inference commonly encountered in business, the social sciences, and natural resource management. Specifically, we will consider methods for actually selecting the sample from an existing population and ways of circumventing various difficulties that arise. Methods for designing surveys that capitalize on characteristics of the population will be presented along with associated estimators to reduce the cost for acquiring an estimate of specified accuracy.

Chapter 2 reviews some of the basic concepts encountered in introductory statistics, including the fundamental role that probability plays in making inferences. Chapter 3 presents some of the basic terminology of sampling, as well as a discussion of problems arising in sample survey design. Simple random sampling, familiar to the beginning student, is carefully presented in Chapter 4; it includes physical procedures for actually selecting the sample. Following chapters cover economical methods for selecting a sample and associated methods for estimating population parameters.

In reading this text, keep in mind that the ultimate objective of each chapter is *inference*. Identify the sampling procedure associated with each chapter, the population parameters of interest, their estimators, and the associated bounds on the errors of estimation. Develop an intuitive understanding of and appreciation for the benefits to be derived from specialized sampling procedures. Focus on the broad concepts and do not become hypnotized by the formulas for estimators and variances that sometimes are unavoidably complicated. In short, focus on the forest rather than the trees. Work some exercises, and the details will fall into place.

CHAPTER 2

A REVIEW OF SOME BASIC CONCEPTS

2.1 INTRODUCTION

Knowledge of the basic concepts of statistics is a prerequisite for a study of sample survey design. Thus, in this chapter, we will review some of these basic concepts.

The ultimate objective of statistics is to make inferences about a population from information contained in a sample. The target of our inference, the population, is a set of measurements, finite or infinite, existing or conceptual. Hence, the first step in statistics is to find a way to phrase an inference about a population or, equivalently, to describe a set of measurements. Thus, frequency distributions and numerical descriptive measures are the first topics of our review.

The second step in statistics is to consider how inferences can be made about the population from information contained in a sample. For this step, we must consider probability distributions of sample quantities, or sampling distributions. Knowledge of probability distributions associated with the sample allows us to choose proper inference-making procedures and to attach measures of goodness to such inferences.

The method of inference primarily employed in business and the social sciences is estimation. We may wish to estimate the total assets of a corporation, the fraction of voters favoring candidate Jones, or the number of campers using a state park during a given period of time. Hence, we must understand the basic concepts underlying the selection of an estimator of a population parameter, the method for evaluating its goodness, and the concepts involved in interval estimation. Because the bias and the variance of estimators determine their goodness, we need to review the basic ideas concerned with the expectation of a random variable and the notions of variance and covariance.

The subsequent sections follow the outline just given. We begin with a review of the primary problem, namely, how to describe a set of measurements. We then rapidly review the probabilistic model for the repetition of an experiment. We explain how the model can be used to infer the characteristics of a population and discuss random variables, probability distributions, and expectations. Finally, we present the basic concepts associated with point and interval estimation.

2.2 DESCRIBING DATA WITH GRAPHS AND TABLES

LINE PLOTS

Data analysis begins by trying to get a good "picture" of the data. Of the many possible graphs, the *line plot* is one of the simplest to construct. This plot merely indicates the location of the data points along a segment of the real number line. The method of construction is as follows:

1. Find the smallest and largest values in the data set.
2. Draw a real number line that encompasses these values.
3. Place a mark (usually an x or a dot) on the real number line corresponding to each data point.

A large data set from a survey of college students will be used throughout this section to illustrate aspects of data analysis. A questionnaire was developed to provide data on certain characteristics of an undergraduate class in statistics geared primarily to students in education and the social sciences. The actual data are shown in the appendix and on the data disk CLASSUR. The variables in the data set are defined as follows:

GENDER (1 = male and 2 = female)
AGE (in years)
GPA (cumulative grade point average)
CLASS (1 = freshman, 2 = sophomore, 3 = junior, 4 = senior, 5 = other)
HEIGHT (in inches)

WEIGHT (in pounds)

STUDYHR (number of hours of study in a typical week)

SLEEPHR (number of hours of sleep in a typical night)

JOB (1 = have a job; 2 = do not have a job, other than student)

TEXTPAY (dollar amount paid for textbooks this term)

RESIDE (permanent residence: 1 = >150 miles from Gainesville; 2 = <150 miles from Gainesville; 3 = outside Florida)

Line plots for such large data sets are best done on a computer, as were the ones shown in Figure 2.1 for AGE, HEIGHT, and STUDYHR. Study the line plots to see what information they may be revealing about these three variables.

FIGURE 2.1 Line plots

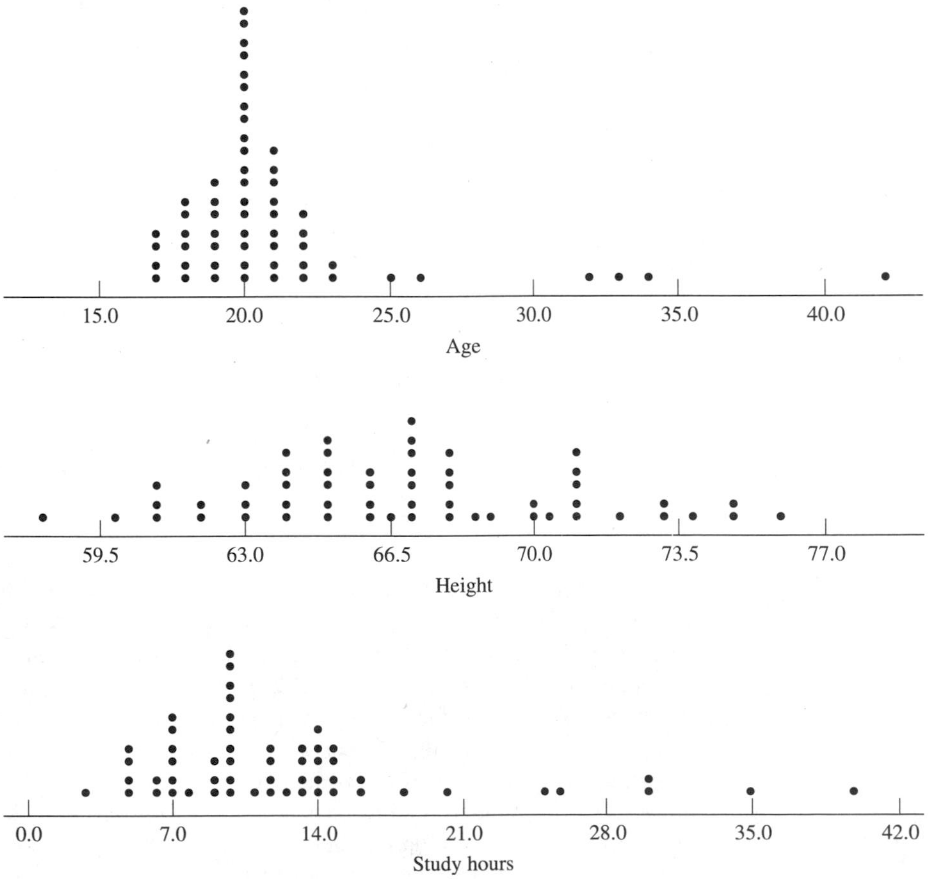

Describe the age distribution of students in the survey.

Describe the height distribution. From the height distribution (do not look at the data set), do you think there are more women than men in the class? Why?

Describe the distribution of typical study hours per week. Do you think that the mode is affected by the fact that this data is self-reported by the students?

Line plots display the shape of the data distributions. Certain terms are commonly used to describe these shapes.

Definition 2.1

A distribution is *symmetric* if one-half of the distribution looks like the mirror image of the other half.

Skewness in a distribution occurs if a long, thin tail stretches out in one direction. The direction of the skewness is the direction in which the tail is pointing.

Clusters are simply groups of points close together, separated from other clusters by *gaps*.

A *mode* is the high point of a data distribution. This represents the data value(s) that occur most often in the data set.

Looking at Figure 2.1, you might say that the distribution of ages is fairly symmetric, except for the four ages that form clusters to the far right of the main body of data. The distribution of study hours is skewed toward the larger values. The distribution of heights is only slightly skewed toward the larger values and seems to have two peaks, which makes it a bimodal distribution. What do you think could account for this?

STEM-AND-LEAF PLOTS

Although the line plots of age, heights, and study hours of students give good pictures of the data distributions, recovering the exact numerical values from these plots is impossible. A *stem-and-leaf plot* (often called a *stem plot*) provides both the picture of the data distribution and the exact values of the data points in a display that is similar to the line plot. Using the variables age, height, study hours, and grade point average (GPA) from the class survey data, computer-drawn (MINITAB) stem plots are shown in Figure 2.2. These are exactly like the printouts of the program.

Look at the stem plot for age and ignore the left-most column for the moment. The second column holds the stems of the data, and the third column (after the space) holds the leaves. The legend for the table tells you that the leaf digit is the units digit of the measurement. Thus, there are four students reporting an age of 17 and two students reporting an age of 23. How old is the oldest student in the survey? How many students are age 19 or younger? Going back to the left-most column, these numbers are the cumulative counts of the data points up to and including that row, counting toward the middle of the data distribution from either extreme.

FIGURE 2.2 Stem plots

Stem and leaf of AGE N = 57
Leaf unit = 1.0

4	1 7777
17	1 8888889999999
(27)	2 000000000000000000111111111
13	2 2222233
6	2 5
5	2 6
4	2
4	3
4	3 23
2	3 4
1	3
1	3
1	4
1	4 2

Stem and leaf of HEIGHT N =
Leaf unit = 1.0 N* =

1	5 8
5	6 0011
10	6 22333
21	6 44444555555
(12)	6 666667777777
22	6 8888889
15	7 00011111
7	7 233
4	7 455
1	7 6

Stem and leaf of STUDYHR N = 56
Leaf unit = 1.0 N* = 1

1	0 3
17	0 5555667777778999
(25)	1 0000000000112222333344444
14	1 5555668
7	2 0
6	2 56
4	3 00
2	3 5
1	4 0

Stem and leaf of GPA N =
Leaf unit = 0.010 N* =

1	23 0
2	24 0
7	25 00006
10	26 005
14	27 0056
20	28 000004
24	29 0368
24	30 0004
20	31 00
18	32 0000459
11	33 4
10	34 00057
5	35 05
3	36 7
2	37 5
1	38

Now, study all the stem plots in Figure 2.2. Use the following questions to guide your investigation:

How does the leaf unit change for the various plots?

Do the first three plots show similar patterns to those observed on the line plots? What are the advantages of the stem plots?

Why does the distribution of GPAs have no long tail toward the lower values?

HISTOGRAMS

A stem plot shows essential distributional features of a data set by selecting a few "stems" to represent groups of data points and then showing each data point within a group as a "leaf." A line plot shows the distribution of a data set over a portion of a real number line by placing an x or dot at the location of each data point. A *histogram* has similarities to each of these. A histogram for the ages of students can be constructed by drawing bars around the rows of leaves, and then erasing the leaf values. The result, as seen in Figure 2.3, is a graph in which the area of a bar is proportional to the number of data points found in the bar.

The steps in constructing a histogram are as follows:

1. Divide the interval covered by the range of the data into a number of segments of equal length.
2. Count the number of data points in each of these segments.
3. Construct a bar over each segment so that the length of the bar is proportional to the number of data points inside the segment.

Thus, a histogram represents frequencies of data points by area, rather than by x's, dots, or a list of the numerical values.

Turning again to the data from the survey of college students, we want to investigate four variables: age, height, study hours, and sleep hours (the reported number of hours sleep during a typical night). Histograms provide a convenient tool for getting a quick but meaningful summary of these data because they particularly show the main distributional features of large data sets. Figure 2.4 shows the histograms for the four variables in question.

FIGURE 2.3 Histogram of student ages

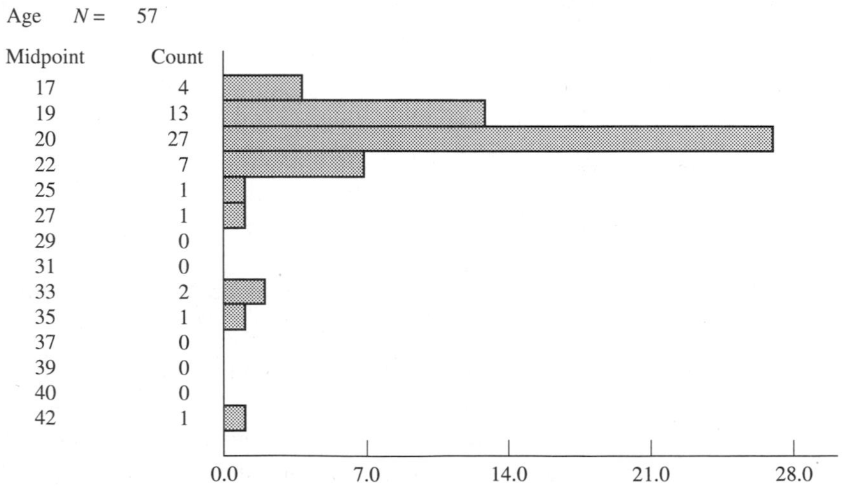

FIGURE 2.4 Histograms for four variables

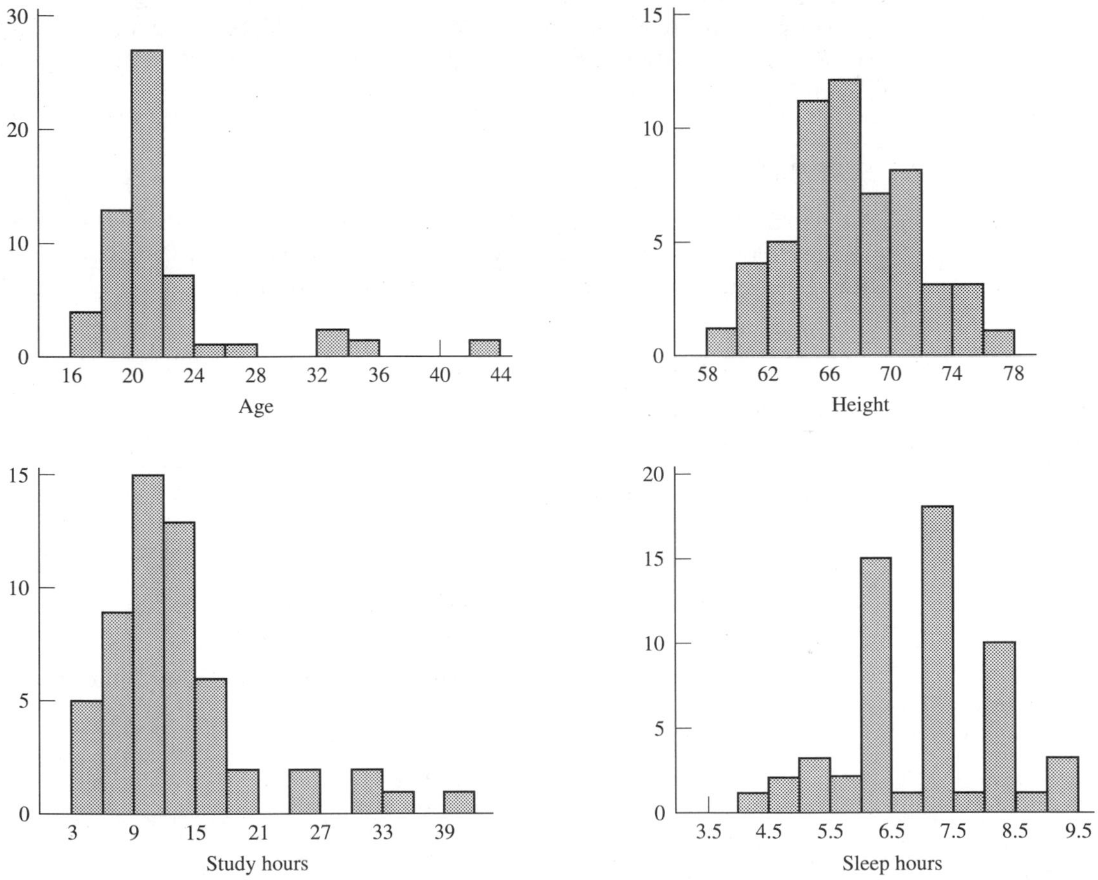

What features of the data are revealed in the histograms? The histogram on age shows a single modal category with a long tail toward the higher ages (skewness toward the larger values), with some definite clusters and gaps. The histogram on height shows a more symmetric, mound-shaped distribution, but there is a hint of a second mode. The histogram on study hours shows a skewness in the direction of the larger values, but the skewness is less extreme than in the case of age. The histogram on sleep hours shows an interesting pattern of high and low bars, sometimes called a "rake" pattern. What could be the cause of this pattern? Notice that the distribution of sleep hours appears to be slightly skewed toward the *smaller* values. Do any of these distributions appear to be unusual for a survey of college students?

The possible second mode on the height distribution can be explored further by separating the data according to the gender of the person responding. The two histograms in Figure 2.5 show heights for females and heights for males on different plots. The female heights have a nearly symmetric, mound-shaped distribution centering around 65 inches. The male heights have a more irregular distribution, primarily due to one extreme data point on the short side, centering around 70 inches. Do you see how these two histograms together could form the bimodal picture seen in Figure 2.4?

BOX PLOTS

All graphs introduced to this point are based on the individual data points themselves. The next plot to be introduced is different in that it depends on certain summaries of the data. The summaries needed are the median and the quartiles.

> **Definition 2.2**
>
> The *median* of a data set is the data value that occurs in the middle of the ordered set of values, or the midpoint between two values that might be tied for the middle position.
>
> The *lower quartile* is the middle value of the half of the data below the median.
>
> The *upper quartile* is the middle value of the half of the data above the median.

A function of the quartiles of some interest in measuring variability is the interquartile range.

FIGURE 2.5 Histograms for heights of (a) females and (b) males

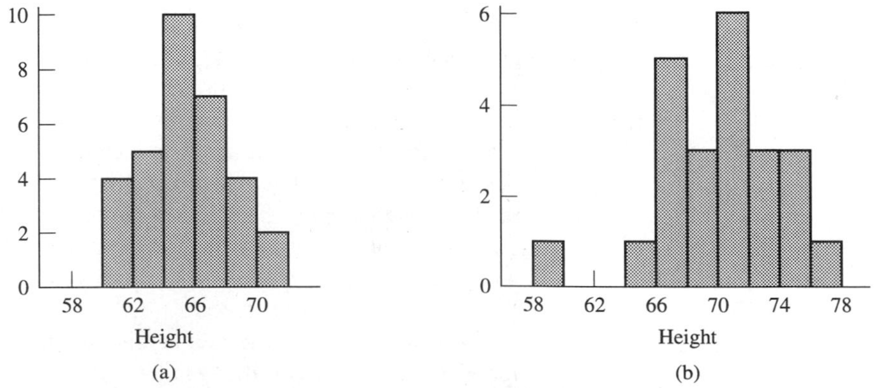

Definition 2.3

The *interquartile range* (IQR) is the difference between the upper and lower quartiles.

The numerical summary that includes the median, quartiles, and *extremes* (largest and smallest values) for a data set is called the *five-number summary.* This turns out to be a very useful way to describe data and to compare data sets. The five-number summary can be placed on a real number line to make a convenient and informative plot for interpreting and comparing data sets. The resulting plot is called a *box-and-whisker plot,* or simply a *box plot.* The steps for constructing a box plot are as follows:

1. Look at the extremes and mark off a section on the real number line that includes these values.
2. Mark the position of the five numbers in the summary a little above the real number line.
3. Draw a narrow box that connects the quartiles, draw a line across the box to mark the median, and connect the extremes to the box by single lines ("whiskers"). The five-number summary is now seen on a plot.

Figure 2.6 shows a box plot representation of the student height data. The extremes are 58 and 76 inches, the quartiles are at 64 and 70 inches, and the median is 67 inches. Among other things, this tells us that about 50% of the data values are between 64 and 70 inches. The symmetry of the box plot reflects the symmetry of the distribution, but the lack of detail in the box plot covers up the fact that this distribution, as we saw earlier, might be bimodal.

Figure 2.7 gives box plots for GPA, study hours (per week), and age. The distribution of GPAs seems reasonably well behaved, but the other two look a bit different. The * and 0 symbols indicate data points that are far enough away from the body of the data to be considered as outliers. A few students study a lot more than most students; a few students are much older than most students.

Outliers are easily visualized in comparison to the summary values of a box plot, and these plots are usually drawn with a formal definition of outlier in mind.

FIGURE 2.6 Box plot of student heights

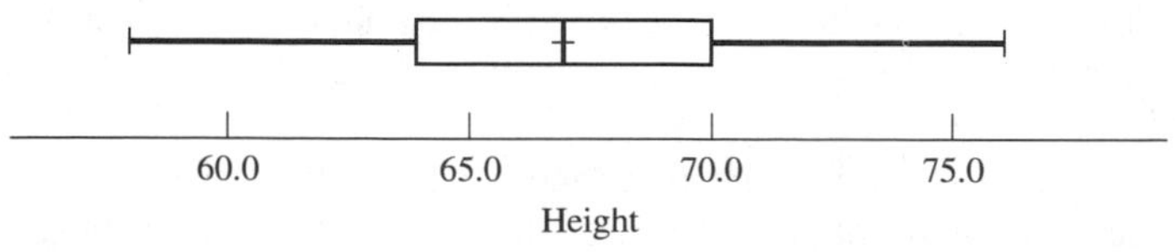

FIGURE 2.7 Box plots of three variables

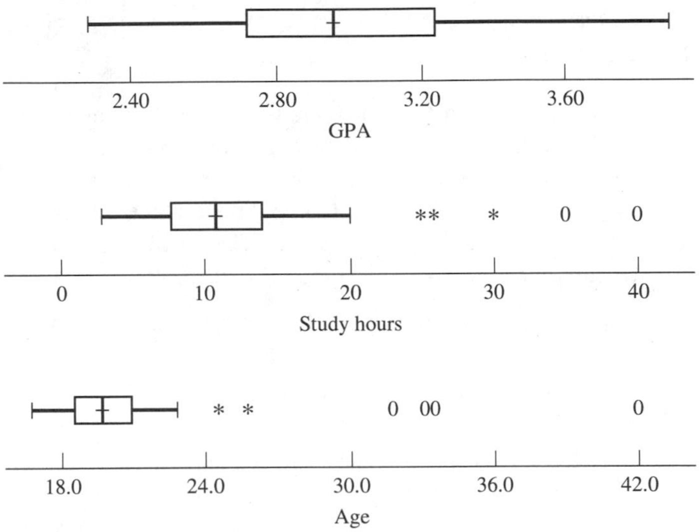

> ***Definition 2.4***
>
> An *outlier* is a data point that lies at a distance of more than 1.5(IQR) above the upper quartile or below the lower quartile. That is, an outlier is a data point that lies more than 1.5(IQR) away from the ends of the box. [MINITAB denotes an outlier by * and an extreme outlier, one that is more than 3(IQR) away from the end of the box, by 0.]

For the data on study hours, the upper end of the box is about 14.5 and the IQR is approximately 6. Thus, the outliers begin at $14.5 + 1.5(6) = 23.5$ on the upper side. Looking back at the stem plot for study hours, we can see that six data values are above 23.5 and should be declared outliers. The * at 30 actually represents two data values. The largest observation that is not an outlier is a value of 20, so we stop the whisker at 20 and leave a gap to emphasize the outliers.

Box plots for larger data sets are most easily drawn by use of a computer, as were the ones in Figure 2.7. Take a few minutes to study the patterns in these box plots. The distribution of GPAs is somewhat symmetric but has a slightly longer tail toward the higher values. What is the median GPA for this class? The distribution of study hours is highly skewed toward the larger values. Notice that age has more extreme outliers than does study hours.

Another useful extension is the construction of parallel box plots to compare data sets. When looking at histograms, it was observed that the heights could be described as a mixture of two distributions, one for males and one for females. This comparison can be made easily by looking at a box plot of heights for males (coded Gender 1) and box plot of heights for females on the same real number line, as shown in Figure 2.8. Notice that the male heights tend to be larger, with the exception of an outlying male height on the low side.

A similar comparison of study hours for those who have jobs (coded 1) and those who do not is shown in Figure 2.8. In general, those who do not have jobs tend to study a little more in a typical week, but there are exceptions. A comparison of sleep hours in a typical night for those who have jobs and those who do not is shown next. How would you compare these two distributions? Do you see anything unusual in the box plots here?

SCATTER PLOTS

Scatter plots are used for plotting measurement data that occur in ordered pairs. In the data from the survey of the undergraduate introductory statistics class, the relationship between height and weight of the students will be used to illustrate the main ideas.

Each student reported a height (in inches) and a weight (in pounds); Figure 2.9 shows a scatter plot of these data. Notice that the plot slants upward as we move from lower

FIGURE 2.8 Parallel box plots

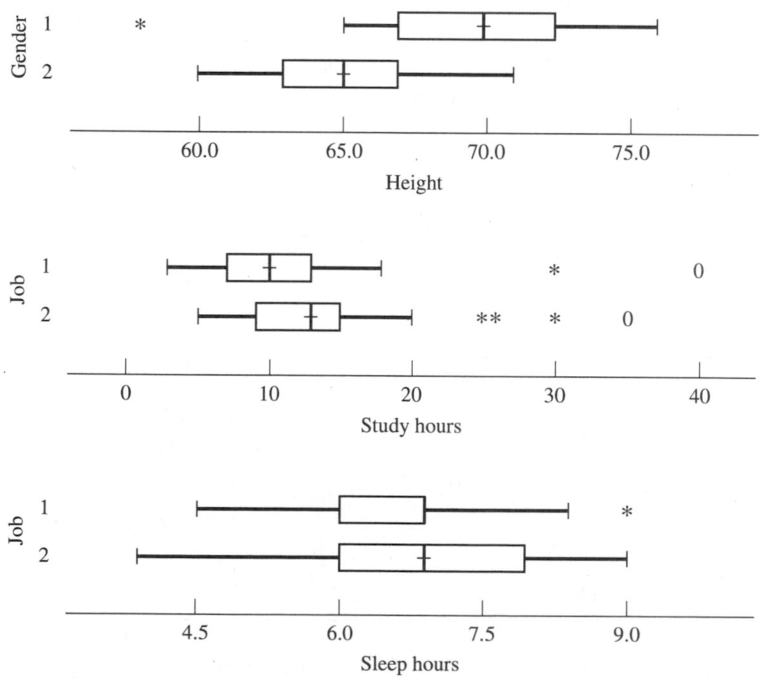

FIGURE 2.9 Scatter plot of height versus weight

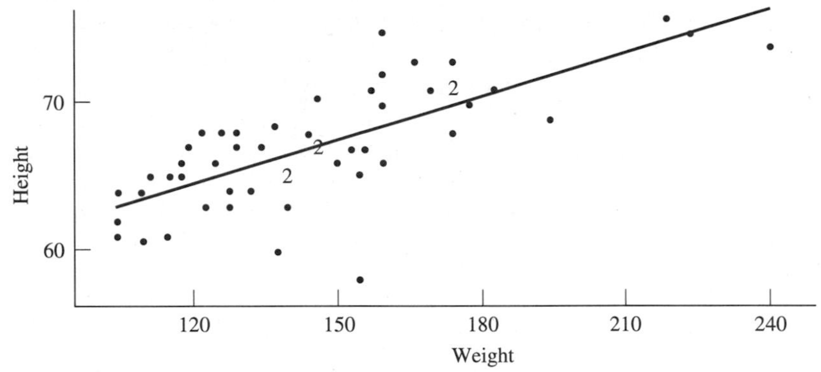

weights to higher weights. (Draw an ellipse around the cloud of data points; the ellipse should be slanted upward to the right.) We summarize this type of pattern by saying that height is positively associated, or *positively correlated*, with weight.

The second scatter plot (Figure 2.10) shows the reported GPAs of the students plotted against study hours during a typical week. This plot shows no tendency to tilt in either direction. In other words, there is *little correlation* between GPA and number of study hours. (Do you find this surprising?)

The third scatter plot (Figure 2.11) shows sleep hours for a typical week plotted against study hours. The correlation here is fairly weak, but if there is any, it is in the negative direction. Why would sleep hours and study hours tend to be negatively correlated?

FIGURE 2.10 Scatter plot of GPA versus study hours

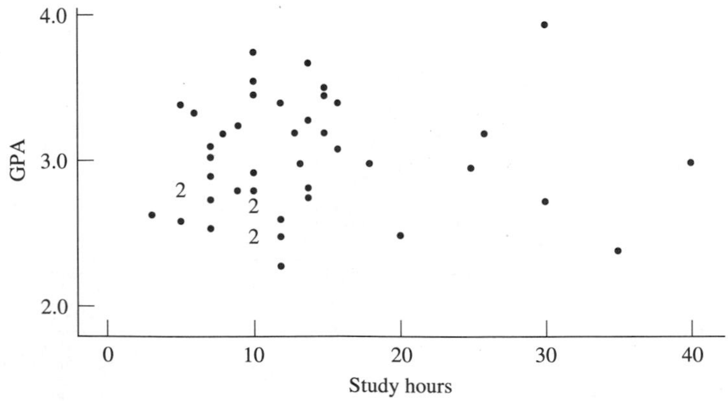

FIGURE 2.11 Scatter plot of sleep hours versus study hours

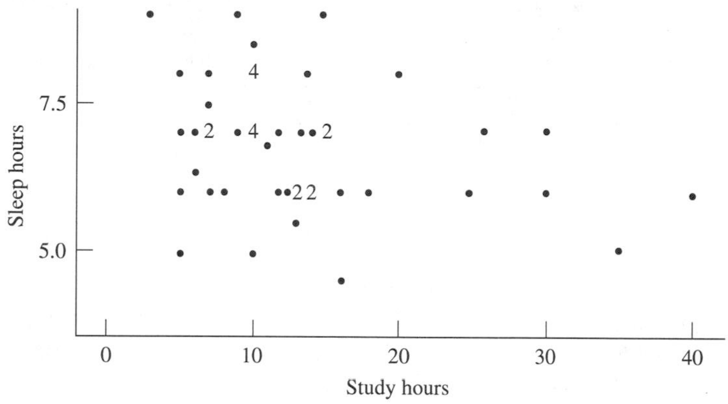

As in all earlier cases, one good way to think about data interpretation is to look for patterns and departures from patterns. In the plot of height versus weight, the overall pattern presented by the cloud of points is one that slopes upward as weight increases. Thus, there is evidence of positive correlation. The straight line captures the center of this increasing trend, much as a mean captures the center of a single set of data. There are interesting departures from the pattern, however. One or two students seem unusually short, although their weights are still in the middle of the pack. The three points forming a cluster of their own in the upper right of the plot might form a group worthy of special investigation. Do you think a scatter plot for men alone and another for women alone would be different from each other? Why? Moving to the plot of sleep hours versus study hours, discuss any interesting features of the pattern seen here. Discuss at least two ways that it differs from the plot of height versus weight.

FREQUENCY TABLES

Much of the data collected in practice comes in the form of frequency counts for categorical outcomes. Students are categorized and counted by gender, grade level, academic level, and ethnic group. Automobiles are categorized and counted with respect to manufacturer, style, color, and size. Library books are categorized and counted according to their types (novel, biography, history, reference, and so on). In fact, most of the data collected in polls and other sample surveys are categorical data summarized by counts. Such count data are best displayed, at least initially, in a *frequency table.*

The data used to introduce frequency tables come from the previous study. Students in an undergraduate statistics class were asked to complete a questionnaire seeking answers about their activities. The variables and the data are listed in the appendix. One variable of interest is the job status of the students (that is, how many of them have jobs outside of being a student). Table 2.1 is a simple table of the outcomes summarized by

TABLE 2.1 **A frequency table**

	Job status of students		
	Frequency	**Proportion**	**Percent**
Job	26	.46	46
No Job	31	.54	54
Total	57	1.00	100

frequencies, *relative frequencies* (or *proportions*), and *percents*. Note that the proportion is simple—the frequency for a category is divided by the total number of students in the study—and the percent is the proportion multiplied by 100.

Another variable of interest is number of hours of sleep during a typical night. The values reported by the students can be rounded off to the closest integer number of hours (4, 5, 6, 7, 8, or 9 in all cases) so that the variable can be treated as categorical with a natural ordering among the categories. (Everyone would agree that the person checking 9 gets more sleep than the person checking 7, and the fractions of hours do not add much to the accuracy of the data.) In such cases, the *cumulative frequencies* and *cumulative percents* can also be used to form interesting summaries of the data (Table 2.2).

Table 2.3 shows that 26 students have jobs. But, the number 26 must be interpreted in light of the number of students completing the survey, namely, 57. Thus, it is more meaningful to say that .46 or 46% of the students in the class have jobs. If this class is representative of undergraduates at the university, then we might say we estimate that approximately 46% of undergraduate students have jobs. The percentage translates directly from the class to the larger group, whereas the count does not. Table 2.2 contains more information and has more room for interpretation. The modal class of 7 hours contains the responses of 19 of the 57 students, or about 33% of the students. The cumulative percents show that about 74% of the students sleep 7 hours or less in a typical night. What percentage of the students sleep at least 6 hours in a typical night?

TABLE 2.2 **A more detailed frequency table**

	Hours of sleep per night		
Hours	**Frequency**	**Cumulative frequency**	**Cumulative percent**
4	1	1	1.75
5	5	6	10.53
6	17	23	40.35
7	19	42	73.68
8	11	53	92.98
9	4	57	100.00

TABLE 2.3 A two-way table

	Job status		
Gender	Yes	No	Total
Male	13	10	23
Female	13	21	34
Total	26	31	57

TWO-WAY TABLES

When two categorical variables are to be compared, a natural way to summarize the frequency counts is in a *two-way table*, with the categories for one variable forming the column headings and categories for the other variable forming the row headings. Students might be jointly categorized by ethnic group and assigned classroom so that the pattern of assignment of ethnic groups across classrooms can be clearly seen.

Does the job status of students differ among men and women? An answer to this question can be found by looking at the two-way table of job status versus gender for the 57 students surveyed (Table 2.3). At first glance, the jobs seem to be distributed equally across both genders because 13 men have jobs and 13 women have jobs. However, there are fewer men than women, and this must be taken into account in the analysis. In total, there are 34 women and 23 men. Therefore, we cannot directly compare the 13 men who have jobs with the 13 women who have jobs. We can compare the *proportion* of men who have jobs ($13/23 = .56$) with the *proportion* of women who have jobs ($13/34 = .38$). The fact that these proportions differ might indicate that more male students than female students tend to have jobs. Another way to say this is that the two categorical variables, gender and job status, seem to be *associated* with each other. The proportions calculated for a particular row or column of a table are called the *conditional proportions*. Here, we looked at the proportion of students who had jobs under the *condition* that they were male and under the *condition* that they were female. Do you see any similarity in the idea of association based on a two-way table of frequency data and association (called *correlation*) from a scatter plot?

Graphical plots and frequency tables provide useful and important ways to display data so that important features of the data can be observed. Careful study of what the data have to say should precede any intensive statistical analysis of the data.

2.3

SUMMARIZING INFORMATION IN POPULATIONS AND SAMPLES

Grasping the essential characteristics of a large set of measurements by looking at a listing of the numbers is not easy, so we usually must summarize the measurements through the use of graphical or numerical techniques. Even though all the measurements in a population for a study are generally not available, we may still be able to assume some reasonable graphical shape for the relative frequency distribution of this population. Of course, we can always construct a frequency or relative frequency histogram for a sample, since the sample measurements are known, and use it to make an empirical assessment of the shape of the population.

Once a relative frequency distribution is established for a population, we can, by using probability arguments, calculate summarizing numerical measures like the mean, variance, and standard deviation. Similar quantities can be calculated directly from sample measurements.

For purposes of illustration, let's assume that a population consists of a large number of integers, 0, 1, 2, ..., 9, in equal proportions. We may think of these integers as written on slips of paper and mixed up in a box, as stored on a table (like a random number table), or as generated in a computer file. Since all integers occur in equal proportions, the *relative frequency histogram*, which shows the distribution of the population measurements, is as shown in Figure 2.12.

These relative frequencies can be thought of in probabilistic terms. If one number is selected *at random* (for example, if someone reaches into the box and blindly pulls out one piece of paper), then the *probability* that the selected number will be a 4 is $\frac{1}{10}$. Suppose one number is to be selected at random from the population under study and let its value be denoted by y. Then the possible values for y (0, 1, 2, ..., 9 in this case) and the probabilities associated with those values ($\frac{1}{10}$ for each in this case) constitute the *probability distribution* for the *random variable* y. The probability associated with y is sometimes denoted by $p(y)$. Thus, for this population,

$$p(0) = p(1) = \cdots = p(8) = p(9) = \frac{1}{10}$$

The numerical measures used to summarize the characteristics of a population are defined as *expected values* of y or a function of y. By definition, the *expected value* of y, $E(y)$, is given by

$$E(y) = \sum_y yp(y)$$

where the summation is over all values of y for which $p(y) > 0$.

FIGURE 2.12 Distribution of population containing integers 0 through 9 with equal frequency

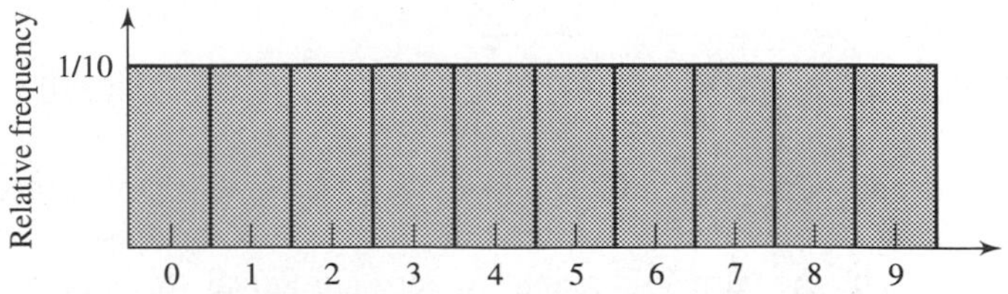

For the population and random variable y under study,

$$\begin{aligned} E(y) &= \sum_y yp(y) \\ &= 0p(0) + 1p(1) + 2p(2) + \cdots + 8p(8) + 9p(9) \\ &= \frac{1}{10}(45) = 4.5 \end{aligned}$$

One can see that $E(y)$ is equal to the average value, or mean value, of all the measurements in our conceptual population. In general, a population mean will be denoted by μ, and it follows that

$$\mu = E(y)$$

where y is the value of a single measurement chosen at random from the population.

The variability of measurements in a population can be measured by the *variance*, which is defined as the expected value, or average value, of the square of the deviation between a randomly selected measurement y and its mean value μ. Thus, the variance of y, $V(y)$, is given by

$$V(y) = E(y - \mu)^2 = \sum_y (y - \mu)^2 p(y)$$

For the population used for illustration in this section,

$$\begin{aligned} V(y) &= E(y - \mu)^2 = \sum_y (y - \mu)^2 p(y) \\ &= (0 - 4.5)^2\left(\frac{1}{10}\right) + (1 - 4.5)^2\left(\frac{1}{10}\right) + \cdots + (9 - 4.5)^2\left(\frac{1}{10}\right) \\ &= \frac{1}{10}[(0 - 4.5)^2 + (1 - 4.5)^2 + \cdots + (9 - 4.5)^2] \\ &= \frac{1}{10}(82.5) = 8.25 \end{aligned}$$

Variance $V(y)$ is commonly denoted by σ^2.

The *standard deviation* (SD) is defined to be the square root of the variance, and it is denoted by $\sigma = \sqrt{\sigma^2}$. For the specific population under discussion,

$$\sigma = \sqrt{8.25} = 2.9$$

In statistical studies, the population of interest consists of unknown measurements; hence, we can only speculate about the nature of the relative frequency histogram or the size of μ and σ. To gain some information about the population, we select a sample of n measurements and study the properties of this sample. We then *infer* characteristics of the population from what we observe in the sample. The sample measurements, in general, will be denoted by $y_1, y_2, \ldots, y_n$.

Following the pattern set for summarizing the information in a population, we can calculate the mean, the variance, and the standard deviation of a sample. These numerical descriptive measures are given, respectively, by

$$\bar{y} = \frac{1}{n}\sum_{i=1}^{n} y_i$$

$$s^2 = \frac{\sum_{i=1}^{n}(y_i - \bar{y})^2}{n-1}$$

$$s = \sqrt{s^2}$$

Note that s^2 has a divisor of $n - 1$ instead of n.

For the population of integers 0, 1, ..., 9 in equal proportion, a sample of $n = 10$ measurements was selected. Each of the 10 measurements was selected at random. (Think of drawing 10 slips of paper out of a box containing a large number of slips, each marked with an integer between 0 and 9.) The sample measurements were

6, 9, 3, 8, 1, 7, 8, 8, 4, 0

For this sample

$$\bar{y} = \frac{1}{n}\sum_{i=1}^{n} y_i$$

$$= \frac{1}{10}(6 + 9 + 3 + \cdots + 4 + 0)$$

$$= \frac{1}{10}(54) = 5.4$$

$$s^2 = \frac{1}{n-1}\sum_{i=1}^{n}(y_i - \bar{y})^2$$

$$= \frac{1}{9}[(6 - 5.4)^2 + (9 - 5.4)^2 + \cdots + (0 - 5.4)^2]$$

$$= \frac{1}{9}(92.4) = 10.27$$

$$s = \sqrt{s^2} = \sqrt{10.17} = 3.2$$

Uses for these sample quantities will be discussed in the next two sections, but we can see that $\bar{y}$ might form a reasonable approximation to μ if μ were unknown. Similarly, s^2 might form a reasonable approximation to σ^2 if σ^2 were unknown, and s might form a reasonable approximation to σ.

2.4

SAMPLING DISTRIBUTIONS

In the following chapters, sample quantities like $\bar{y}$ will be used extensively for making inferences about unknown population quantities; hence, we must study the properties of certain functions of sample observations. This study begins with a numerical illustration, which will then be generalized to cover a wide variety of sampling situations.

Consider the population discussed in Section 2.3 in which the integers 0, 1, . . . , 9 were represented in equal proportions. Fifty samples each of size $n = 10$ were selected from this population, each sample chosen in a manner similar to the method used for the one sample selected at the end of Section 2.2. Table 2.4 lists the sample means $\bar{y}$ for these 50 samples in increasing numerical order.

Figure 2.13 shows a frequency histogram for the 50 sample means. This distribution is an approximation to the *theoretical sampling distribution of* $\bar{y}$, since it shows how the $\bar{y}$'s tend to be distributed when repeated samples are taken. The sampling distribution can be thought of as a probability distribution for $\bar{y}$. Note that the distribution of $\bar{y}$'s tends to center close to the population mean of $\mu = 4.5$, has much less spread (or variability) than the original population measurements, and has a mound shape rather than the flat shape of the population distribution.

From an elementary statistics course (see Mendenhall and Beaver, 1994), we know that the sampling distribution of $\bar{y}$ for random samples should have a mean of μ, a standard deviation of $\sigma/\sqrt{n}$, and a shape like that of a *normal curve* (a symmetric, bell-shaped curve). This display of 50 sample means has an average of 4.22 (which is close to $\mu = 4.5$) and a standard deviation of .79 (which is close to $\sigma/\sqrt{n} = 2.9/\sqrt{10} = .92$). Also, the frequency histogram has an approximate bell shape, although it is not quite symmetric. These facts concerning the behavior of sample means will be important in the development of inference procedures.

From known properties of the normal curve, it follows that approximately 68% of the values of $\bar{y}$, in repeated sampling, should fall within 1 SD of the mean of the sampling distribution of $\bar{y}$'s. Approximately 95% of the values of $\bar{y}$, in repeated sampling, should fall within 2 SD of the mean. Checking these statements for the observed sample of 50 $\bar{y}$'s, we see that

$$4.22 \pm .79 \qquad \text{or} \qquad (3.43, 5.01)$$

contains 39 out of 50 (78%) of the $\bar{y}$ values in the sample, and

$$4.22 \pm 2(.79) \qquad \text{or} \qquad (2.64, 5.80)$$

TABLE 2.4 **Sample means for 50 samples, each of size $n = 10$**

2.3	3.6	4.1	4.3	4.8
2.6	3.7	4.1	4.3	4.8
2.6	3.7	4.1	4.4	4.8
3.2	3.7	4.1	4.5	4.9
3.3	3.8	4.2	4.7	5.0
3.4	3.9	4.3	4.7	5.1
3.5	4.0	4.3	4.7	5.3
3.5	4.1	4.3	4.8	5.5
3.6	4.1	4.3	4.8	6.0
3.6	4.1	4.3	4.8	6.6

FIGURE 2.13 Distribution of 50 sample means, with $n = 10$ for each sample

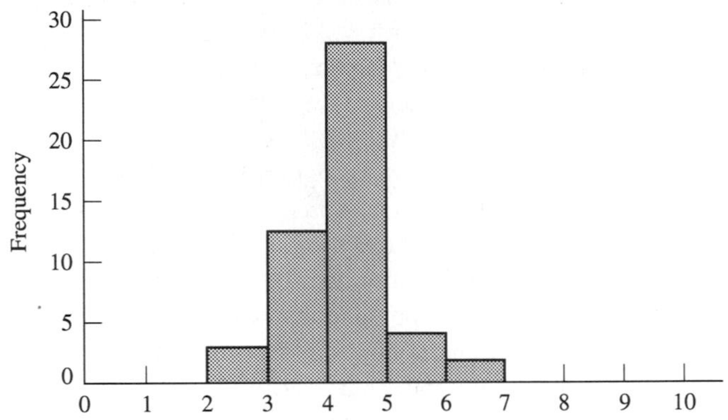

contains 45 out of 50 (90%) of the $\bar{y}$'s. These percentages are reasonably close to the theoretical values of 68% and 95%. (Remember, we've only seen an approximation, based on 50 samples, to the true sampling distribution of $\bar{y}$.)

If the sampling distribution of some sample quantity does not follow a normal distribution, at least approximately, then relative frequency interpretations can still be obtained from Tchebysheff's theorem. This theorem states that, for any $k \geq 1$, *at least* $(1 - 1/k^2)$ of the measurements in any set must lie within k SD of their mean. For example, setting $k = 2$ yields that at least $[1 - (\frac{1}{2})^2] = (1 - \frac{1}{4}) = \frac{3}{4}$ of any set of measurements must lie within 2 SD of their mean. Usually, this fraction is much greater than $\frac{3}{4}$.

The high percentage of measurements falling within 2 SD of the mean, from either the normal distribution of Tchebysheff's theorem, suggests that the range of any set of measurements usually encompasses a little more than 4 SD. Put another way, the standard deviation of a set of measurements can be approximated as one-fourth of the range of that set of measurements.

The discussion of sampling distributions thus far has assumed that the population from which the samples were selected was essentially infinite. But we may want to work with populations of N measurements, where N may be relatively small. Does the approximate normality of the sampling distribution of $\bar{y}$ still hold? Results of two empirical investigations into this question follow.

A population of $N = 100$ measurements was generated by computer and yielded a distribution like the one shown in Figure 2.14. A sample of $n = 20$ measurements was drawn from this population in such a way that every possible sample of size 20 had an equal chance of being selected (see Chapter 4). This process was repeated until 50 such samples were selected. The sample mean $\bar{y}$ was calculated for each sample, and the distribution of these sample means is shown in Figure 2.15. Notice the tendency of this histogram to be bell shaped, although not perfectly symmetric.

FIGURE 2.14 Distribution of population for $N = 100$

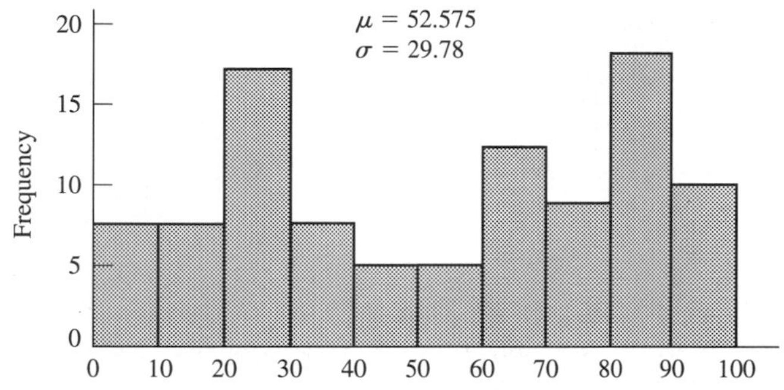

A population of $N = 20$ measurements was then generated, yielding the distribution shown in Figure 2.16. The methodology outlined above was used to select 50 samples, each of size $n = 15$. The distribution of the 50 sample means is shown in Figure 2.17. Note once again a tendency toward a bell-shaped, somewhat symmetric distribution. However, this distribution differs considerably from that of Figure 2.15. Here the sample means are grouped very tightly around the population mean, as should be expected since the sample size of 15 is nearly equal to the population size of 20. One might suggest that the distribution in Figure 2.17 does not resemble the normal distribution nearly as much as does the one in Figure 2.15.

In summary, the sampling distribution of $\bar{y}$ should closely approximate a normal distribution if n is reasonably large but no more than 20% of N. This observation is

FIGURE 2.15 Distribution of sample means for $N = 100$ and $n = 20$

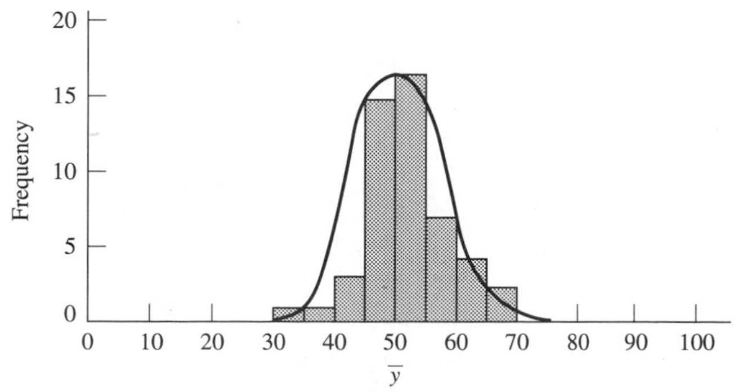

FIGURE 2.16 Distribution of population for $N = 20$

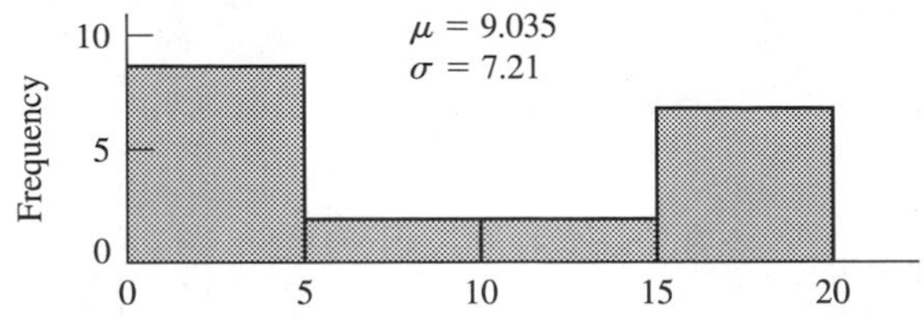

especially true if $n \geq 30$. When n is a very high percentage of N, the sampling distribution of $\bar{y}$ is grouped even more tightly around the population mean.

2.5

COVARIANCE AND CORRELATION

Often an experiment yields more than one random variable of interest. For example, a psychologist measures more than one characteristic per individual in a study of human behavior. Typical variables might be a measure of intelligence y_1, a personality measure y_2, and other variables representing test scores or measures of physical characteristics. Often we are interested in the simple dependence of pairs of variables, such as the relationship between personality and intelligence or between college achievement and college board scores. Particularly, we ask whether data representing paired observations of y_1 and y_2 on a number of people imply a dependence between the two variables. If so, how strong is the dependence?

FIGURE 2.17 Distribution of sample means for $N = 20$ and $n = 15$

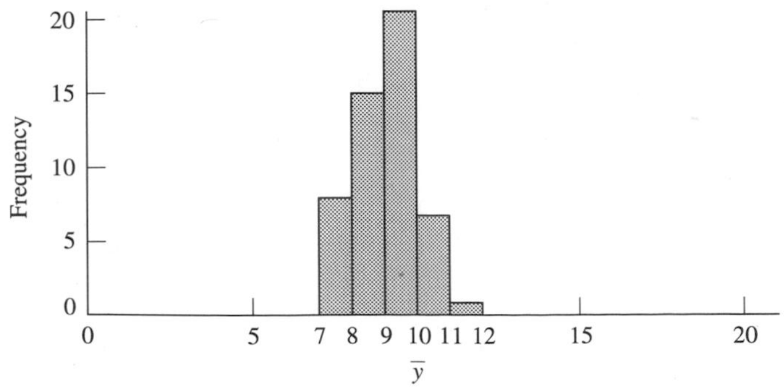

Intuitively, we think of dependence of two random variables, y_1 and y_2, as implying that one—say, y_1—either increases or decreases as y_2 changes. We will confine our attention to two measures of dependence, the *covariance* and the *simple coefficient of linear correlation* and will use Figure 2.18(a) and 2.18(b) to justify choosing them as measures of dependence. These graphs represent plotted points for two (random) samples of $n = 10$ experimental units drawn from a population. Measurements of y_1 and y_2 were made on each experimental unit. If all of the points lie on a straight line, as indicated in Figure 2.18(a), y_1 and y_2 are obviously dependent. In contrast, Figure 2.18(b) indicates little or no dependence between y_1 and y_2.

Suppose we actually know μ_1 and μ_2, the means of y_1 and y_2, respectively, and locate this point on the graphs of Figure 2.18. Now locate a plotted point on Figure 2.18(a) and measure the deviations, $(y_1 - \mu_1)$ and $(y_2 - \mu_2)$. Note that both deviations assume the same algebraic sign for a particular point; hence, their product, $(y_1 - \mu_1)(y_2 - \mu_2)$, is positive. This result will be true for all plotted points on Figure 2.18(a). Points to the right of (μ_1, μ_2) will yield pairs of positive deviations, points to the left will produce pairs of negative deviations, and the average of the product of the deviations $(y_1 - \mu_1)(y_2 - \mu_2)$ will be "large" and positive. If the linear relation indicated in Figure 2.18(a) had sloped downward to the right, all corresponding pairs of deviations would have been of the opposite sign, and the average value of $(y_1 - \mu_1)(y_2 - \mu_2)$ would have been a large negative number.

The situation just described will not occur for Figure 2.18(b), where little or no dependence exists between y_1 and y_2. Corresponding deviations, $(y_1 - \mu_1)$ and $(y_2 - \mu_2)$, will assume the same algebraic sign for some points and opposite signs for others. Thus, the product $(y_1 - \mu_1)(y_2 - \mu_2)$ will be positive for some points, negative for others, and will average to some value near zero.

Clearly, then, the expected (average) value of $(y_1 - \mu_1)(y_2 - \mu_2)$ provides a measure of the linear dependence of y_1 and y_2. This quantity, defined over the two corresponding populations associated with y_1 and y_2, is called the *covariance* of y_1 and y_2. We denote the covariance between y_1 and y_2 thus:

$$\operatorname{cov}(y_1, y_2) = E[(y_1 - \mu_1)(y_2 - \mu_2)]$$

FIGURE 2.18 Plotted points of two samples

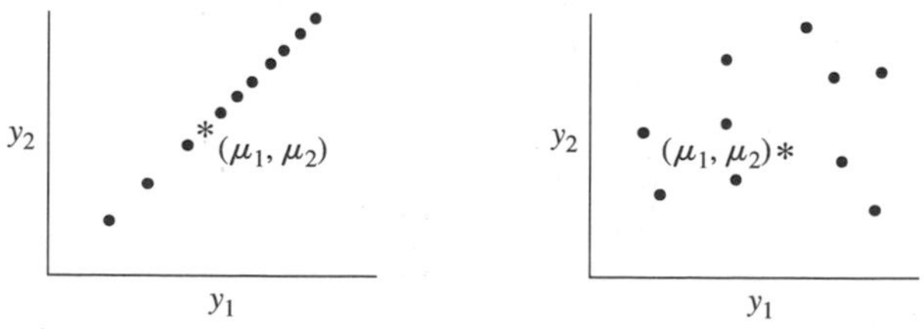

(a) (b)

The larger the absolute value of the covariance of y_1 and y_2, the greater will be the linear dependence between y_1 and y_2. Positive values indicate that y_1 increases as y_2 increases; negative values indicate that y_1 decreases as y_2 increases. A zero value of the covariance indicates no linear dependence between y_1 and y_2.

Unfortunately, to use the covariance as an absolute measure of dependence is difficult because its value depends on the scale of measurement. Consequently, to determine, at first glance, whether a particular covariance is "large" is difficult. We can eliminate this difficulty by standardizing its value, using the simple coefficient of linear correlation. Thus, the population linear coefficient of correlation,

$$\rho = \frac{\text{cov}(y_1, y_2)}{\sigma_1 \sigma_2}$$

(where σ_1 and σ_2 are the standard deviations of y_1 and y_2, respectively) is related to the covariance and can assume values in the interval $-1 \leq \rho \leq 1$. The sample coefficient of correlation is used as an estimator of ρ and is discussed in most introductory courses. Further information on this subject can be found in Mendenhall and Beaver (1994, Chapter 10).

2.6 ESTIMATION

The objective of any sample survey is to make inferences about a population of interest on the basis of information obtained in a sample from that population. Inferences in sample surveys are usually aimed at the *estimation* of certain numerical characteristics of the population, such as the mean, total, or variance. These numerical descriptive measures of the population are called *parameters*.

An *estimator* is a function of observable random variables, and perhaps other known constants, used to estimate a parameter. For example, the sample mean $\bar{y}$ can be used as an estimator of the population mean μ. The sample mean $\bar{y}$ is an estimator since it is a function of sample observations. Note that $\bar{y}$, however, is a random variable and has a probability distribution, or sampling distribution, that depends on the sampling mechanism, as pointed out in Section 2.3. Some of the possible values that $\bar{y}$ can take on will be close to μ, and others may be quite far from μ on either the positive or the negative side. If we are to take a sample and calculate a specific value as our best estimate of μ, we would like to know that on the average $\bar{y}$ generates values that center about μ and are in general quite close to μ. Thus, we will want to select a sampling plan that ensures us that $E(\bar{y}) = \mu$ and $V(\bar{y})$ is "small."

In general, suppose $\hat{\theta}$ is an estimator of the parameter θ. Two properties that we would like $\hat{\theta}$ to possess are as follows:

1. $E(\hat{\theta}) = \theta$.
2. $V(\hat{\theta}) = \sigma^2_{\hat{\theta}}$ is small.

An estimator possessing property 1 is said to be *unbiased*. As for property 2, we will not discuss minimum-variance unbiased estimators in this text, but we will compare unbiased estimators on the basis of their variances. If two unbiased estimators are available for θ, we will generally give preference to the one with the smaller variance.

To summarize, this text will investigate a number of combinations of sampling plans and estimators that give rise to unbiased estimators with small variance.

Although the probability distributions of $\bar{y}$, a common estimator, will depend on the sampling mechanism and the sizes of the sample and the population, in many instances the sample mean tends to have a bell-shaped symmetric distribution known as the *normal distribution*.

Once we know which estimator $\hat{\theta}$ we are using in a situation and something about its probability distribution, we can assess the magnitude of the error of estimation. We define the *error of estimation* to be $|\hat{\theta} - \theta|$. How good will a single estimate be? We cannot state that an observed estimate will be within a specified distance of θ, but we can, at least approximately, find a bound B such that

$$P(|\hat{\theta} - \theta| \leq B) = 1 - \alpha$$

for any desired probability $1 - \alpha$, where $0 < \alpha < 1$. If $\hat{\theta}$ has a normal distribution, then $B = z_{\alpha/2}\sigma_{\hat{\theta}}$, where $z_{\alpha/2}$ is the value cutting off an area of $(\alpha/2)$ in the right-hand tail of the standard normal distribution. The table values of $z_{\alpha/2}$ are given in Table 1 of the appendix. If $1 - \alpha = .95$, then $z_{.025} = 1.96$, or approximately 2. Since many estimators we use throughout the text will not be precisely normally distributed for many values of n and N and since Tchebysheff's theorem states that at least 75% of the observations for any probability distribution will be within 2 SD of their mean, we will use $2\sigma_{\hat{\theta}}$ as a bound on the error of estimation. This value gives $P(|\hat{\theta} - \theta| \leq B) \doteq .95$ for the approximately normal cases and $P(|\hat{\theta} - \theta| \leq B) \geq .75$ in any case. Note that for a normally distributed $\hat{\theta}$, a bound to satisfy any desired probability $(1 - \alpha)$ can be found through use of Appendix Table 1.

If $P(|\hat{\theta} - \theta| \leq B) = 1 - \alpha$, then $P(\hat{\theta} - B \leq \theta \leq \hat{\theta} + B) = 1 - \alpha$. In this form, $(\hat{\theta} - B, \hat{\theta} + B)$ is called a *confidence interval* for θ with *confidence coefficient* $(1 - \alpha)$. The quantity $\hat{\theta} - B$ is called the lower confidence limit (LCL), and $\hat{\theta} + B$ is called the upper confidence limit (UCL).

2.7 SUMMARY

This chapter presented a capsule review of the basic concepts of statistics. Making an inference about a population requires a method for describing a set of measurements and, consequently, requires a discussion of frequency histograms and numerical descriptive measures. Two very useful numerical descriptive measures are the mean and the standard deviation. Although the mean is an easily understood measure of central tendency, the standard deviation acquires meaning as a measure of variation only when interpreted by using Tchebysheff's theorem or some specific distribution such as the normal.

Another important concept is the role that probability plays in making inferences about a population. The probabilist reasons from a known population to a sample. In contrast, the statistician uses probability as the vehicle to make inferences about a population from information contained in a sample. Although a good background in probability is desirable, knowledge of the basic concepts of probability and the use of probability in inference making provides a sufficient background for understanding this text.

Random variables and their probability distributions are presented to provide a background for describing the properties of estimators of population parameters. The notions of expectations, covariance, and correlation assist in evaluating the properties of estimators.

The estimation of population parameters is the primary method of inference making used in sample survey methods. The concept of a point estimator with its corresponding measure of goodness (bound on the error of estimation) is presented and is used as the method of inference in all subsequent chapters.

EXERCISES

2.1 Give a definition of *statistics* and discuss the role of statistics in modern society.

2.2 What are some unique features of sample survey design within the broad framework of statistics?

2.3 What features would you look for in a data set? How might you go about describing these features?

2.4 What is the difference between a *statistic* and a *parameter*?

2.5 Show that the sample variance is equivalent to

$$s^2 = \frac{1}{n-1}\left[\sum_{i=1}^{n} y_i^2 - n\bar{y}^2\right]$$

2.6 The factors that affect our lives and the lives of those around us are variable, but they tend to be characterized by a typical value. Most people have about 13 years of formal schooling. Most workers are on the job for about 40 years before they retire. There are about 4 people in a typical family, and that family most likely has 2 cars. On a summer day, the temperature will be around 80 degrees Fahrenheit, on a winter day around 35 degrees Fahrenheit. The winning football team may score around 21 points, the winning basketball team around 90 points, and the winning baseball team around 5 runs. The newborn baby weighs around 8 pounds. So, you see, all of us have a set of "typical" values by which we make judgments every day. Discuss a variety of typical values that are useful in your life. Where did you learn these values? Do you think the values are correct, or nearly so?

2.7 How can I choose one brand of sports drink over another? Perhaps calories and price will help. The data below show this information for the liquid leading sports drinks. One 8-ounce serving is the basic unit for both calories and cost.

Brand	Calories (C)	Cost in dollars (D)
1. 10-K	60	.22
2. All Sport	70	.24
3. Daily's 1st Ade	60	.26
4. Exceed	70	.34
5. Gatorade	50	.26
6. Hydra Fuel	66	.52
7. Nautilus Plus	60	.22
8. PowerAde	67	.24
9. Snapple Snap-Up	80	.35

Source: Consumer Reports, August, 1993.

(a) What is a good summary number for typical calories per serving for these drinks? What is a good summary number for the variation in the calories per serving?
(b) What is a good summary number for typical cost per serving for these drinks? What is a good summary for the variation in costs per serving?
(c) Does the total of the calories column provide useful information, for practical purposes? How about the total of the cost column?
(d) Suppose Hydra Fuel is eliminated from the list. What impact does that have on the average calories per serving? On the standard deviation of calories per serving? On the average cost per serving? On the standard deviation of the cost per serving?
(e) Which drink has most influence on the average calories per serving? What reason did you use in making this choice?

2.8 Some of the sports drinks come in powdered form or in "light" versions.
(a) Exceed powder comes in a 32-serving size for $9.43 and has 70 calories/serving. Gatorade powder has a 32-serving size for $3.59 and has 60 calories/serving. Is it fair to include the powdered drinks on the same list with the liquid drinks and figure their values into the averages? Why, or why not?
(b) All Sport Lite has 2 calories/serving, at a cost of 24¢ a serving. Gatorade Light has 25 calories/serving, at a cost of 26¢ a serving. Will adding the light varieties to the list have much of an effect on the average cost per serving? Will adding the light varieties have much of an effect on the standard deviation of cost per serving?
(c) Will adding the light varieties have much of an effect on the average calories per serving? Will adding the light varieties have much of an effect on the standard deviation of calories per serving? Describe the nature of this effect.
(d) Can you think of a way to choose a typical value for the calories that is less affected by the two low values for the light varieties?

2.9 Many animals are in danger of extinction. One way to see the extent of the problem is to study the numbers of animals on the endangered species list. The following data show the numbers of endangered species for various groups of animals. The count is the total number of endangered species within the group. Thus, there are 37 mammals on the endangered species list within the United States and 249 others in the rest of the world.

Group	U.S.	U.S. and foreign	Foreign only
Mammals	37	19	249
Birds	57	16	153
Reptiles	8	8	64
Amphibians	6	0	8
Fishes	55	3	11
Snails	12	0	1
Clams	50	0	2
Crustaceans	10	0	0
Insects	13	2	4
Arachnids	3	0	0

Source: The World Almanac, 1994.

(a) If you wanted to summarize these data for the United States in a single number, what number do you think would be the most meaningful? Why?
(b) If you wanted to summarize the situation for endangered mammals worldwide, including the United States, what number do you think would be the most meaningful? Why?
(c) Does the average of the numbers in the United States column have a useful interpretation? Explain.
(d) Write a paragraph summarizing the information in the data set. Make use of the summary numbers you chose in parts (a) and (b).

2.10 The table below shows the number of days a year, over a 10-year period, that the Pollution Standard Index (PSI) was over 100, indicating poor air quality, for various test sites. Describe some of the key features of these data by working through the following problems.

	Year									
Sites	**1981**	**1982**	**1983**	**1984**	**1985**	**1986**	**1987**	**1988**	**1989**	**1990**
Atlanta	9	3	23	8	9	17	19	15	3	16
Boston	0	3	10	6	2	0	4	11	1	1
Chicago	3	3	12	6	6	2	9	15	1	0
Dallas	12	11	17	10	11	5	6	3	3	5
Denver	4	4	11	1	0	1	4	3	0	0
Detroit	12	17	16	4	1	3	6	16	10	3
Houston	33	22	43	30	30	26	29	31	19	35
Kansas City	4	0	4	11	3	3	2	3	1	2
Los Angeles	152	133	142	154	153	159	146	165	137	116
New York	21	20	33	16	13	6	14	30	5	8
Philadelphia	22	33	52	22	25	19	32	34	17	11
Pittsburgh	9	4	15	0	2	2	7	21	4	0
San Francisco	1	0	2	0	1	0	0	0	0	0
Seattle	1	0	0	0	0	1	0	1	0	2
Washington	12	19	38	12	12	9	18	33	4	5
Total	295	272	418	280	268	253	296	381	205	204

Source: U.S. Environmental Protection Agency, Office of Air Quality and Standards.

(a) Compare 1990 with 1981 by constructing parallel box plots. If there are any outliers, remove them and construct new box plots. Does it appear that air quality improved over this period?

(b) Compare the PSI days for Atlanta, Boston, Chicago, Houston, and New York, using parallel box plots. How do these cities compare?
(c) Compare the data for Atlanta and Boston by constructing a scatter plot. Compare the data for Chicago and Boston by constructing a scatter plot. Is Boston more like Chicago or Atlanta? Explain.
(d) Compare the 1990 data with the 1981 data by constructing a scatter plot. What pattern do you see? Can you improve the detail of the scatter plot by removing one of the cities?
(e) Write an overall summary of the changes in PSI days across this period for these sites.

2.11 The following data set shows the home run production of four great players for the New York Yankees. Based on statistical evidence, order these players from best to worst. Make appropriate plots and calculate appropriate summary statistics to defend your answer.

Babe Ruth		***Lou Gehrig***		***Mickey Mantle***		***Roger Maris***	
Year	**Home Runs**	**Year**	**Home Runs**	**Year**	**Home Runs**	**Year**	**Home Runs**
1920	54	1923	1	1951	13	1960	39
1921	59	1924	0	1952	23	1961	61
1922	35	1925	20	1953	21	1962	33
1923	41	1926	16	1954	27	1963	23
1924	46	1927	47	1955	37	1964	26
1925	25	1928	27	1956	52	1965	8
1926	47	1929	35	1957	34	1966	13
1927	60	1930	41	1958	42		
1928	54	1931	46	1959	31		
1929	46	1932	34	1960	40		
1930	49	1933	32	1961	54		
1931	46	1934	49	1962	30		
1932	41	1935	30	1963	15		
1933	34	1936	49	1964	35		
1934	22	1937	37	1965	19		
		1938	29	1966	23		
		1939	0	1967	22		
				1968	18		

Source: The Baseball Encyclopedia, 4th ed., Joseph L. Reicher, ed., 1978.

2.12 When a few data points are repeated in a data set, the results are often arrayed in a frequency table. For example, a quiz given to each of 25 students was graded on a 4-point scale (0, 1, 2, 3) with 3 being a perfect score. Here are the results:

Score *(X)*	***Frequency*** *(F)*	***Proportion*** *(P)*
3	16	.64
2	4	.16
1	2	.08
0	3	.12

(a) Show how the average score can be calculated by using the frequencies.
(b) Show how the average score can be calculated by using the proportions.
(c) Calculate the standard deviation of these scores.

2.13 Leasing new cars instead of purchasing them is now a popular option for individuals as well as businesses. A lease agreement usually requires the payment of an extra fee if the car is driven over 15,000 miles a year. The figure of 15,000 is chosen to be fair to most drivers because lease companies say that a typical driver drives around 10,000 miles a year. How do they know? Recorded below are data on motor vehicle registrations and vehicle miles driven for a selection of five states in 1991. These data will be used to investigate the question of average driving per vehicle. The notation is

$R=$ Number of motor vehicle registrations
$M=$ Vehicle miles of travel

State	*R* (thousands)	*M* (billions)
Alaska	471	4
California	22,253	258
Florida	9,980	114
New York	9,771	108
Rhode Island	628	7

(a) Calculate the average miles per vehicle for each state. Order the five states by this criterion.
(b) Combine the data to calculate the average miles per vehicle across the five states.
(c) Does the 15,000-mile figure used by the lease companies seem reasonable? Explain.

2.14 Explain what is meant by the term *sampling distribution* for a random variable.

2.15 The following data show the average SAT scores per state, along with the percentage of students taking the SAT within each state. Plot these data and summarize any patterns you see, with regard to the relationship between these two variables.

State	SAT	Percent	State	SAT	Percent
AL	991.000	8.000	MD	904.000	64.000
AK	920.000	41.000	MA	896.000	79.000
AZ	932.000	26.000	MI	980.000	11.000
AR	1005.000	6.000	MN	1023.000	12.000
CA	897.000	47.000	MS	997.000	4.000
CO	959.000	29.000	MO	1002.00	12.000
CT	897.000	81.000	MT	982.000	22.000
DE	892.000	61.000	NE	1024.000	10.000
DC	840.000	71.000	NV	919.000	25.000
FL	882.000	48.000	NH	921.000	75.000
GA	844.000	62.000	NJ	886.000	74.000
HI	883.000	55.000	NM	996.000	12.000
ID	968.000	18.000	NY	881.000	75.000
IL	1006.000	16.000	NC	844.000	57.000
IN	865.000	57.000	ND	1073.000	6.000
IA	1093.000	5.000	OH	946.000	22.000
KS	1039.000	10.000	OK	997.000	9.000
KY	993.000	11.000	OR	922.000	54.000
LA	994.000	9.000	PA	876.000	67.000
ME	879.000	64.000	RI	880.000	67.000

(continued)

State	SAT	Percent	State	SAT	Percent
SC	832.000	58.000	VA	890.000	60.000
SD	1047.000	5.000	WA	913.000	49.000
TN	1015.000	12.000	WV	926.000	17.000
TX	874.000	44.000	WI	1023.000	11.000
UT	1031.000	5.000	WY	980.000	13.000
VT	890.000	68.000			

2.16 According to the U.S. Census Bureau, the distribution of family sizes in the United States are as shown on the accompanying table:

Number of persons	1	2	3	4	5	6	7
Proportion of families	.25	.32	.17	.16	.07	.02	.01

That is, 32% of the families in the United States have 2 members, while only 2% have 6 members. (Families with more than 7 children are very rare.)
(a) Find the mean family size, approximately, from this distribution of family sizes. Will this approximation be too large or too small? Explain.
(b) Find the approximate standard deviation of family sizes.
(c) Suppose Nielsen randomly selects 400 families from this population. Describe, as closely as you can, the shape, center, and spread of the 400 data values that might occur in the sample.
(d) Nielsen is actually interested in the mean number of persons per family in samples of 400 families. Describe, as closely as you can, the shape, center, and spread of the distribution of possible values of the sample mean in random samples of 400 families.

2.17 Does aspirin really help prevent heart attacks? During the 1980s, approximately 22,000 physicians over the age of 40 agreed to participate in a long-term health study for which one important question was to determine whether asprin helps lower the rate of heart attacks (myocardial infarctions). The treatments of this part of the study were aspirin or placebo, and the physicians were randomly assigned to one treatment or the other as they entered the study. The method of assignment was equivalent to tossing a coin and sending the physician to the aspirin arm of the study if a head appeared on the coin. After the assignment, neither the participating physicians nor the medical personnel who treated them knew who was taking aspirin and who was taking placebo. This is called a double-blind experiment. (Why is the double blinding important in a study such as this?) The physicians were observed carefully for an extended period of time, and all heart attacks, as well as other problems, that might have occurred were recorded. All data can be summarized on two-way tables.

Other than aspirin, many variables could have an effect on the rate of heart attacks for the two groups of physicians. For example, the amount of exercise they get and whether they smoke are two prime examples of variables that should be controlled in the study so that the true effect of aspirin can be measured. The table below shows how the subjects eventually divided according to exercise and to cigarette smoking.

	Aspirin	Placebo
Exercise vigorously		
Yes	7910	7861
No	2997	3060
Cigarette smoking		
Never	5431	5488
Past	4373	4301
Current	1213	1225

Source: "The Final Report on the Aspirin Component of the Ongoing Physicians' Health Study," *The New England Journal of Medicine* 231, 3 (1989): 129–135.

(a) Do you think the randomization scheme did a good job in controlling these variables? Explain.
(b) Would you be concerned about the results for aspirin being unduly influenced by the fact that most of the aspirin takers were also nonsmokers? Explain.
(c) Would you be concerned about the placebo group possibly having too many members who did not exercise? Explain.

2.18 This study (see Exercise 2.17) reports that 139 heart attacks developed among the aspirin users and 239 heart attacks developed in the placebo group. This was said to be a significant result in favor of aspirin as a possible prevention for heart attacks. To demonstrate this difference, place the data on heart attacks in an appropriate two-way table. (Remember, the 22,000 participants were about evenly split between aspirin and placebo.) What are the appropriate conditional proportions to study if we want to compare the rates of heart attacks for the two treatment groups? Do these proportions turn out to be different?

2.19 Heart attacks aren't the only cause for concern in the Physician's Health Study. Another is that too much aspirin can cause an increase in strokes. Among the aspirin users in the study, 119 had strokes during the observation period. Within the placebo group, only 98 had strokes. Place these data on an appropriate two-way table and comment on the association between aspirin use and strokes, as compared with the association between aspirin use and heart attacks.

2.20 What about smoking as it relates to heart attacks and the use of aspirin? The table below shows the number of heart attacks for each treatment group, separated out according to whether the participant was a current smoker or had never smoked.

		Current smokers			***Never smoked***	
		Aspirin	**Placebo**		**Aspirin**	**Placebo**
Heart attack	Yes	21	37	Yes	55	96
	No	1192	1188	No	5376	5392

Is aspirin as effective a preventive among current smokers as it is among those who never smoked? What can we say about the rate of heart attacks among the current smokers as compared with those who never smoked?

2.21 What is an *estimator*?

2.22 How does one evaluate the goodness of an estimator?

2.23 Describe two desirable properties of an estimator.

2.24 What is an *unbiased estimator*?

2.25 What is the *error of estimation*?

2.26 What is a *reasonable bound* on the error of estimation?

2.27 Suppose a population consists of measurements denoted by $u_1, u_2, \ldots, u_N$. A single observation y is chosen at random from this population. Show that

$$\sigma^2 = V(y) = \frac{1}{N}\sum_{i=1}^{N}(u_i - \mu)^2$$

where μ is the population mean.

CHAPTER 3

ELEMENTS OF THE SAMPLING PROBLEM

3.1 INTRODUCTION

You will recall that the objective of statistics is to make inferences about a population from information contained in a sample. This same objective motivates our discussion of the sampling problem. We will consider the particular problem of sampling from a finite collection of measurements (population). We will refer occasionally to populations composed of an infinite number of measurements. In most cases, the inference will be in the form of an estimate of a population parameter, such as a mean, total, or proportion, with a bound on the error of estimation. For those more interested in methodology than theory, intuitive arguments will be given whenever possible to justify the use of estimators.

The first part of our discussion of the sampling problem introduces certain technical terms common to sample surveys. Next, we discuss how to select a sample from the population.

Each observation, or item, taken from the population contains a certain amount of information about the population parameter or parameters of interest. Since information costs money, the experimenter must determine how much information he or she should buy. Too little information prevents the experimenter from making good estimates, while too much information results in a waste of money. The quantity of information obtained in the sample depends on the number of items sampled and on the amount of variation in the data. This latter factor can be controlled by the method of selecting the sample, called the *design of the sample survey*. The design of the survey and the sample size determine the quantity of information in the sample pertinent to a population parameter, provided that accurate measurements are obtained on each sampled element. Several sample survey designs are introduced in Section 3.3.

If accurate measurements are not obtained on each element of the survey, then other errors are introduced. These errors are discussed in Section 3.4. Accuracy of measurements can be enhanced by proper methods of data collection, discussed in Section 3.5, and by good questionnaire construction, discussed in Section 3.6. Section 3.7 presents the main elements one should be careful to check when planning a survey.

3.2 TECHNICAL TERMS

Technical terminology is kept to a minimum in this text; however, certain common terms must be defined. Let us introduce these terms by way of an example. In a certain community, an opinion poll was conducted to determine public sentiment toward a bond issue in an upcoming election. The objective of the survey was to estimate the proportion of voters in the community who favored the bond issue.

> **Definition 3.1**
>
> An *element* is an object on which a measurement is taken.

In our example, an element is a registered voter in the community. The measurement taken on an element is the voter's preference on the bond issue. Since measurements are usually considered to be numbers, the experimenter can obtain numerical data by recording a 1 for a voter in favor of the bond issue and a 0 for a voter not in favor.

> **Definition 3.2**
>
> A *population* is a collection of elements about which we wish to make an inference.

The population in our example is the collection of voters in the community. The characteristic (numerical measurement) of interest for each member of this population is his or her preference on the bond issue.

An important task for the investigator is to carefully and completely define the population *before* collecting a sample. The definition must include a description of the elements to be included and a specification of the measurements to be taken, since these two components are interrelated. For example, if the population in the bond issue study consists of registered voters, then one may want to collect information on whether or not each sampled person plans to vote in the upcoming election. Sampling the desired target population is not always possible, and the investigator may have to collect additional information so that answers can be provided for questions of interest. If in our example the only population available for sampling is a list of residents of the community, then information about whether each sampled person is, in fact, a registered voter should be collected.

Definition 3.3

Sampling units are nonoverlapping collections of elements from the population that cover the entire population.

In the bond issue example, a sampling unit may be a registered voter in the community. However, a more efficient process may be to sample households, which are collections of elements, in order to obtain information on voter preferences. If households are the sampling units, they must be defined so that no voter in the population can be sampled more than once and so that each voter has a chance of being selected in the sample.

As the definition states, sampling units should be nonoverlapping. However, situations do arise in which the nonoverlapping condition is virtually impossible to achieve. Field plot samples taken, for example, in studies of animal habitat are often circular. The circular pattern is a convenient one to lay out and has advantages in terms of the amount of walking necessary to study the plot. Obviously, circular plots cannot cover a field without some overlap. The intent here is to suggest that the overlap should be as small as possible for efficient sampling.

If each sampling unit contains one and only one element of the population, then a sampling unit and an element from the population are identical. This situation arises if we sample individual voters rather than households within the community.

Definition 3.4

A *frame* is a list of sampling units.

If we specify the individual voter as the sampling unit, a list of all registered voters may serve as a frame for a public opinion poll. Note that this frame will not include all the elements in the population because updating the list daily is impossible. If we take the household as the sampling unit, then a telephone directory, a city directory, or a list of household heads obtained from census data can serve as a frame.

All these frames have some inadequacies. The lists will not be up to date. They will contain many names of unregistered household heads, and hence a sample drawn from the lists will contain many units that are not in the population. Also, some registered voters may not appear on any of these lists. It is hoped, however, that the gap between the frame and the population is small enough to permit inferences to be made about the population on the basis of a sample drawn from the frame.

Some sampling schemes may involve multiple frames. In sampling voters, one could start by sampling housing units or city blocks and then sampling voters within the selected housing units or blocks. One frame, then, is a list of housing units or a list of city blocks, and the second frame is a list of voters within those larger units. The second frame may not be available until the housing units or blocks are selected and studied in some detail. As another example, estimation of crop yields in a state may involve sampling from a list of growers to be inverviewed and a list of fields to be objectively measured.

Definition 3.5

A *sample* is a collection of sampling units drawn from a frame or frames.

Data are obtained from the elements of the sample and used in describing the population. Let the individual voter be our sampling unit and the list of registered voters be our frame. In the public opinion poll, a number of voters (the sample) will be contacted to determine their preference for the upcoming bond issue. We then use the information obtained from these voters to make an inference about voter preference throughout the community.

3.3

HOW TO SELECT THE SAMPLE: THE DESIGN OF THE SAMPLE SURVEY

The objective of sampling is to estimate population parameters, such as the mean or the total, from information contained in a sample. As stated previously, the experimenter controls the quantity of information contained in the sample by the number of sampling units he or she includes in the sample and by the method used to select the sample data. How do we determine which procedure to use and the number of observations (sampling units) to include in the sample? The answer depends on how much information we want to buy. If θ is the parameter of interest and $\hat{\theta}$ is an estimator of θ, we should specify a bound on our error of estimation; that is, we should specify that θ and $\hat{\theta}$ differ in absolute value by less than some value B. Stated symbolically,

$$\text{error of estimation} = |\theta - \hat{\theta}| < B$$

We also must state a probability, $(1 - \alpha)$, that specified the fraction of times in repeated sampling we require the error of estimation to be less than B. This condition can be stated as

$$P[\text{error of estimation} < B] = 1 - \alpha$$

We will usually select $B = 2\sigma_{\hat{\theta}}$, and hence $(1 - \alpha)$ will be approximately .95 for bell-shaped distributions. Most estimators used in this book will exhibit bell-shaped distributions for reasonably large sample sizes, even when the parent population is skewed.

After we obtain a specified bound with its associated probability $(1 - \alpha)$, we can compare different designs (methods of selecting the sample) to determine which procedure yields the desired precision at minimum cost. The problem of selecting the sample size to achieve a certain bound on error is discussed in Mendenhall and Beaver (1994, Chapter 8).

PROBABILITY SAMPLING

The classical formulation of a statistical estimation problem, as described above and in Chapter 2, requires that randomness be built into the sampling design so that properties of the estimators can be assessed probabilistically. With proper randomness in the sampling, one can make statements such as "Our estimate is unbiased and we are 95% confident that our estimate will be within 2 percentage points of the true proportion." Sample designs that are based on planned randomness are called *probability samples*. Virtually all of the remainder of this book will deal with probability samples; the main types of probability sampling designs are outlined here.

The basic design (*simple random sampling*) consists of selecting a group of n sampling units in such a way that each sample of size n has the same chance of being selected. Thus, we can obtain a random sample of n eligible voters in the bond issue poll by drawing names from the list of registered voters in such a way that each sample of size n has the same probability of selection. The details of simple random sampling will be discussed in Chapter 4. At this point, we merely state that a simple random sample will contain as much information on the community preference as any other sample survey design, provided all voters in the community have similar socioeconomic backgrounds.

Suppose, however, that the community consists of people in two distinct income brackets, high and low. Voters in the high bracket may have opinions on the bond issue that are quite different from the opinions of voters in the low bracket. Therefore, to obtain accurate information about the population, we want to sample voters from each bracket. We can divide the population elements into two groups, or strata, according to income and select a simple random sample from each group. The resulting sample is called a *stratified random sample.*

Note that stratification is accomplished by using knowledge of an auxiliary variable, namely, personal income. By stratifying on high and low values of income, we increase the accuracy of our estimator. *Ratio estimation* is a second method for using the information contained in an auxiliary variable. Ratio estimators not only use measurements on the response of interest but also incorporate measurements on an auxiliary variable. If the goal is to estimate the average yearly amount spent on entertainment by households in a community, it might be best first to estimate the ratio of entertainment expenses to household income, and then to multiply this result by the annual household income total for the community. Ratio estimation can also be used with stratified random sampling.

Although individual preferences are desired in the survey, a more economical procedure, especially in urban areas, may be to sample specific families, apartment buildings, or city blocks rather than individual voters. Individual preferences can then be obtained from each eligible voter within the unit sampled. This technique is called *cluster sampling.* Although we divide the population into groups for both cluster sampling and stratified random sampling, the techniques differ. In stratified random sampling, we take a simple random sample within each group; in cluster sampling, we take a simple random sample of groups and then sample items within the selected groups (clusters).

Sometimes, the names of persons in the population of interest are available in a list, such as a registration list, or on file cards stored in a drawer. For this situation, an economical technique is to draw the sample by selecting one name near the beginning of the list and then selecting every tenth or fifteenth name thereafter. If the sampling is conducted in this manner, we obtain a *systematic sample.* As you might expect, systematic sampling offers a convenient means of obtaining sample information; unfortunately, we do not necessarily obtain the most information for a specified amount of money.

We know that observations cost money. Note that the cost of an observation may vary from design to design, and even within a design, depending on the method of data collection. The experimenter should choose the design that gives the desired bound on error with the smallest number of observations (assuming the same cost per observation). However, if the cost per observation varies from design to design, the experimenter should choose the design that gives the desired bound on the error of estimation at a minimum cost.

QUOTA SAMPLING

Probability sampling has a certain amount of randomness built in so that estimator bias or unbiasedness can be established and probability statements can be made about the accuracy of the methods; but could not we do better if we did not randomize at all? Suppose we know that our school's student body contains 60% men and 40% women. Rather than take a simple random sample that almost certainly will not contain exactly 60% men, why not select students one at a time until we get exactly 60% men? Samples of the latter type are called quota samples. We now will discuss the performance of *quota samples* as compared to probability samples.

One of the best examples on which to compare probability sampling and quota sampling is the U.S. presidential election polls, because the true outcome is eventually known (the accuracy of the poll can be seen) and both types of sampling have been used over the years. During and before 1948, quota sampling was the favored method of political pollsters. The thinking was that results would be more accurate if the sample contained ratios of sex, age, income, education, and other factors related to political persuasion that perfectly mirrored the ratios for those same factors in the population as a whole. That thinking was first seriously questioned after the famous presidential election of 1948, in which the underdog Harry Truman defeated the heavily favored Thomas E. Dewey. The results of three famous polls of the time (Crossley, Gallup, and Roper) are shown in Table 3.1. We see here that all three polls seriously overestimated the popular vote for Dewey; yet the quota samples had been designed well, as we see in Table 3.2. The polls achieved ratios of sex, age, education, color, and veteran status quite close to those found in the U.S. population. (A notable exception may be education, in which the "grade school or less" category is seriously underrepresented in the polls.)

Quota sampling was not the only reason for the failure of the polls in 1948, but it certainly was a major contributor. Why? A major reason for poor results from quota sampling is that the final selection of the respondent is left up to the subjective judgment of the interviewer rather than being determined objectively (as by a random number generator). Interviewers in quota sampling tend to be given general instructions ("Find two men and three women in your block, and make sure four are over 25 years of age and one is under 25."). Interviewers in probability sampling are given names or addresses already selected by a randomization device, without human subjectivity. Let us speculate, then, why Republicans were overrepresented in the quota samples of 1948. Perhaps Republicans tended to be better educated or to have higher incomes than Democrats. Thus, there might have been a preponderance of them in the better sections

TABLE 3.1 **The election polls of 1948**

	Percentage of total presidential vote				
	Dewey	Truman	Thurmond	Wallace	Total*
National vote	45.1	49.5	2.4	2.4	99.4
Crossley	49.9	44.8	1.6	3.3	99.6
Gallup	49.5	44.5	2.0	4.0	100.0
Roper	52.2	37.1	5.2	4.3	98.8

* Exclusive of percentages for minor candidates. Gallup percentages calculated on total vote for four principal candidates

Source: F. Mosteller, *The Pre-Election Polls of 1948*. Copyright © 1949, Social Sciences Research Council, New York.

TABLE 3.2 **Comparison of Gallup and Roper samples with U.S. population**

	Percentage		
	Population estimate	**Gallup Oct. 14 sample**	**Roper Oct. 25 sample**
Sex			
Male	49.1*	50.5	50.1
Female	50.9	49.5	49.9
Age			
21–34	34.3	29.6	36.3
35–49	30.9	34.4	30.8
50 and over	34.8	35.9	32.9
Education (last school attended)			
Grade school or less	43.5	35.3	27.5
High school	43.4	46.8	48.8
College	13.0	17.9	23.7
Color			
White	96.1	95.0	97.4
Negro	3.9	5.0	2.6
Veteran status			
Male veterans	14.2	13.3	18.3
All others	85.8	86.7	81.6
Labor union membership of males			
Member		17.5	23.1
Nonmember		82.5	76.9
Number of respondents		2972	3501

*Population aged 21 years and over as of November 1948

Source: See Table 3.1.

of town, and they might have been the more attractive persons to seek out in a crowd. Even unintentionally, inverviewers under the quota sampling rules could be easily drawn toward choosing too many Republicans.

There is a more subtle reason for the failure of quota samples. Table 3.2 shows that Gallup and Roper tried to set quotas for six variables. But there are many more variables (perhaps hundreds more) that might affect voter preference in a presidential election. What about attitudes toward foreign powers, health care, taxes, or even sports? What about national origin, marital status, and income level? It is impossible to control all such factors in quota, or any, sampling designs. The randomization inherent in probability sampling helps balance out factors that cannot be controlled or possibly even measured directly. A random selection of students from your school may result in approximately 60% men and, at the same time, show proportions of fraternity members, in-state residents, and married students about equal to the corresponding population ratios, even though no one controlled for these factors, whereas setting a quota for 60% men might disturb the balance on the other equally important factors. Randomization thus provides balance in uncontrolled (and uncontrollable) factors to a much greater extent than does quota sampling.

Quota sampling failed in 1948. But what about other years? Table 3.3 shows the results of Gallup presidential election polls from 1936 (when the poll began) to 1984. Quota sampling was used prior to and including 1948, while probability sampling has been used since 1952. Note that all of the polls from 1936 to 1948 produced an underestimate of the Democratic vote, but the outcomes were so clear that the correct winner could still be predicted until the very close election of 1948. Since 1948, no clear pattern emerges between Democratic and Republican vote percentages, as compared to the winners. However, the errors involved (the differences between the estimated and true percentages) are generally smaller. In fact, the Gallup poll reports that their average error in 24 national election polls prior to 1950 was 2.3%, whereas their average error in 17 national election polls since 1950 was only 1.5%. This improved accuracy is being accomplished with *smaller* sample sizes. The 1948 Gallup poll had a sample size of 3250, whereas most modern polls aim for 1500 respondents.

To complete the saga of the 1948 election, one other example is pertinent. The Washington State Public Opinion Laboratory conducted two polls prior to the election, one using quota and one using probability sampling. The results are shown in Table 3.4. Are you surprised?

How does the Gallup poll work today? A statement from the American Institute of Public Opinion (the Gallup organization) is given below. In it, you see an emphasis on randomness and objectivity that was sadly lacking prior to 1948. Whatever else you might learn from a book on sampling methods, keep in mind the necessity for randomization in the sampling design. The Gallup poll statement is followed by a briefer one from the *New York Times*.

TABLE 3.3 **Gallup poll accuracy**

Year	Gallup final survey	Election result
1936	55.7% Roosevelt (D)	62.5% Roosevelt
1940	52.0% Roosevelt (D)	55.0% Roosevelt
1944	51.5% Roosevelt (D)	52.3% Roosevelt
1948	44.5% Truman (D)	49.9% Truman
1952	51.0% Eisenhower (R)	55.4% Eisenhower
1956	59.5% Eisenhower (R)	57.8% Eisenhower
1960	51.0% Kennedy (D)	50.1% Kennedy
1964	64.0% Johnson (D)	61.3% Johnson
1968	43.0% Nixon (R)	43.5% Nixon
1972	62.0% Nixon (R)	61.8% Nixon
1976	48.0% Carter (D)	50.0% Carter
1980	47.0% Reagan (R)	50.8% Reagan
1984	59.0% Reagan (R)	59.2% Reagan

The figure shown is the winner's percentage of the Democratic–Republican vote, except in the elections of 1948, 1968, and 1976.

Source: G. Gallup, Jr., *The Gallup Poll, Public Opinion 1984.* Copyright © 1985, Scholarly Resources Inc., Wilmington, DE.

TABLE 3.4 **The Washington state poll of 1948**

	Actual Washington state vote	Probability sample	Quota sample
Dewey	42.7	46.0	52.0
Truman	52.6	50.5	45.3
Wallace	3.5	2.9	2.5

Source: See Table 3.1.

The Gallup Poll

Although most Gallup poll findings are based on telephone interviews, a significant proportion is based on interviews conducted in person in the home. The majority of the findings reported in Gallup Poll surveys is based on samples consisting of a minimum of 1,000 interviews. The total number, however, may exceed 1,000, or even 1,500, interviews, where the survey specifications call for reporting the responses of low-incident population groups such as young public-school parents or Hispanics.

Design of the Sample for Telephone Surveys

The findings from the telephone surveys are based on Gallup's standard national telephone samples, consisting of unclustered directory-assisted, random-digit telephone samples utilizing a proportionate, stratified sampling design. The random-digit aspect of the sample is used to avoid "listing" bias. Numerous studies have shown that households with unlisted telephone numbers are different from listed households. "Unlistedness" is due to household mobility or to customer requests to prevent publication of the telephone number. To avoid this source of bias, a random-digit procedure designed to provide representation of both listed and unlisted (including not-yet-listed) numbers is used.

Telephone numbers for the continental United States are stratified into four regions of the country and, within each region, further arranged into three size-of-community strata. The sample of telephone numbers produced by the described method is representative of all telephone households within the continental United States.

Only working banks of telephone numbers are selected. Eliminating nonworking banks from the sample increases the likelihood that any sampled telephone number will be associated with a residence.

Within each contacted household, an interview is sought with the youngest man 18 years of age or older who is at home. If no man is home, an interview is sought with the oldest woman at home. This method of respondent selection within households produces an age distribution by sex that closely approximates the age distribution by sex of the total population.

Up to three calls are made to each selected telephone number to complete an interview. The time of day and the day of the week for callbacks are varied to maximize the chances of finding a respondent at home. All interviews are conducted on weekends or weekday evenings in order to contact potential respondents among the working population.

The final sample is weighted so that the distribution of the sample matches current estimates derived from the U.S. Census Bureau's Current Population Survey (CPS) for the adult population living in telephone households in the continental United States.

Design of the Sample for Personal Surveys

The design of the sample for personal (face-to-face) surveys is that of a replicated area of probability sample down to the block level in the case of urban areas and to segments of townships in the case of rural areas.

After stratifying the nation geographically and by size of community according to information derived from the most recent census, over 350 different sampling locations are selected on a mathematically random basis from within cities, towns, and counties that, in turn, have been selected on a mathematically random basis.

The interviewers are given no leeway in selecting the areas in which they are to conduct their interviews. Each interviewer is given a map on which a specific starting point is marked and is instructed to contact households according to a predetermined travel pattern. At each occupied dwelling unit, the interviewer selects respondents by following a systematic procedure that is repeated until the assigned number of interviews has been completed.

The New York Times

The latest New York Times/CBS News Poll is based on telephone interviews conducted from Sept. 8 to 11 with 1,161 adults around the country, excluding Alaska and Hawaii.

The sample of telephone exchanges called was selected by a computer from a complete list of exchanges in the United States. The exchanges were chosen to assure that each region of the country was represented in proportion to its population. For each exchange, the telephone numbers were formed by random digits, thus permitting access to both listed and unlisted numbers. Within each household, one adult was designated by a random procedure to be the respondent for the survey.

The results have been weighted to take account of the household size and the number of telephone lines into the residence, and to adjust for variations in the sample relating to region, race, sex, age and education.

In theory, in 19 cases out of 20 the results based on such samples will differ by no more than three percentage points in either direction from what would have been obtained by seeking out all American adults. For smaller subgroups the potential sampling error is larger. For example, for blacks it is plus or minus 10 percentage points.

In addition to sampling error, the practical difficulties of conducting any survey of public opinion may introduce other sources of error into the poll. Variations in question wording or the order of questions, for example, can lead to somewhat different results.

3.4 SOURCES OF ERRORS IN SURVEYS

Sample surveys are afflicted with many types of errors, some arise because only a sample from the population is intended for measurement and because, even for the sampled elements, data may be incomplete or incorrect. Experts in sample survey design and practice have classified survey errors in a variety of ways, but one of the most complete and most informative comes from Groves (1989). His scheme will be modified for use here.

Survey errors can be divided into two major groups: *errors of nonobservation*, where the sampled elements comprise only part of the target population, and *errors of observation*, where recorded data deviate from the truth. Errors of nonobservation can be attributed to sampling, coverage, or nonresponse. Errors of observation can be attributed to the interviewer (data collector), respondent, instrument, or method of data collection.

ERRORS OF NONOBSERVATION

Generally, the data observed in a sample will not precisely mirror the data in the population from which that sample was selected, even if the sampling and measuring are done with extreme care and accuracy. This deviation between an estimate from an ideal sample and the true population value is the *sampling error* that is produced simply because this is a sample and not a census. Sampling error can be measured theoretically and estimated from the sample data for probability samples. The error of estimation discussed in Section 3.3 and in Chapter 2 is one way of assessing the size of this error. It is important to note that sampling error can be reduced by good survey designs and appropriate choice of sample size. Thus, the investigator has some control over this component of error; methods to control it are the subject of most of the remainder of this book.

In almost all surveys, the sampling frame does not match up perfectly with the target population, leading to errors of *coverage*. For telephone surveys, telephone directories are inadequate because of unlisted numbers. For mail surveys of property owners, the most recent list of addresses available at the county court house will be out of date because some nonresident owners have moved and some have sold their property recently. For surveys of hunters or anglers, lists of license purchases are inadequate because children are not required to purchase a license. This lack of coverage introduces an error in the sampling process, an error that is not easily measured or corrected in many surveys. This coverage problem should be clearly elucidated in the report on the data analysis so that those using the results of the study can see clearly how the sampled population differs from the target population.

Probably the most serious of all the nonobservational errors, however, is *nonresponse*. This is a particularly difficult and important problem in surveys that attempt to collect information directly from people through some form of interview. Nonresponse rates are easily obtained because the investigator knows both the sample size and the number of responses to the survey, and sometimes these rates are used to judge the quality of a survey. This is a mistake because a small nonresponse rate could still cause a survey to miss an important part of the population, say, all people over age 70. Data from a survey with a high nonresponse rate could still be informative if the nonrespondents looked like the respondents in all important characteristics. The important consideration here is the nature of the nonrespondents. A good survey must attempt to obtain some information on this group in order to measure how far from the respondent group it may be.

Nonresponse arises in one of three ways: the inability to contact the sampled element (person or household, for example), the inability of the person responding to come up with the answer to the question of interest, or refusal to answer. Data must be collected from precisely those elements that were selected by the randomization scheme used in the design of the survey. An interviewer must not substitute a next-door neighbor, who just happens to be home at 3:00 P.M., for the person actually selected for the sample. This type of substitution might lead to a survey that is biased because of too many families with children or too many retired persons or too many people who work nights are being interviewed. Besides these obvious biases, haphazard substitutions alter the probabilistic structure of the design and may make it impossible to estimate the sampling error.

The inability of the interviewed person to answer the question of interest is a serious problem only in questions dealing with fact. A question on opinion can easily have a "don't know" option and the survey design can account for a certain percentage being in this category. A survey on the economic impact of businesses on a community, however, can be seriously biased if a few of the larger businesses do not know how much they spend on transportation. This is the type of question, however, for which an answer can be found by deeper checking.

The most serious aspect of the nonresponse problem today is refusal to answer. Perhaps because of the proliferation of surveys, perhaps because of fear related to increases in crime, and, no doubt, for a variety of other reasons, people are refusing to answer survey questions in ever-increasing numbers. Actually, many surveys report that their response rates are as good as ever and have not decreased in recent years. On closer scrutiny, though, this maintenance of response rates is often due to an increased effort to replace the refusals with others who will respond. Figure 3.1 shows the nonresponse rates for the National Health Interview Survey, an annual survey

FIGURE 3.1 Nonresponse rates for the National Health Interview Survey, 1967–1985

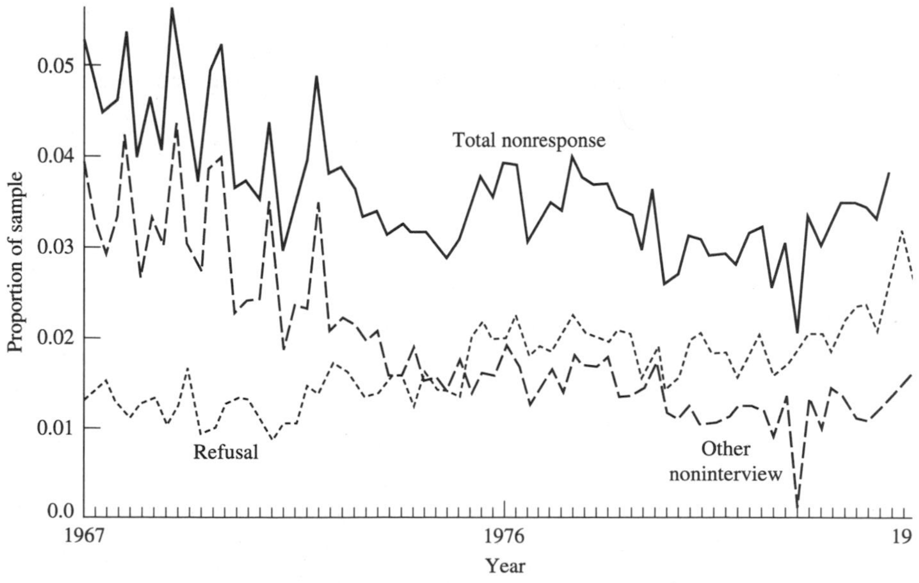

conducted for the federal government, from 1967 through 1985. The overall nonresponse rate appears to be fairly steady, but the refusals, when broken down by type, show a marked increase over the years. Figure 3.2 shows a similar pattern for the Current Population Survey. (The cyclic effect of the nonresponse rates for the CPS is due to some months having much more detailed questionnaires than others.) These increases in nonresponse rates are for government surveys done on a regular basis; the situation is worse for commercial surveys and for those done sporadically.

What do survey designers and analysts know about those who tend to refuse to answer surveys? The highest refusal rates occur among the elderly and the poorly educated, although this is not uniformly true for all surveys. This pattern seems to exist over *all* ethnic and salary groups. Single-person households are more likely to refuse an interview than are multiple-person households, but this issue is confounded with the elderly issue because the elderly belong to many single-person households. These groups, the poorly educated and the elderly, often feel that surveys suggest someone else (often the government) is attempting to gain more power over them and they are relatively powerless. Thus, they will not give those in "authority" any more ammunition. Of course, proliferation of surveys is causing a tremendous intrusion on privacy for all groups, especially when most people include the sales calls (which may begin with a comment about conducting a survey) with the serious surveys. If a survey produces a high refusal rate, it behooves the investigator to find some information on those refusing to answer in order to reduce a potentially great bias.

Careful planning can lower refusal rates. Alerting the respondents in advance, by a letter or a telephone call, that they have been selected for a survey may help improve the response rate. This is especially true if the letter is from a prestigious organization (in the eyes of the potential respondents), if the letter explains that the survey can be beneficial to them and to others, and if the letter explains why it is important that the very person selected must respond in order to make the survey valid. In general, a potential respondent does not see why his or her next-door neighbor cannot be substituted. (After all, he is home all the time and loves to talk.) Explaining the nature

FIGURE 3.2 Nonresponse rates for the Current Population Survey, 1955–1986

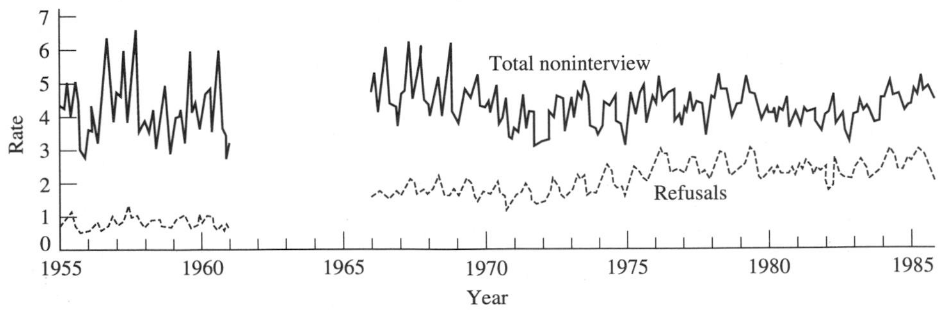

From T. DeMaio et al., "Cognitive and Motivational Bases of Census and Survey Response," *Source*: *Proceedings of Second Annual Research Conference*, U.S. Bureau of Census, 1986, Figure 5a.

of random sampling in nontechnical language sometimes helps. Long introductions about the technical merits of the survey and its outcomes, however, are not seen as beneficial. (Direct rewards to the respondent and the effect of the interviewer will be discussed later.)

ERRORS OF OBSERVATION

Once a person (or object) is in place and ready to be "measured," there are still more errors that can creep into the survey. These errors can be classified as due to the interviewer, the respondent, the *measurement instrument*, or the method of data collection.

Interviewers have a direct and dramatic effect on the way a person responds to a question. Reading a question with inappropriate emphasis or intonation can force a response in one direction or another. Most people who agree to an interview do not want to appear disagreeable and will tend to side with the view apparently favored by the interviewer, especially on questions for which the respondent does not have a strong opinion. Friendly interviewers have more success, of course, than the overtly forceful ones. How gender issues affect interviews is not clear, but male interviewers get a higher rate of cooperation from male respondents than do female interviewers. In general, interviewers of the same gender, racial, and ethnic groups as those being interviewed are slightly more successful.

Respondents differ greatly in motivation to answer correctly and in ability to do so. Each respondent must understand the entire question and be clear about the options for the answer. Sometimes, flashcards showing the question in written form helps this process in personal interviews. That means that questions must be clearly phrased and the questionnaire should not be too long because people will quickly tire of the interview. (Section 3.5 is devoted to questionnaire design.) Obtaining an honest response to sensitive questions, such as questions on business or sexual practices, is particularly difficult and may require special techniques (see Chapter 11). An attempt to place response errors into categories suggests that most are due to either recall bias (the respondent simply does not remember correctly), prestige bias (the respondent exaggerates a little on income or hunting success), intentional deception (the respondent will not admit breaking a law or has a particular gripe against an agency), or incorrect measurement (the respondent did not understand the units and reported feet instead of inches or did not understand the definition of *children* and reported grandchildren as well).

The incorrect measurement issue is related to the *measurement instrument* as a source of error. In any measurement question, the unit of measurement must be clearly defined, whether it be inches on a tape measure, pounds on a scale, or glasses of water (where a "glass" could be any standard size such as 12 ounces). Inaccurate responses are often caused by errors of definition in survey questions. Some examples are the following: (1) The word *children* must be clearly defined. (2) What does the term *unemployed* mean? Should the unemployed include those who have given up looking for work, teenagers who cannot find summer jobs, and those who lost part-time jobs?

(3) Does *education* include only formal schooling or technical training, on-the-job classes and summer institutes as well? Items to be measured must be precisely defined and be unambiguously measurable.

The interviewer, the respondent, and the instrument are brought together in various ways, depending on the *method of data collection*. The most commonly used methods of data collection in sample surveys are personal interviews and telephone interviews. These methods, with appropriately trained interviewers and carefully planned callbacks, commonly achieve response rates of 60% to 75%; sometimes, these rates can be even higher. A mailed questionnaire sent to a specific group of interested persons can achieve good results, but, response rates for this type of data collection are generally so low that all reported results are suspect. Frequently, objective information can be found from direct observation rather than from an interview or mailed questionnaire. These four types of data collection are discussed next.

Personal Interviews

Data are frequently obtained by *personal interviews*. For example, we can use personal interviews with eligible voters to obtain a sample of the public sentiments toward a community bond issue. The procedure usually requires the interviewer to ask prepared questions and to record the respondent's answers. The primary advantage of these interviews is that people will usually respond when confronted in person. In addition, the interviewer can note specific reactions and eliminate misunderstandings about the questions asked. The major limitations of the personal interview (aside from the cost involved) concern the interviewers. If they are not thoroughly trained, they may deviate from the required protocol, thus introducing a bias into the sample data. Any movement, facial expression, or statement by the interviewer can affect the response obtained. For example, a leading question such as "Are you also in favor of the bond issue?" may tend to elicit a positive response. Finally, errors in recording the response can also lead to erroneous results.

Telephone Interviews

Information can also be obtained from persons in the sample through *telephone interviews*. Surveys conducted through telephone interviews are less expensive than personal interviews, owing to the elimination of travel expenses. The investigator can also monitor the interviews to be certain the specified interview procedure is being followed.

A major problem with telephone surveys is the establishment of a frame that closely corresponds to the population. Telephone directories have many numbers that do not belong to households, and many households have unlisted numbers. A few households have no phone service, although lack of phone service is now only a minor problem for most surveys in the United States. A technique that avoids the problem of unlisted numbers is random digit dialing. In this method, a telephone exchange number (the first three digits of the seven-digit number) is selected, and then the last four digits are dialed randomly until a fixed number of households of a specified type are reached. This technique seems to produce unbiased samples of households in selected target populations and avoids many of the problems of trying to sample a telephone directory.

With random digit dialing in a residential survey, only about 20% of the numbers will lie within the frame of interest. Most of the remaining 80% will be unused numbers or numbers belonging to businesses and institutions. The rate of usable numbers can be improved by using random digit dialing to locate clusters (blocks of numbers). Once a residential number is identified, more residences can be selected from the same cluster by leaving the first eight digits the same and randomizing only the last two. This improves the proportion of usable responses because telephone companies assign numbers in blocks.

A study of postelection attitudes and voting behavior (Bergstein, 1979) used this clustered technique. It found that only about 23% of first-stage calls resulted in usable residential numbers; with the clustering technique, however, the percentage of usable residential numbers rose to approximately 57%. This technique therefore can pay big dividends in savings of time and money. Incidentally, in this same study, trained interviewers (those with over six months of interviewing experience) produced a 77% response rate, while those with less training produced only a 67% response rate.

Telephone interviews generally must be kept shorter than personal interviews because respondents tend to get impatient more easily when talking over the telephone. With appropriately designed questionnaires and trained interviewers, telephone interviews can be as successful as personal interviews. [See Schuman and Presser (1981) for more details.]

Self-Administered Questionnaires

Another useful method of data collection is the *self-administered questionnaire*, to be completed by the respondent. These questionnaires usually are mailed to the individuals included in the sample, although other distribution methods can be used. The questionnaire must be carefully constructed if it is to encourage participation by the respondents.

The self-administered questionnaire does not require interviewers, and thus its use results in a savings in the survey cost. This savings in cost is usually brought at the expense of a lower response rate. Nonresponse can be a problem in any form of data collection, but since we have the least contact with respondents in a mailed questionnaire, we frequently have the lowest rate of response. The low response rate can introduce a bias into the sample because the people who answer questionnaires may not be representative of the population of interest. To eliminate some of this bias, investigators frequently contact the nonrespondents through follow-up letters, telephone interviews, or personal interviews.

Direct Observation

The fourth method for collecting data is *direct observation*. For example, if we were interested in estimating the number of trucks that use a particular road during the

4–6 P.M. rush hours, we could assign a person to count the number of trucks passing a specified point during this period. Possibly, electronic-counting equipment could also be used. The disadvantage in using an observer is the possibility of errors in observation.

Direct observation is used in many surveys that do not involve measurements on people. The U.S. Department of Agriculture, for instance, measures certain variables on crops in sections of fields in order to produce estimates of crop yields. Wildlife biologists may count animals, animal tracks, eggs, or nests in order to estimate the size of animal populations.

A closely related notion is that of getting data from objective sources that are not affected by the respondents themselves. Health information can sometimes be obtained from hospital records, income information from employer's records (especially for state and federal government workers). This approach may take more time but may yield large rewards in important surveys.

REDUCING ERRORS IN SURVEYS

Both errors of nonobservation and errors of observation can seriously affect the accuracy of a survey. Errors cannot be eliminated from a survey, but their effects can be reduced by careful adherence to a good sampling plan. Some of the major points in reducing survey errors are presented next.

Callbacks

Nonresponse can be reduced by having a carefully prepared plan for callbacks on sampled elements. A fixed number of callbacks should be required for each sampled element, and these callbacks should be on different days of the week and at different hours of the day. A specific method for determining an appropriate number of callbacks will be given in Chapter 11. That some responses be obtained on at least a subset of the original nonrespondents is important so that large biasing factors can be eliminated. For example, in surveying opinions on gun control legislation, one would want to make sure that all the nonrespondents were not people who favor gun control but not strongly enough to bother responding to a questionnaire.

If a survey requires interviews, the interview times for the original contact and the callbacks should be planned carefully so as to maximize the response rate with a minimum number of callbacks. Studies of optimum interview times have been made, and the results of one such study are presented in Table 3.5. Note that the highest proportions are in the early evening hours, Sunday through Tuesday. (Have you ever been called by a pollster during your dinner hour?)

It is no secret that it is now getting more difficult to find anyone at home to respond to surveys, due to the open workforce and mobile society in which we live. Figure 3.3 shows the results of three studies, completed in 1960, 1971, and 1976, that compare at-home rates. Note that the 1976 curve is much lower than the others at *all* hours of the day.

TABLE 3.5 **Proportion of households in which at least one person aged 14 or older was at home**

	Proportion by day of week						
Time of day	Sun.	Mon.	Tues.	Wed.	Thurs.	Fri.	Sat.
8:00–8:59 A.M.	(B)	(B)	(B)	(B)	(B)	(B)	(B)
9:00–9:59 A.M.	(B)	(B)	(B)	.55	.28	.45	(B)
10:00–10:59 A.M.	(B)	.47	.42	.38	.45	.40	.55
11:00–11.59 A.M.	.35	.41	.49	.46	.43	.50	.62
12:00–12.59 P.M.	.42	.53	.49	.56	.45	.55	.60
1:00–1:59 P.M.	.49	.44	.50	.48	.43	.51	.63
2:00–2:59 P.M.	.49	.50	.52	.47	.45	.45	.59
3:00–3:59 P.M.	.54	.47	.49	.54	.50	.50	.65
4:00–4:59 P.M.	.52	.58	.55	.57	.57	.56	.53
5:00–5:59 P.M.	.61	.67	.65	.67	.59	.57	.56
6:00–6.59 P.M.	.75	.73	.72	.68	.65	.64	.59
7:00–7:59 P.M.	.73	.74	.75	.64	.61	.57	.66
8:00–8:59 P.M.	(B)	.51	.51	.59	.74	.52	(B)
9:00–9:59 P.M.	(B)	(B)	(B)	.64	(B)	(B)	(B)

(B) = base less than 20

Source: M. F. Weeks et al. (1980).

What response rates can be expected for a survey of the general public? Studies reported by Stephan and McCarthy (1958) give the information in Table 3.6. In this table, the first group pertains to surveys in which a particular randomly selected adult was to be the respondent. (For example, Mary Smith's name was chosen from a directory, and the interviewer *must* locate Mary Smith for the interview.) The second

FIGURE 3.3 Proportion of households in which at least one person aged 14 or older was at home (weekdays).

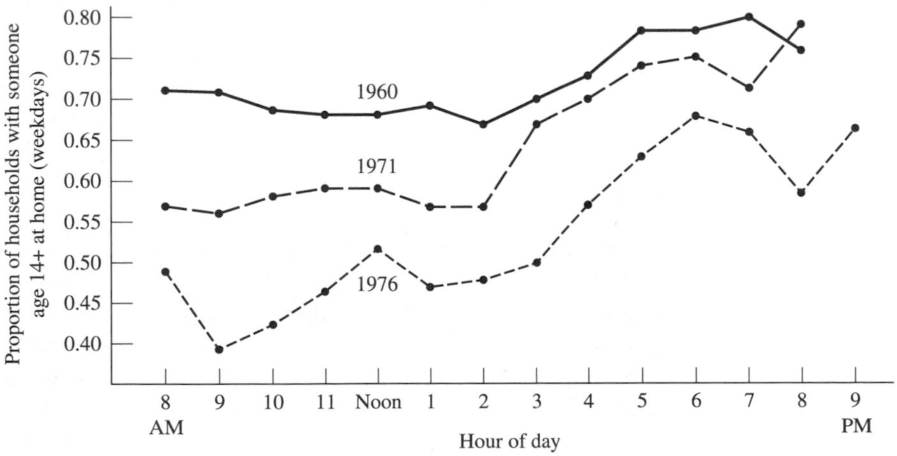

Source: See Table 3.5.

TABLE 3.6 **Number of calls required to complete interviews**

Survey number	Respondent	*Percent of those interviewed contacted on*		
		First call	Second call	Third or later call
1	Random adult	38.0	32.9	29.1
2	Random adult	43.8	35.4	20.8
3	Random adult	36.9	30.8	32.3
4	Any adult	63.5	22.2	14.3
5	Any adult	72.0	19.4	8.6
6	Any adult	83.5	10.3	6.2
7	Housewife	77.6	16.2	6.2
8	Farm operator	72.2	20.8	7.0

Source: Stephan and McCarthy, *Sampling Opinions*. Copyright © 1958, John Wiley & Sons, Inc., New York.

group of surveys would accept responses from any responsible adult who happened to be at home when the interviewer called. The third group was targeted for specific titles (not specific individuals) such as the housewife or the farm operator.

As you see, specific targeted individuals are difficult to find on the first call, and an interviewer should expect to be successful only about 40% of the time. The less specificity about who can respond for the dwelling unit, the better the response rate on first call. If information is required only for the dwelling unit, then there is a good chance of finding someone who can respond on first call. If the information must be obtained on or from a specific individual (Mary Smith), obviously the response rate on first call will decrease.

Rewards and Incentives

Sometimes, an appropriate tactic for encouraging responses is to offer an award for responding. This reward may be a cash payment to a person who agrees to participate in a study. In studies of consumer products, a participant may be given a supply of the product. The rewards should be offered to potential participants in a study only *after* they have been selected for the sample by some objective procedure. To take as a sample those who respond to an advertised reward is usually not appropriate, since those who respond under such inducements may not be representative of the target population.

Incentives to respond are particularly helpful for samples from groups that have a particular interest in the problem under study. Insured motorists may be more willing to respond to a questionnaire on automobile insurance if a cover letter from the state insurance commissioner's office states that the results may help to promote lower rates. Hunters will respond to a questionnaire on game management practices if they are assured that the results may improve hunting conditions. Many similar examples can be given, but the important point is that people are more likely to respond to a survey if they see some potential benefit coming from the results.

Trained Interviewers

The skill of the interviewer is directly related to the quantity and quality of data resulting from a survey, whether the interview is in person or over the telephone. Good interviewers can ask questions in such a way as to encourage honest responses and can tell the difference between those who really don't know the answer and those who are simply reluctant to answer. Newly recruited interviewers should practice on typical respondents like those they might meet in the field. These practice sessions should be under the watchful eye of experienced interviewers, who can then evaluate the interview and suggest improvements in interview technique.

Data Checks

Completed questionnaires should be scrutinized carefully by someone other than the interviewer to see that the form is filled out correctly. At this stage, and again later if data is entered into a computer, a predesigned system of data checks should be made to spot obvious errors in information.

The ranges of measurements can be checked to sort out the cases in which, say, the age of a person is listed as 1040, or a married adult is listed as 9 years old, or a family is reported to have 53 children under the age of 12. Data can be cross-checked in a well-designed questionnaire to see, for example, whether the respondent's reported age agrees with the reported year of birth. Simple arithmetic facts—for instance, proportions must be between 0 and 1 and the hours per day assigned to different work tasks cannot sum to more than 24—can be included in these data checks. Checking data quickly, so that questionable responses can be corrected while the repondent is still available, is very important to the success of a sample survey.

After all responses are collected and the data are being analyzed, additional data checks can be employed. The survey results should be representative of the population, and, sometimes, sample data can be checked against known facts for the population to see whether potential problem areas exist. For example, if the population is 50% female but the sample is only 10% female, there may be serious errors in summary measurements that average over males and females. If average income for survey respondents is well below the reported average from other sources for the target population, then large errors may show up in summary measurements on variables related to income. Some of these potential problems may be solved by augmenting the sample or by changing the form of analysis, but even if they cannot be solved, any inconsistencies should be pointed out in the final analysis.

Questionnaire Construction

After sample selection, the most important component of a well-run, informative, and accurate sample survey is a properly designed questionnaire. This subject is the topic of Section 3.5.

One of the classical errors in opinion-polling history came about because of nonsampling errors related to nonresponse and poor selection of a frame. The *Literary*

Digest attempted to predict the outcome of the 1936 presidential election by sending postcard questionnaires to 10 million people selected, evidently, from subscribers to the *Digest*, telephone directories, and automobile owners. The 2,376,523 returned cards showed Landon a winner over Roosevelt by 57% to 43%. However, Roosevelt won the election by 62.5% to 37.5%. The large error may have been partly due to the frame being weighted toward higher-income people, but, certainly, the high nonresponse rate was a significant factor.

According to the account in his book *The Sophisticated Poll Watcher's Guide*, George Gallup polled a subsample of 3000 of the 10 million polled by the Digest, and he *predicted* that the *Digest* poll would show 56% for Landon and 44% for Roosevelt. Furthermore, another Gallup poll predicted that Roosevelt would win with 56% of the vote. The error in the Gallup poll was still sizable but much smaller than the error in the *Digest* poll, even though the *Digest* had many more respondents. For further details on this interesting case, read the article by Maurice C. Bryson listed in the references in the appendix.

3.5 DESIGNING A QUESTIONNAIRE

As stated earlier, one objective of any survey design is to minimize the nonsampling errors that may occur. If a survey is to obtain information from people, then many potential nonsampling errors should be considered and, it is hoped, controlled by the careful design of the questionnaire. We briefly discuss questionnaire construction in this section, but it is a very important topic and should be investigated further by those attempting to design complex questionnaires for surveys. An excellent reference, and the one on which we rely extensively for the discussion that follows, is Schuman and Presser (1981). Some major concerns in questionnaire construction are outlined in the following sections.

QUESTION ORDERING

Respondents to questionnaires generally try to be consistent in their responses to questions. Respondent consistency may cause the ordering of the questions to affect the responses, sometimes in ways that seem unpredictable to the inexperienced investigator. An example discussed in Schuman and Presser (1981) illustrates the point.

An experiment was conducted with the following two questions:

A. Do you think the United States should let Communist newspaper reporters from other countries come in here and send back to their papers the news as they see it?

B. Do you think a Communist country like Russia should let American newspaper reporters come in and send back to America the news as they see it?

For surveys in 1980 in which the questions appeared in the order (A, B), 54.7% of the respondents answered yes to A and 63.7% answered yes to B. For surveys in which the questions appeared in the order (B, A), 74.6% answered yes to A and 81.9% answered yes to B. So the evidence suggests that asking question B first puts the respondents in a more lenient frame of mind toward allowing Communist reporters into the United States. In other words, those who answered yes to B, when it was asked first, tried to be consistent and also answer yes to a similar question, A. Thus, the context in which a question is asked is very important and should be understood and explained in the analysis of questionnaire data.

Order is also important in the relative positioning of specific versus general questions. Respondents may be asked the following questions:

A. Will you support an increase in state taxes for education?

B. Will you support an increase in state taxes?

It would not be unusual to find more people supporting B if asked in the order (B, A) than if asked in the order (A, B). If question A is asked first, persons who support taxes for education and answer A affirmatively may think that B implies an increase in taxes not necessarily going to education, and they may then say no to this question. If B is presented first, the same people who support more taxes for education may answer affirmatively since they have not yet seen a specific question on taxes for education.

Attitude toward a question in a survey is very often set, or changed, by preceding questions that bear on the same topic. Schuman and Presser report that more crime victimization was reported by respondents when the question on victimization occurred *after* a series of questions on crime than when it occurred by itself. Evidently, the questions on crime helped the person responding to remember small incidents when he or she was a victim of crime, incidents that might otherwise have been forgotten. Attitudes toward government can be quite negative after a series of questions emphasizing government waste and inefficiency, and they can be much more positive after a series of questions emphasizing the necessary and timely functions government performs.

In a series of questions involving ratings, the first question is often considered in a different light from those that follow, and it tends to receive more extreme ratings. For example, suppose a person is to rate a number of possible vacation sites, with each one receiving a numerical rating from 1 to 10, 10 being very good. If the first site looks good to the respondent, it will tend to be rated close to 10 and the others will tend to be rated lower. If the first site looks unattractive, it will tend to be rated close to 1 and the others will tend to be rated higher. Thus, among the group of good sites, each will tend to receive its highest ratings when it appears first on the list. Similarly, each bad site will tend to receive its lowest rating when it appears first on the list. Evidently, the first item on the list is used as a reference point, and other items are rated up or down relative to the first item.

For many survey questions, the order of the possible responses (or choices) to a particular question is as important as the position of the question on the questionnaire. If a person being interviewed is presented with a long list of possible choices or if each possible choice is wordy or difficult to interpret, then the person is likely to respond with the most recent choice (the last one on the list). If a respondent must choose items from a long written list, then the items appearing toward the top of the list have a selection advantage. For example, consider the election of candidates for office from a long slate: Those toward the top of the list tend to get elected. In a list of simple choices, such as strongly agree, agree, disagree, and strongly disagree in an attitude survey, alternatives tend to receive their highest frequency of response when listed first. That is, the proportion who strongly agree will tend to be higher when that option is a first choice rather than a fourth choice.

Researchers attempting to design a questionnaire should be aware of the common ordering problems for question and response. They should attempt to counter potential difficulties by considering the following techniques:

1. Printing questionnaires with different orderings for different subsets of the sample
2. Using show cards or repeating the question as often as necessary in an interview so that the question and possible answers are clearly understood
3. Carefully explaining the context in which a question was asked in the analysis of the survey data

OPEN-VERSUS-CLOSED QUESTIONS

Since questionnaires today are often designed to be electronically scored after completion, with the data in a form for computer handling, most questions will be *closed questions*. That is, each question will have either a single numerical answer (like age of the respondent) or fixed number of predetermined choices, one of which is to be selected by the respondent.

Even though closed questions allow for easy data coding and analysis, some thought should be given to *open questions*, in which the respondent is allowed to freely state an unstructured answer. The open question allows the respondent to express some depth and shades of meaning in the answer. But it can cause great difficulties in analysis because answers may not be easily quantified and may be nearly impossible to compare across questionnaires. In contrast, the closed question may not always provide the appropriate alternatives, and the alternatives listed may, themselves, influence the opinion of the person responding. Once a questionnaire is completed, however, the data handling is fairly routine, and valid statistical summaries of reported answers are easily constructed.

A typical open question, similar to ones actually used in Gallup polls, is as follows:

> What is the most important problem facing the United States today?

This question can provide meaningful results as it is, since many people will choose similar problems as being most important. However, their choices could be forced into predetermined categories by the following closed question:

> The most important problem facing the United States today is (check one)
>
> A. National security.
>
> B. Crime.
>
> C. Inflation.
>
> D. Unemployment.
>
> E. Budget deficits.

One can see that any closed form of this question will limit the alternatives and may force a respondent into an answer that would not necessarily be a first choice.

A good plan for designing a closed question with appropriate alternatives is to use a similar open question on a pretest; then choose as the fixed alternatives those that most nearly represent the choices expressed in the open answers. To come up with a short list of alternatives from the open-ended answers will not always be easy, but this approach will provide more realistic alternatives than could be obtained from mere speculation.

RESPONSE OPTIONS

On almost any question that can be posted, someone being interviewed will want to say that he doesn't know or has no opinion. Since such responses give no useful information about the question and essentially reduce the sample size, typical survey practice is to avoid using these options. The respondent is forced to make a choice from among the listed informative answers, unless the interviewer decides that such a choice simply cannot be made.

However, to force people to make decisions on questions they know nothing about seems inappropriate. Thus, a good questionnaire will provide screening questions to determine whether the respondent has enough information to form an opinion on certain issues. If so, the main question is asked without a no-opinion option. If not, the question may be skipped.

In other words, questions about which nearly everyone has enough information to form some opinion, such as questions on stricter enforcement of speed limit laws for drivers, should be stated without a no-opinion option. Questions of a specific, narrow, or detailed nature, such as questions on a recently passed city ordinance, should be prefaced by screening questions to see whether the respondent has any information on the subject.

Even after the no-opinion option is eliminated from a question, there remains the problem of deciding how many options to allow. Frequently, questionnaires attempt to polarize opinion on one side or the other, as in the following question:

> Do you think the enforcement of traffic laws in our city is too strict or too lenient?

Here no middle ground is offered. One reason for not allowing a middle choice, such as "just right the way things are," is that respondents may take this choice far too often as an easy way out. The two-choice option forces the person responding to think about the direction of the response, but the interviewer should explain that various degrees of strictness or leniency can be taken into account. "Which pole am I closest to?" is the point that the respondent is urged to consider. Of course, if one wants to categorize the degree of strictness or leniency in this question, then more than two options can be presented. However, questionnaire designers usually wish to keep the number of options as small as possible.

WORDING OF QUESTIONS

Even for questions in which the number of response options is clearly determined, the designer should be concerned about the phrasing of the main body of the question. Yes–no questions like

> Do you favor the use of capital punishment?

should be asked in a more balanced form, such as

> Do you favor or oppose the use of capital punishment?

Some questions have strong arguments and counterarguments woven into them. Schuman and Presser (1981, p. 186) show results for a comparison of the following questions:

A. If there is a union at a particular company or business, do you think that all the workers there should be required to be union members, or are you opposed to this?

B. If there is a union at a particular company or business, do you think that all the workers there should be required to be union members, or should it be left to the individual to decide whether or not he wants to be in the union?

Among persons presented with question A, 32.1% responded that workers should be required to be union members; but among those presented with question B, only 23.0% responded in this way. Question B has a stronger counterargument in the second phase of the question. People with no strong feelings either way are particularly susceptible to strong arguments or counterarguments within the body of the question. Again, questions should be asked in a balanced form, with little argument or counterargument within the text of the question.

> Do you agree that courts are too lenient with criminals?

will receive many more yes responses than it should simply because that response seems to agree with the interviewer's notion of the correct response. Leading questions should be rephrased in a balanced form, as discussed earlier in this subsection.

Responses to many questions can be drastically altered just by an appropriate, or inappropriate, choice of words. Schuman and Presser (1981, p. 277) report on studies of the following questions:

A. Do you think the United States should forbid public speeches against democracy?

B. Do you think the United States should allow public speeches against democracy?

In one study, those presented with question A gave 21.4% yes responses, while those presented with question B gave 47.8% no responses. People are somewhat reluctant to *forbid* public speeches against democracy, but they are much more willing to *not allow* such speeches. *Forbid* is a strong word and elicits a negative feeling that many cannot favor. *Allow* is a much milder word and doesn't elicit strong feelings. The important point to remember is that the tone of the question, set by the words employed, can have a significant impact on the responses.

Questions also must be stated in clearly defined terms in order to minimize response errors. A question like

How much water do you drink?

is unnecessarily vague. It may be reworded as follows:

Here is an eight-ounce glass. (Hold one up.) How many eight-ounce glasses of water do you drink each day?

If total water intake is important, the interviewer must remind the person that coffee, tea, and other drinks are mostly water.

Similarly, a question like

How many children are in your family?

is too ambiguous. It may be restated as follows:

How many persons under the age of 21 live in your household and receive more than one-half of their financial support from you?

Again, the question must be specific, with all components well defined.

In designing a questionnaire, we must always remember that people do not remember factual information very well. An interesting study in this area is reported by Bradburn et al. (1987). Three main points in the article are as follows:

1. Do not count on people to remember even the simplest facts. One study reports that only 31% of respondents correctly recalled their savings account balance, and only 47% got it correct *when allowed to consult their records.*
2. People do not generally determine frequencies of events by simple counting. If asked "How many times have you visited a doctor in the past year?," they will tend to establish a rate for a shorter period of time and then multiply. For example, a certain respondent may think she visits a doctor about once a month and then multiply by 12 to get an annual figure. If asked "How many times have you eaten at a restaurant in the past month?," an interviewee may decompose the event into breakfast, lunch, and dinner, and approximate an answer for each meal before adding them back together.
3. People tend to telescope events that they remember well into a shorter time frame. Thus, an automobile accident or a reward on the job may seem to

be more recent than it actually was. Similarly, events that are not recalled easily may seem to be longer ago than they actually were.

Knowledge of these facets of human behavior can be helpful in designing a good questionnaire. We can, for example

1. Ask questions about facts in more than one way, seek out more than one source, or use direct observations as much as possible.
2. Help with the decomposition process by decomposing the questions we ask (such as asking about water, soft drink, beer, and coffee consumption rather than simply asking about drink consumption).
3. Relate questions about events in relationship to important milestones in life (such as "Was the hospital visit before or after you moved to this address?"; "Was it before or after your daughter left for college?") to compensate for the telescoping.

Responses will always contain some errors, but careful questioning can reduce these errors to a point at which the results are still useful.

Many more items could be discussed on the topic of questionnaire construction. But the items presented here are the most important ones, and each should be considered very carefully before sampling is begun.

3.6

PLANNING A SURVEY

We will now review and extend some of the ideas presented in previous sections in the form of a checklist. Each item on the checklist should be carefully considered in the planning of any survey.

1. *Statement of objectives*. State the objectives of the survey clearly and concisely and refer to these objectives regularly as the design and the implementation of the survey progress. Keep the objectives simple enough to be understood by those working on the survey and to be met successfully when the survey is completed.
2. *Target population*. Carefully define the population to be sampled. If adults are to be sampled, then define what is meant by *adult* (all those over the age of 18, for example) and state what group of adults are included (all permanent residents of a city, for example). Keep in mind that a sample must be selected from this population and define the population so that sample selection is possible.
3. *The frame*. Select the frame (or frames) so that the list of sampling units and

the target population show close agreement. Keep in mind that multiple frames may make the sampling more efficient. For example, residents of a city can be sampled from a list of city blocks coupled with a list of residents within blocks.

4. *Sample design*. Choose the design of the sample, including the number of sample elements, so that the sample provides sufficient information for the objectives of the survey. Many surveys have produced little or no useful information because they were not properly designed.
5. *Method of measurement*. Decide on the method of measurement, usually one or more of the following methods: personal interviews, telephone interviews, mailed questionnaires, or direct observations.
6. *Measurement instrument*. In conjunction with step 5, carefully specify how and what measurements are to be obtained. If a questionnaire is to be used, plan the questions so that they minimize nonresponse and incorrect response bias.
7. *Selection and training of field-workers*. Carefully select and train field-workers. After the sampling plan is clearly and completely set up, someone must collect the data. Those collecting data, the field-workers, must be carefully taught what measurements to make and how to make them. Training is especially important if interviews, either personal or telephone, are used because the rate of response and the accuracy of responses are affected by the interviewer's personal style and tone of voice.
8. *The pretest*. Select a small sample for a pretest. The pretest is crucial because it allows you to field-test the questionnaire or other measurement device, to screen interviewers, and to check on the management of field operations. The results of the pretest usually suggest that some modifications must be made before a full-scale sampling is undertaken.
9. *Organization of fieldwork*. Plan the fieldwork in detail. Any large-scale survey involves numerous people working as interviewers, coordinators, or data managers. The various jobs should be carefully organized and lines of authority clearly established before the survey is begun.
10. *Organization of data management*. Outline how each piece of datum is to be handled for all stages of the survey. Large surveys generate huge amounts of data. Hence, a well-prepared data management plan is of the utmost importance. This plan should include the steps for processing data from the time a measurement is taken in the field until the final analysis is completed. A quality control scheme should also be included in the plan in order to check for agreement between processed data and data gathered in the field.
11. *Data analysis*. Outline the anlayses that are to be completed. Closely related to step 10, this step involves the detailed specification of what analyses are to be performed. It may also list the topics to be included in the

final report. If you think about the final report before a survey is run, you may be more careful in selecting items to be measured in the survey.

If these steps are followed diligently, the survey is off to a good start and should provide useful information for the investigator.

3.7 SUMMARY

The objective of a sample survey is to make inferences about the population of interest from information contained in a sample. The population consists of the body of data about which we wish to make an inference and is composed of elements or bits of information. Nonoverlapping collections of elements from the population are called sampling units. The frame is a list of sampling units that we use to represent the population. The sample is a collection of sampling units drawn from the frame. Using the sample data, we will estimate certain population parameters and place bounds on our error of estimation.

The quantity of information obtained from the sample can be controlled by the number of sampling units drawn and the sample design or method of data collection used. Some of the designs introduced were simple random sampling, stratified random sampling, cluster sampling, and systematic sampling. Each is discussed in detail in a later chapter. The best design for a given problem is the one that provides the necessary precision in terms of a bound on the error of estimation for a minimum cost.

After the design has been selected, there are various methods of collecting the sample data. Personal interviews, telephone interviews, direct observations, and questionnaires were discussed and assessed as means of collecting the sample data. Each method has its advantages and limitations.

We discussed the actual construction of questionnaires. Again, we emphasize the importance of obtaining information in the sample that is representative of the population. This problem is of prime significance when we consider methods of data collection.

EXERCISES

3.1 An experimenter wants to estimate the average water consumption per family in a city. Discuss the relative merits of choosing individual families, dwelling units (single-family houses, apartment buildings, and so on), and city blocks as sampling units. What would you use as a frame in each case?

3.2 A forester wants to estimate the total number of trees on a tree farm that possess diameters exceeding 12 inches. A map of the farm is available. Discuss the problem of choosing appropriate sampling units and an appropriate frame.

3.3 A safety expert is interested in estimating the proportion of automobile tires with unsafe tread. Should he use individual cars or collections of cars, such as those in parking lots, as sampling units? What could he use as a frame?

3.4 An industry is composed of many small plants located throughout the United States. An executive wants to survey the opinions of the employees on the vacation policy of the industry. What would you suggest she use as sampling units? As a frame?

3.5 A state department of agriculture desires to estimate the number of acres planted in corn within the state. Suggest possible sampling units and frames.

3.6 A political scientist wants to estimate the proportion of adult residents of a state who favor a unicameral legislature. Discuss possible units and frames. Also, discuss the relative merits of personal interviews, telephone interviews, and mailed questionnaires as methods of data collection.

3.7 Discuss the relative merits of using personal interviews, telephone interviews, and mailed questionnaires as methods of data collection for each of the following situations:
(a) A television executive wants to estimate the proportion of viewers in the country who are watching her network at a certain hour.
(b) A newspaper editor wants to survey the attitudes of the public toward the type of news coverage offered by his paper.
(c) A city commissioner is interested in determining how homeowners feel about a proposed zoning change.
(d) A country health department wants to estimate the proportion of dogs that have had rabies shots within the last year.

3.8 Discuss problems associated with question ordering. Give a list of two or three questions for which you think order is important and explain why.

3.9 Discuss the use of open-versus-closed questions. Give an example of an appropriate open question. Give an example of how a similar question could be closed. What are the advantages of closed questions?

3.10 Give an example of a question that contains a weak counterargument. Give an example of a question that contains a strong counterargument.

3.11 Discuss the use of a no-opinion option in a closed question.

3.12 Give an example of a question that could force a response in a certain direction because of its strong wording.

3.13 Discuss the importance of proper data management techniques and quality control in a survey.

3.14 Discuss the importance of having a pretest.

3.15 Why is the response rate an important consideration in surveys? Discuss methods for reducing the nonresponse rate.

3.16 Respondents commonly receive telephone calls from people taking surveys during the evening dinner hour. Those planning the survey probably think that many potential respondents will be home at that time. Discuss the pros and cons of this approach.

3.17 You are hired to estimate the proportion of registered Republicans in your county who favor an increase in the number of nuclear weapons owned by the United States. How would you plan the survey? (Go through the eleven steps of Section 3.6, realizing that steps 4–11 cannot be answered completely at this time.)

3.18 A Yankelovich, Skelley, and White poll taken in the fall of 1984 showed that a fifth of the 2207 people surveyed admitted to having cheated on their federal income taxes. Do you think that this fraction is close to the actual proportion who cheated? Why? (Discuss the difficulties of obtaining accurate information on a question of this type.)

3.19 In a Gallup youth survey (*Gainesville Sun*, Feb. 13, 1985), 414 high school juniors and seniors were asked:

> What course or subject that you have studied in high school has been the best for preparing you for your future education or career?

In their responses to this question, 25% of the students chose mathematics, and 25% chose English. Do you think this question is a good question with informative results?

3.20 A survey by Group Attitudes, Inc., was said to measure American attitudes toward college (*Gainesville Sun*, Sept. 9, 1982). The polling firm mailed questionnaires to 4200 people across the United States and received 1188 responses. About 55% of those polled said they had major concerns about being able to pay for their child's college education. Would you regard this figure as highly reliable and representative of the true proportion of Americans with this concern? (What groups of people are likely to respond to such a question?)

3.21 Readers of the magazine *Popular Science* (Aug. 1990) were asked to phone in (on a 900 number) their responses to the following question: "Should the United States build more fossil-fuel generating plants or the new so-called safe nuclear generators to meet the energy crisis of the 90s?" Of the total call-ins, 86% chose the nuclear option. Would do you think about the way the poll was conducted? What do you think about the way the question was worded? Do you think the results are a good estimate of the prevailing mood of the country?

3.22 "Food Survey Data All Wrong?" This was the headline of a newspaper article (*Gainesville Sun*, Sept. 11, 1991) on a report from the General Accounting Office of the U.S. government related to the Nationwide Food Consumption Survey. The survey of 6000 households of all incomes and 3600 low-income households is intended to be the leading authority on who consumes what foods. Even though the original households were randomly selected, the GAO said the results were questionable because only 34% of the sampled households responded. Do you agree? What is the nature of the biases that could be caused by the low response rate?

3.23 "Why did they take my favorite show off the air?" The answer lies, no doubt, in low Nielsen ratings. What is this powerful rating system, anyway? Of the 95.1 million households in America, Nielsen Media Research randomly samples 4000 on which to base their ratings. The sampling design is rather complex, but at the last two stages it involves randomly selecting city blocks (or equivalent units in rural areas) and then randomly selecting one household per block to be the Nielsen household. The *rating* for a program is the percentage of the sampled households that have a TV set on and tuned to the program. The *share* for a program is the percentage of the *viewing households* that have a TV set tuned to the program, where a viewing household is a household that has at least one TV set turned on.

(a) How many households are equivalent to 1 rating point?
(b) Is 1 share going to be larger or smaller than 1 rating point?
(c) For the week of April 19, 1992, "60 Minutes" was the top-rated show, with a rating of 21.7. Explain what this rating means.
(d) Discuss potential biases in the Nielsen ratings, even with the randomization in the selection of households carefully built in.

3.24 How does Nielsen determine who is watching what show? The determination comes from the data recorded in a journal by members of a Nielsen household. When a person begins to watch a show, he or she is supposed to log on. Computer vision researchers at the University of Florida are developing a peoplemeter that uses computer image recognition to passively, silently, and automatically record who is watching each show. Discuss the potential for this electronic device to reduce bias in the Nielsen ratings. Are there any new problems that might be caused by this device?

3.25 A serious quality-of-life issue in today's world is the threat of HIV infection and AIDS. Good decisions on how to battle this threat must come from sound data, but data on such a personal issue are difficult to obtain. Improvements are being made in data collection, however, as is seen in a study reported in *Science* (J. Catania et al., "Prevalence of AIDS-Related Risk Factors and Condom Use in the United States," Nov. 13, 1992). In this study, a random sample of 2673 U.S. residents between the ages of 18 and 75 was selected by random digit dialing. A larger sample of 8263 residents was randomly selected from high-risk cities. Some of the demographic characteristics of the resulting samples are shown below:

	High-risk cities		*National*	
	Percentage	Number	Percentage	Number
Women	57.9	4785	58.4	1561
Men	42.1	3478	41.6	1112
African-American	33.8	2795	13.5	360
Hispanic	20.7	1711	8.3	222
White	42.7	3525	75.9	2030
Other	2.8	230	2.3	61

(a) One of the main purposes of using randomization is to ensure that the sample is representative of the target population, of which there are two in this study. For the target population of the United States as a whole, census data tells us that 51% are female, 12% are black, and 9% are Hispanic. How well did the randomized survey do in fairly representing these groups?
(b) Why are the percentages for the target population of high-risk cities much different from the census figures?

3.26 The agreement between the percentages for the randomized study and the actual population percentages for various groups of people may allow us to have more confidence in the main results of the survey, which are the percentages of various HIV-related risk groups as shown below:

	National		High-risk cities	
Risk group	**Percentage**	**Number**	**Percentage**	**Number**
Multiple partners*	7.0	170	9.5	651
Risky partner	3.2	76	3.7	258
Transfusion recipient	2.3	55	2.1	144
Multiple partner and risky partner	1.7	41	3.0	209
Multiple partner and transfusion recipient	0.0	1	0.3	20
Risky partner and transfusion recipient	0.2	4	0.3	19
All others	0.7	16	0.7	51
No risk	84.9	2045	80.4	5539

* Past 12 months.

(a) For the nation as a whole, what percentage of those between the ages of 18 and 75 are at risk for HIV infection? What is the percentage for the high-risk cities? (Actually, these percentages go much higher if sexual practices for more than the past year are taken into account.)
(b) The percentages can be used as *estimates* of probabilities for anticipating what might happen in certain situations. For example, if a national firm is to hire 1000 workers across the country, how many would they expect to be at risk for HIV infection? What would happen to this expected value if the firm is to hire all 1000 workers in high-risk cities?

3.27 *Bias* is the tendency for a whole set of responses to read high or low because of some inherent difficulty with the measurement process. (A chipped die may have a bias toward 6's, in that it comes up 6 much more often than we would expect.) In the study of HIV-related risk groups (Exercise 3.25), as part of a telephone interview, people in the survey were asked intimate questions about their personal lives. Is there a possibility of bias in the responses? If so, in which direction? Will randomization in selecting the respondents help reduce potential bias related to the sensitive questions? Will randomization in selecting the respondents help reduce any potential bias?

3.28 The Teenage Attitudes and Practices Survey obtained completed questionnaires, either by telephone or mail, from a randomly selected group of 9965 12–18 year olds living in households across the country. One type of question asked the teenagers being interviewed about the perceived behavior of their peers. The teenagers answering the questions are classified as

NS = Never smoked
EX = Experimenter with smoking
FS = Former smoker
CS = Current smoker

One question was "Do your peers care about keeping their weight down?" The data shown are the population projections (in thousands) calculated from the sample responses.

	Care about keeping weight down			
	NS	**EX**	**FS**	**CS**
A lot	6297	3613	197	2114
Somewhat	2882	1677	90	793
A little	1441	625	16	354
Don't care	1709	822	33	377

(a) Why would a question be formed in terms of peer behavior rather than as a direct question to the person being interviewed?
(b) Approximately how many teenagers in the United States never smoked (as of 1989)?
(c) What proportion of teenagers think their peers care a lot about keeping their weight down?
(d) Approximately how many teenagers are current smokers in the United States? What proportion of them think their peers care a lot about keeping their weight down?
(e) Among those that care a lot about keeping their weight down, what proportion have never smoked? What proportion are current smokers?
(f) Do you think perceived peer attitudes toward keeping weight down are associated with smoking status of the teenager? Calculate appropriate proportions to justify your answer.

3.29 Another question on the survey of teenagers was "Do your peers care about staying away from marijuana?" The population projected frequencies (in thousands) are as follows:

	Care about staying away from marijuana			
	NS	**EX**	**FS**	**CS**
A lot	7213	2693	75	857
Somewhat	2482	1861	109	1102
A little	744	542	27	298
Don't care	1878	1550	119	1312

(a) What proportion of those who never smoked think their peers care a lot about staying away from marijuana?
(b) What proportion of the current smokers think their peers care a lot about staying away from marijuana?
(c) Among those who think their peers care a lot about staying away from marijuana, what proportion have never smoked?
(d) Among those who think their peers don't care about staying away from marijuana, what proportion have never smoked?
(e) Do you think perceived peer attitudes toward staying away from marijuana are associated with smoking status of the teenager? What proportions help justify your answer?

3.30 Additional questions on the Teenage Attitudes and Practices Survey were asked directly of those being interviewed. Some of the data are reported in the form of percentages rather than frequencies. Two examples are the following:

	Do you believe cigarette smoking helps reduce stress?			
	NS	**EX**	**FS**	**CS**
Yes	12.0	18.7	29.8	46.5
No	84.9	78.5	68.9	51.7
Don't know	3.0	2.5	1.6	1.6

	Do you believe almost all doctors are strongly against smoking?			
	NS	**EX**	**FS**	**CS**
Yes	80.1	78.8	80.1	80.5
No	17.3	18.8	17.3	16.7
Don't know	2.5	2.3	2.6	2.6

(a) How were these percentages calculated? What do they mean? Are these percentages joint, marginal, or conditional?
(b) Do the opinions on whether or not cigarette smoking helps reduce stress appear to be associated with the smoking status of the person responding? Write a paragraph justifying your answer.
(c) Do the opinions on physicians being against smoking appear to be associated with the smoking status of the person responding? Write a paragraph justifying your answer.

EXPERIENCES WITH REAL DATA

How do you rate the place you live, or the place you would like to live? Each year *MONEY* magazine publishes a ranking of "best" places to live. The article below, from the September 1994 issue of the magazine, explains their methodology, and the data following that show some of the results for the 1994 ranking.* Read the article carefully and review the data. Then, write a critique of the article, pointing out the strengths and weaknesses of this method. Would you move to Raleigh, North Carolina, on the basis of this ranking?

> Our basic survey methodology remains unchanged. We hired Beta Research, a Syosset, N.Y. polling firm, to ask a statistically valid sample of MONEY subscribers (median age: 48; median household income: $75,320) what qualities they value in a place to live. Specifically, this year, in rating the importance of 43 factors on a scale of 1 to 10, they said their top priorities were a low crime rate, clean water, clean air and plentiful doctors. (For the complete list, see page 132.)
>
> Next with assistance from Fast Forward, a Portland, Ore, demographic consulting firm, we collected the most timely data available on each of the 300 largest U.S. metropolitan statistical areas, as defined by the U.S. Office of Management and Budget. The data come from the government as well as from private firms (housing permits from the *U.S. Housing Markets* report; cost-of-living figures from the American Chamber of Commerce Researchers Association; hospitals with low mortality rates from the *Consumers' Guide to Hospitals*; environmental rankings from *The Green Index*; top restaurants from *The Mobile Guide*; and the abundance of orchestras from *Symphony* magazine). Century 21 again provided an exclusive list of the current prices and property taxes for a typical three-bedroom home in each area, plus price trends from a year ago. Thirteen of this year's metro areas, all Boston and New York City suburbs, weren't ranked last year because OMB no longer classified them as stand-alone metro areas but folded them into the major city nearby. This year they returned as separate entities to our list.
>
> Finally, we awarded points to metro areas based on how well they delivered the attributes subscribers value the most. The results were then consolidated into nine broad categories: health, crime, economy, housing, education, transportation, weather, leisure, and arts and culture. With the completed rankings in hand. MONEY reporters visited the top 10 and bottom five places to personally assess attractions and drawbacks our statistics might have missed.

*Reprinted from the September 1994 issue of *Money* by special permission, copyright 1994, Time Inc.

Readers' Poll: You now rate low crime over a clean environment

Our readers have a new No. 1 concern when choosing a place to live. Reflecting the public's growing outcry for improved safety, the desire for a low crime rate edged out clean water, the leading factor for the past four years. Clean air came in third again. In all we asked readers to rate the importance of 43 factors on a scale of 1 to 10, that they might value in a community. Two we added this year scored strongly: future job growth (7.7) and a racially diverse population (6.3). Also, you were more concerned this year than last about sunny weather and the prospects for higher local taxes. But good schools, nearby skiing, low house prices and proximity to major league sports teams mattered less to you compared with a year ago.

Characteristic	Mean score	Characteristic	Mean score
Low crime rate	9.1	Low unemployment rate	7.0
Clean water	9.0	Short commutes	6.9
Clean air	8.8	Sunny weather	6.5
Plentiful doctors	8.6	Close to a big airport	6.5
Many hospitals	8.5	Low risk of natural disasters	6.5
Strong state government	8.0	Close to relatives	6.5
Low income taxes	8.0	Near national forests or parks	6.5
Low property taxes	7.9	Low house prices	6.3
Housing appreciation	7.8	Racially diverse population	6.3
Affordable medical care	7.7	Near places of worship	6.3
Future job growth	7.7	Near a big city	5.9
Good schools	7.6	Museums nearby	5.4
Low risk of state tax rise	7.6	Good public transportation	5.3
Inexpensive living	7.5	Proximity to major league sports	5.0
Recession resistance	7.5	Local symphony orchestras	5.0
Low sales taxes	7.5	Zoos or aquariums	4.6
Recent job growth	7.3	Near amusement parks	4.2
Close to colleges	7.2	Close to skiing area	3.6
Affordable car insurance	7.2	Proximity to minor league sports	3.6
High civic involvement	7.2	Far from nuclear reactors	2.7
Near lakes or ocean	7.1	Lack of hazardous waste	1.6
High marks from ecologists	7.1		

Source: Beta Research, Syosset, N.Y.

Comparing our Top 10

Although no metro area sweeps top honors across the board, No. 1 ranked Raleigh/Durham/Chapel Hill, N.C. scores extremely well in four of nine broad categories. The table awards 100 points for the best score of all 300 places in each category. Rochester, Minn. came in second by excelling in the health, crime, education and transit categories, and No. 6 Stamford/Norwalk achieves its perfect arts score because of proximity to New York City. No. 9 Sioux Falls, Utah's No. 3 Provo/Orem and No. 4 Salt Lake City/Ogden boast humming economies. So does No. 7 Gainesville, though its crime rate is a drag. No. 5 San Jose shines in weather, health, leisure and arts, owing to nearby San Francisco's impressive medical facilities. No. 10 Albuquerque is blowing the roof off its housing market, and No. 8 Seattle nosed out San Jose in leisure.

	Health	Crime	Economy	Housing	Education	Transit	Weather	Leisure	Arts
1. Raleigh/Durham/ Chapel Hill, N.C.	88	20	93	88	95	41	38	4	28
2. Rochester, Minn.	96	63	71	43	97	81	14	25	26
3. Provo/Orem, Utah	59	58	79	61	41	58	29	37	22
4. Salt Lake City/ Ogden	72	27	81	75	51	40	26	35	22
5. San Jose	82	35	37	25	40	27	83	93	91
6. Stamford/ Norwalk, Conn.	86	53	74	40	14	18	28	88	100
7. Gainesville, Fla.	45	4	92	49	45	45	79	5	22
8. Seattle	79	20	67	28	62	28	47	94	52
9. Sioux Falls, S.D.	71	54	96	43	6	75	9	2	14
10. Albuquerque	61	13	63	94	36	42	43	23	17

C A S E   S T U D Y

ARE AMERICANS DRINKING LESS ALCOHOL?

A Gallup poll of approximately 1200 Americans conducted in early 1994 was directed toward determining if the consumption of alcoholic beverages was declining. In addition, gender differences in drinking habits were to be noted. Two of the questions and the resulting percentages are shown below:

I have a few questions about alcoholic beverages . . . Do you have occasion to use alcoholic beverages such as liquor, wine or beer, or are you a total abstainer?

Percent Who Drink, by Gender—Trend

	Total	Men	Women
1994 June 3–6	65%	70%	61%
1992	64	72	57
1990	64	64	51

During the past five years, has your consumption of alcoholic beverages increased, decreased, or stayed about the same?

Consumption Changes Over Past 5 Years—Trend

	June 1984	June 1994
Increased	11%	7%
Decreased	29	41
Stayed the same	51	51
No opinion	5	1
	100%	100%

Has the proportion of men who drink decreased significantly since 1992? Has the proportion of women who drink increased significantly since 1992? Has the overall proportion who have decreased drinking in the past five years changed since 1984? Is the proportion who have stayed the same in amount over the past five years significantly greater than the proportion who have decreased their amount of alcohol consumed? The methods presented in this chapter will help us find answers to these questions. Solutions are given at the end of the chapter.

CHAPTER 4

SIMPLE RANDOM SAMPLING

4.1 INTRODUCTION

The objective of a sample survey is to make an inference about the population from information contained in a sample. Two factors affect the quantity of information contained in the sample and hence the precision of our inference-making procedure. The first is the size of the sample selected from the population. The second is the amount of variation in the data; variation can frequently be controlled by the method of selecting the sample. The procedure for selecting the sample is called the *sample survey design*. For a fixed sample size n, we will consider various designs, or sampling procedures, for obtaining the n observations in the sample. Since observations cost money, a design that provides a precise estimator of the parameter for a fixed sample size yields a savings in cost to the experimenter. The basic design, or sampling technique, called simple random sampling is discussed in this chapter.

Definition 4.1

If a sample of size n is drawn from a population of size N in such a way that every possible sample of size n has the same chance of being selected, the sampling procedure is called *simple random sampling*. The sample thus obtained is called a *simple random sample*.

It is a consequence of this definition that all individual elements in a population have the same chance of being selected; however, this statement cannot be taken as a definition of simple random sampling because it does not imply that all samples of size n have the same chance of being selected.

We will use simple random sampling to obtain estimators for population means, totals, and proportions.

Consider the following problem. A federal auditor is to examine the accounts for a city hospital. The hospital records obtained from a computer data file show a particular accounts receivable total, and the auditor must verify this total. If there are 28,000 open accounts in the hospital, the auditor cannot afford the time to examine every patient record to obtain a total accounts receivable figure. Hence, the auditor must choose some sampling scheme for obtaining a representative sample of patient records. After examining the patient accounts in the sample, the auditor can then estimate the accounts receivable total for the entire hospital. If the computer figure lies within a specified distance of the auditor's estimate, the computer figure is accepted as valid. Otherwise, more hospital records must be examined for possible discrepancies between the computer figure and the sample data.

Suppose that all $N = 28{,}000$ patient records are recorded on computer cards and a sample size $n = 100$ is to be drawn. The sample is called a simple random sample if every possible sample of $n = 100$ records has the same chance of being selected.

Simple random sampling, which forms the bases of most sampling designs discussed in this book, is the foundation of most scientific surveys. Based on a sample of around 4000 metered households geographically dispersed across the United States, the Nielsen Television Index (NTI) is the most widely used audience-measurement service in the world. This sample is actually selected by a complex, multistage stratified design, but at the last two levels, blocks are randomly selected from Census Bureau Block Groups and households are randomly selected from within the blocks. The meter records whether a television set is turned on, what channel is being viewed, and channel changes. In addition, members of the household keep a diary on who is watching various shows. The NTI reports the number of households in the viewing audience, the number of television sets actually tuned to a particular show, and information on the characteristics of the audience.

The Gallup polls that are done by telephone surveys begin by stratifying banks of telephone numbers into various geographic regions but then randomly select phone numbers within these regions by random digit dialing. Gallup polls that are done by face-to-face interviews also begin with geographic stratification but then randomly sample 350 locations across the country from within randomly selected cities, towns, and counties. Again, the ideas of simple random sampling play a key role, even though the final sample design is much more complex.

Auditors study simple random samples of accounts in order to check for compliance with audit controls set up by the firm or to verify the actual dollar value of the accounts. Thus, they may wish to estimate the proportion of accounts not in compliance with controls or the total value of, say, accounts receivable.

Marketing research often involves a simple random sample of potential users of a product. The researcher may want to estimate the proportion of potential buyers who prefer a certain color of car or flavor of food.

A forester may estimate the volume of timber or proportion of diseased trees in a forest by selecting geographic points in the area covered by the forest and then attaching a plot of fixed size and shape (such as a circle of 10-meter radius) to that point. All the trees within the sample plots may be studied, but, again, the basic design is a simple random sample.

Two problems now face the experimenter: (1) How does he or she draw the simple random sample, and (2) how can he or she estimate the various population parameters of interest? These topics are discussed in the following sections.

4.2

HOW TO DRAW A SIMPLE RANDOM SAMPLE

To draw a simple random sample from the population of interest is not as trivial as it may first appear. How can we draw a sample from a population in such a way that every possible sample of size n has the same chance of being selected? We might use our own judgment to "randomly" select the sample. This technique is frequently called haphazard sampling. A second technique, representative sampling, involves choosing a sample that we consider to be typical or representative of the population. Both haphazard and representative sampling are subject to investigator bias, and, more importantly, they lead to estimators whose properties cannot be evaluated. Thus, neither technique leads to a simple random sample.

Simple random samples can be selected by using tables of random numbers. A table of random numbers is shown in Appendix Table 2.

A random number table is a set of integers generated so that in the long run the table will contain all ten integers (0, 1, . . . , 9) in approximately equal proportions, with no trends in the pattern in which the digits were generated. Thus, if one number is selected from a random point in the table, it is equally likely to be any of the digits 0 through 9.

Choosing numbers from the table is analogous to drawing numbers out of a hat containing those numbers on thoroughly mixed pieces of paper. Suppose we want a simple random sample of three persons to be selected from seven. We could number the people from 1 to 7, put slips of paper containing these numbers (one number to a slip) into a hat, mix them, and draw out three, *without replacing* drawn numbers. Analogously, we could drop a pencil point on a random starting point in Appendix Table 2. Suppose the point falls on the 15th line of column 9 and we decide to use the right-most digit (a 5, in this case). This procedure is like drawing a 5 from the hat. We may now proceed in any direction to obtain the remaining numbers in the sample. Suppose we decide before starting to proceed down the page. The number immediately below the 5 is a 2, so our second sampled person is number 2. Proceeding, we next come to an 8, but there are only seven people in our population; hence, the 8 must be ignored. Two more 5s then appear, but both must be ignored since person 5 has already been selected. (The 5 has been removed from the hat.) Finally, we come to a 1, and our sample of three is completed with persons numbered 5, 2, and 1.

Note that any starting point can be used and one can move in any predetermined direction. If more than one sample is to be used in any problem, each should have its own unique starting point. Many computer programs, such as MINITAB, can be used to generate random numbers.

A more realistic illustration is given in Example 4.1.

Example 4.1

For simplicity, assume there are $N = 1000$ patient records from which a simple random sample of $n = 20$ is to be drawn. We know that a simple random sample will be obtained if every possible sample of $n = 20$ records has the same chance of being selected. The digits in Appendix Table 2, and in any other table of random numbers, are generated to satisfy the conditions of simple random sampling. Determine which records are to be included in a sample of size $n = 20$.

Solution

We can think of the accounts as being numbers 001, 002, ..., 999, 000. That is, we have 1000 three-digit numbers, where 001 represents the first record, 999 the 999th patient record, and 000 the 1000th.

Refer to Appendix Table 2 and use the first column; if we drop the last two digits of each number, we see that the first three-digit number formed is 104, the second is 223, the third is 241, and so on. Taking a random sample of 20 digits, we obtain the numbers shown in Table 4.1.

TABLE 4.1 **Patient records to be included in the sample**

104	779	289	510
223	995	635	023
241	963	094	010
421	895	103	521
375	854	071	070

If the records are actually numbered, we merely choose the records with the corresponding numbers, and these records represent a simple random sample of $n = 20$ from $N = 1000$. If the patient accounts are not numbered, we can refer to a list of the accounts and count from the 1st to the 10th, 23rd, 70th, and so on, until the desired numbers are reached. If a random number occurs twice, the second occurrence is omitted, and another number is selected as its replacement.

4.3

ESTIMATION OF A POPULATION MEAN AND TOTAL

We stated previously that the objective of survey sampling is to draw inferences about a population from information contained in a sample. One way to make inferences is to estimate certain population parameters by utilizing the sample information. The objective of a sample survey is often to estimate a population mean, denoted by μ, or a population total, denoted by τ. Thus, the auditor of Example 4.1 might be interested in the mean dollar value for the accounts receivable or the total dollar amount in these accounts. Hence, we consider estimation of the two population parameters, μ and τ, in this section.

Suppose that a simple random sample of n accounts is drawn, and we are to estimate the mean value per account for the total population of hospital records. Intuitively, we would employ the sample average

$$\bar{y} = \frac{\sum_{i=1}^{n} y_i}{n}$$

to estimate μ.

Of course, a single value of $\bar{y}$ tells us very little about the population mean μ, unless we are able to evaluate the goodness of our estimator. Hence, in addition to estimating μ, we would like to place a bound on the error of estimation. It can be shown that $\bar{y}$ possesses many desirable properties for estimating μ. In particular, $\bar{y}$ is an unbiased estimator of μ and has a variance that decreases as the sample size n increases. More precisely, for a simple random sample chosen without replacement from a population of size N,

$$E(\bar{y}) = \mu$$

and

$$V(\bar{y}) = \frac{\sigma^2}{n}\left(\frac{N-n}{N-1}\right) \tag{4.1}$$

These properties are formally derived in the appendix, but at this point we will show that they hold for a particular case. Suppose we have a population of $N=4$ measurements given by $\{1, 2, 3, 4\}$. If a single observation y is selected at random from this population, then y can take on any of the four possible values, each with probability $\frac{1}{4}$. Thus,

$$\mu = E(y) = \sum yp(y) = 1\left(\frac{1}{4}\right) + 2\left(\frac{1}{4}\right) + 3\left(\frac{1}{4}\right) + 4\left(\frac{1}{4}\right)$$
$$= \left(\frac{1}{4}\right)(1+2+3+4) = \frac{1}{4}(10) = 2.50$$

and

$$\begin{aligned}\sigma^2 = V(y) = E(y-\mu)^2 &= \sum (y-\mu)^2 p(y) \\ &= (1-2.5)^2\left(\frac{1}{4}\right) + (2-2.5)^2\left(\frac{1}{4}\right) + (3-2.5)^2\left(\frac{1}{4}\right) + (4-2.5)^2\left(\frac{1}{4}\right) \\ &= \frac{5}{4}\end{aligned}$$

Now suppose we are to select a random sample of size $n = 2$ (without replacement) from this population of four measurements. There are six possible samples:

$$\{1, 2\}, \{1, 3\}, \{1, 4\}, \{2, 3\}, \{2, 4\}, \{3, 4\}$$

All of these samples should be equally likely, and hence a probability of $\frac{1}{6}$ should be attached to the occurrence of any one sample. The six sample means $\bar{y}$, are, respectively,

$$1.5, 2.0, 2.5, 2.5, 3.0, 3.5$$

Since each of these sample means can occur with probability $\frac{1}{6}$, we know the sampling distribution of $\bar{y}$ and can compute $E(\bar{y})$ and $V(\bar{y})$. From our definition of expected value,

$$\begin{aligned}E(\bar{y}) &= \sum \bar{y}p(\bar{y}) \qquad \text{(summed over all values of } \bar{y}) \\ &= (1.5)\left(\frac{1}{6}\right) + (2.0)\left(\frac{1}{6}\right) + (2.5)\left(\frac{1}{6}\right) + (2.5)\left(\frac{1}{6}\right) \\ &\quad + (3.0)\left(\frac{1}{6}\right) + (3.5)\left(\frac{1}{6}\right) \\ &= 2.50 = \mu\end{aligned}$$

Also,

$$\begin{aligned}V(\bar{y}) = E(\bar{y}-\mu)^2 &= \sum (\bar{y}-\mu)^2 p(\bar{y}) \\ &= (1.5-2.5)^2\left(\frac{1}{6}\right) + (2.0-2.5)^2\left(\frac{1}{6}\right) + (2.5-2.5)^2\left(\frac{1}{6}\right) \\ &\quad + (2.5-2.5)^2\left(\frac{1}{6}\right) + (3.0-2.5)^2\left(\frac{1}{6}\right) + (3.5-2.5)^2\left(\frac{1}{6}\right) \\ &= (2.5)\left(\frac{1}{6}\right) = \frac{5}{12}\end{aligned}$$

Recalling that for this example $\sigma^2 = \frac{5}{4}$, $N = 4$, and $n = 2$, we have

$$\begin{aligned}\frac{\sigma^2}{n}\left(\frac{N-n}{N-1}\right) &= \frac{(5/4)}{2}\left(\frac{4-2}{4-1}\right) \\ &= \frac{5}{8}\left(\frac{2}{3}\right) = \frac{5}{12}\end{aligned}$$

Thus, we have demonstrated that

$$E(\bar{y}) = \mu \qquad \text{and} \qquad V(\bar{y}) = \frac{\sigma^2}{n}\left(\frac{N-n}{N-1}\right)$$

Consider the sample variance

$$s^2 = \frac{1}{n-1}\sum_{i=1}^{n}(y_i - \bar{y})^2$$

It will also be shown in the appendix that

$$E(s^2) = \frac{N}{N-1}\sigma^2$$

so that $V(\bar{y})$ can be unbiasedly estimated from the sample by

$$\hat{V}(\bar{y}) = \frac{s^2}{n}\left(\frac{N-n}{N}\right)$$

The variance of the estimator $\bar{y}$ is the same as that given in an introductory course except that it is multiplied by a correction factor to adjust for sampling from a finite population. The correction factor takes into account the fact that an estimate based on a sample $n = 10$ from a population of $N = 20$ items contains more information about the population than a sample of $n = 10$ from a population of $N = 20{,}000$.

Estimator of the population mean μ:

$$\hat{\mu} = \bar{y} = \frac{\sum_{i=1}^{n} y_i}{n} \tag{4.2}$$

Estimated variance of $\bar{y}$:

$$\hat{V}(\bar{y}) = \frac{s^2}{n}\left(\frac{N-n}{N}\right) \tag{4.3}$$

where

$$s^2 = \frac{\sum_{i=1}^{n}(y_i - \bar{y})^2}{n-1} = \frac{\sum_{i=1}^{n} y_i^2 - n\bar{y}^2}{n-1}$$

Bound on the error of estimation:

$$2\sqrt{\hat{V}(\bar{y})} = 2\sqrt{\frac{s^2}{n}\left(\frac{N-n}{N}\right)} \tag{4.4}$$

The quantity $(N-n)/N$ is called the finite population correction (fpc). Note that this correction factor differs slightly from the one encountered in the true variance of $\bar{y}$. When n remains small relative to the population size N, the fpc is close to unity. Practically speaking, the fpc can be ignored if $(N-n)/N \geq .95$, or, equivalently, $n \leq (\frac{1}{20})N$. In that case, the estimated variance of $\bar{y}$ is the more familiar quantity s^2/n.

In many cases, the population size is not clearly defined or is unknown. Suppose very small laboratory specimens are selected from a large bulk tank of raw sugar in order to measure pure sugar content. How N will be determined is unclear, but it can generally be assumed to be quite large. Hence, the fpc can be ignored. If a sample of voters is selected from the population of a state, to obtain a precise N for that point in time is generally impossible. Again, N is assumed large, and the fpc is ignored.

To show the behavior of confidence intervals for the mean, we selected 50 random samples of size $n=20$ from the population of $N=100$ elements graphed in Figure 2.14. An approximate 95% confidence interval was constructed for each sample, with the results shown in Table 4.2. Note that four (or 8%) of the observed intervals fail to cover the true population mean. This result is quite close to the normal value of 5%.

Table 4.3 shows the results of a similar experiment drawn from the data of Figure 2.16. Here only two (4%) of the intervals fail to cover the true mean.

Notice that in both cases the confidence intervals change in both location and length as we move from sample to sample. Remember also that the intervals are *random*. In repeated sampling, roughly 95% of the intervals include μ, but any one interval may or may not include μ.

Example 4.2

Refer to the hospital audit of Example 4.1 and suppose that a random sample of $n=200$ accounts is selected from the total of $N=1000$. The sample mean of the accounts is found to be $\bar{y}=\$94.22$, and the sample variance is $s^2=445.21$. Estimate μ, the average due for all 1000 hospital accounts, and place a bound on the error of estimation.

Solution

We use $\bar{y}=\$94.22$ to estimate μ. A bound on the error of estimation can be found by using Equation (4.4):

$$2\sqrt{\hat{V}(\bar{y})} = 2\sqrt{\frac{s^2}{n}\left(\frac{N-n}{N}\right)} = 2\sqrt{\frac{445.21}{200}\left(\frac{1000-200}{1000}\right)}$$

$$= 2\sqrt{1.7808} = \$2.67$$

Thus, we estimate the mean value per account, μ, to be $\bar{y}=\$94.22$. Since n is large, the sample mean should possess approximately a normal distribution, so that $\$94.22 \pm \$2.67$ is approximately a 95% confidence interval for the population mean.

TABLE 4.2 **Confidence intervals for $N = 100$ and $n = 20$**

$\bar{y}$	s^2	LCL	UCL	$\mu = 52.575$
56.020	1047.629	43.332	68.708	
53.650	973.679	41.418	65.882	
60.052	1044.769	47.381	72.722	
49.350	606.324	39.697	59.002	
49.082	994.433	36.721	61.444	
49.038	1058.878	36.282	61.794	
42.857	937.009	30.858	54.856	
46.682	901.619	34.911	58.453	
42.694	677.978	32.487	52.901	
52.922	1086.781	39.999	65.844	
47.778	926.727	35.845	59.712	
48.950	705.443	38.539	59.362	
52.200	1227.258	38.467	65.933	
50.395	714.205	39.919	60.871	
54.384	845.914	42.982	65.785	
49.296	968.221	37.099	61.494	
50.167	957.080	38.040	62.295	
50.082	948.243	38.010	62.153	
58.146	840.061	46.785	69.508	
51.010	1144.449	37.749	64.271	
54.947	1021.469	42.418	67.476	
51.596	907.564	39.787	63.405	
60.053	612.693	50.350	69.756	
61.360	730.304	50.767	71.954	
37.612	642.730	27.674	47.550	
45.641	788.646	34.632	56.640	
47.266	678.076	37.059	57.474	
51.645	815.394	40.452	62.839	
48.601	760.584	37.790	59.412	
49.368	1003.110	36.953	61.784	
52.723	874.174	41.133	64.313	
43.005	622.081	33.228	52.782	
33.760	586.996	24.262	43.257	
57.683	656.446	47.639	67.726	
68.100	750.229	57.363	78.837	
59.298	695.199	48.962	69.634	
47.474	1021.986	34.942	60.006	
47.749	962.295	35.588	59.909	
50.098	785.590	39.111	61.085	
51.697	893.741	39.978	63.416	
45.989	731.062	35.390	56.588	
54.382	735.614	42.392	66.373	
56.294	898.002	44.547	68.041	
52.548	1333.015	38.236	66.860	
53.236	1147.398	39.958	66.514	
57.694	766.730	46.840	68.548	
63.771	860.750	52.270	75.271	
48.835	875.848	37.234	60.437	
66.375	645.377	56.416	76.333	
56.731	1070.385	43.906	69.556	

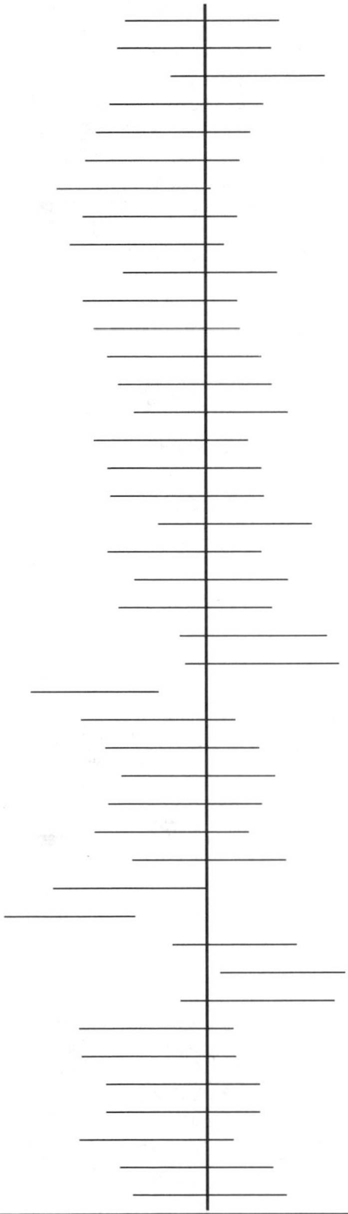

TABLE 4.3 **Confidence intervals for $N=20$ and $n=15$**

$\bar{y}$	s^2	LCL	UCL
10.172	62.698	8.168	12.175
10.312	53.296	8.465	12.160
10.435	58.390	8.501	12.368
8.198	15.953	6.340	10.057
7.410	46.677	5.681	9.139
10.455	49.063	8.682	12.227
9.133	64.951	7.094	11.172
9.255	51.192	7.445	11.066
9.392	54.933	7.516	11.267
10.386	59.018	8.442	12.330
8.700	62.707	6.696	10.703
8.126	46.516	6.401	9.852
8.869	53.483	7.018	10.719
9.074	58.610	7.137	11.012
7.719	52.275	5.889	9.548
7.815	49.723	6.031	9.600
8.794	54.153	6.932	10.656
8.778	53.764	6.923	10.634
11.350	50.345	9.554	13.145
8.205	55.182	6.326	10.085
8.371	56.787	6.464	10.278
9.556	57.116	7.644	11.468
9.442	58.971	7.499	11.385
7.865	53.681	6.011	9.719
9.323	67.261	7.248	11.398
9.300	48.507	7.538	11.062
9.400	47.895	7.667	11.133
9.738	56.042	7.844	11.632
10.100	50.812	8.297	11.904
9.540	58.265	7.608	11.471
9.204	55.947	7.311	11.096
9.057	55.205	7.177	11.937
9.514	48.569	7.750	11.277
10.533	50.279	8.738	12.327
9.076	51.918	7.252	10.899
8.159	52.775	6.321	9.997
7.453	55.096	5.575	9.331
8.323	58.580	6.386	10.260
8.704	57.380	6.787	10.620
9.146	57.078	7.234	11.057
9.301	53.757	7.446	11.156
8.908	59.789	6.951	10.864
7.418	55.680	5.530	9.306
7.335	49.494	5.555	9.115
9.601	58.002	7.674	11.528
8.175	55.629	6.288	10.063
8.634	52.570	6.799	10.468
9.200	55.837	7.310	11.091
7.136	41.977	5.496	8.775
8.032	50.013	6.243	9.822

$\mu = 9.035$

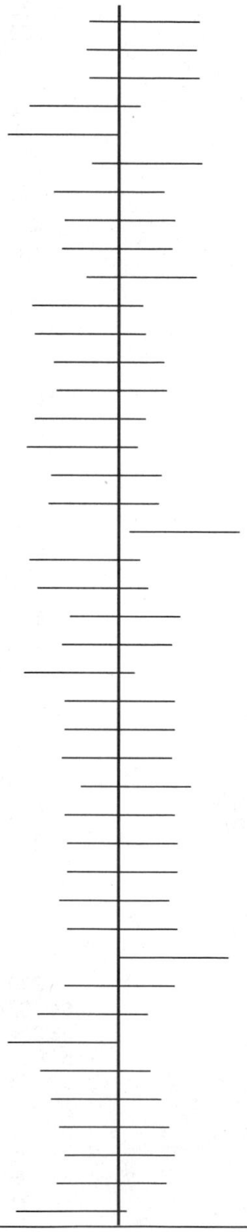

Example 4.3 A simple random sample of $n = 9$ hospital records is drawn to estimate the average amount of money due on $N = 484$ open accounts. The sample values for these nine records are listed in Table 4.4. Estimate μ, the average amount outstanding, and place a bound on your error of estimation.

TABLE 4.4 **Amount of money owed**

y_1	33.50
y_2	32.00
y_3	52.00
y_4	43.00
y_5	40.00
y_6	41.00
y_7	45.00
y_8	42.50
y_9	39.00

Solution

As good data analysis suggests, it is always wise to make an appropriate plot of the data to see if anything unusual appears. Figure 4.1 shows the box plot of these data, which suggest that two of the observations are a bit unusual, compared to the rest, one on the high side and one on the low side. Perhaps the investigator should look at these two accounts again to see if they represent something different from the rest of the accounts in the sample.

Our estimate of μ is

$$\bar{y} = \frac{\sum_{i=1}^{9} y_i}{9} = \frac{368.00}{9} = \$40.89$$

To find a bound on the error of estimation, we must compute

$$s^2 = \frac{\sum_{i=1}^{9}(y_i - \bar{y})^2}{n-1} = \frac{\sum_{i=1}^{9} y_i^2 - \left(\sum_{i=1}^{9} y_i\right)^2 \Big/ 9}{8}$$

$$= \frac{1}{8}\left[15{,}332.50 - \frac{(368)^2}{9}\right] = \frac{1}{8}[15{,}332.50 - 15{,}047.11]$$

$$= 35.67$$

Using Equation (4.4), we obtain the bound on the error of estimation,

$$2\sqrt{\hat{V}(\bar{y})} = 2\sqrt{\frac{s^2}{n}\left(\frac{N-n}{N}\right)} = 2\sqrt{\frac{35.67}{9}\left(\frac{484-9}{484}\right)}$$

$$= 2\sqrt{3.890} = 3.944 = \$3.94$$

FIGURE 4.1 Box plot of the hospital account data

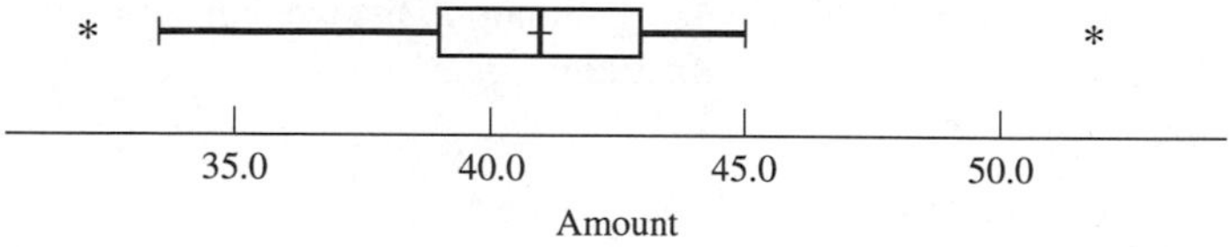

To summarize, the estimate of the mean amount of money owed per account, μ, is $\bar{y} = \$40.89$. Although we cannot be certain how close $\bar{y}$ is to μ, we are reasonably confident that the error of estimation is less than \$3.94.

MINITAB MACRO FOR ESTIMATING A MEAN; SIMPLE RANDOM SAMPLING

To minimize the time spent on computational details, small programs called *macros* can perform the standard computations used in analyzing data from surveys. These macros are available for both MINITAB and SAS statistical software, and similar macros can be developed for other systems. The MINITAB macro for estimating a population mean is called 'srsmean.MTB'. It is described in the appendix and can be called up from the data disk accompanying this book. Its use is illustrated below for the data in Example 4.3.

After issuing the Execute Macro command, select 'srsmean.MTB' as the macro to execute. Other data will be inserted as the program asks for it. In this case, there is only one column of data, and it is in column 1 of MINITAB's spreadsheet. The results are as follows: (The final summary will not be displayed unless your version of MINITAB has a MATRIX command.)

```
MTB > Execute 'srsmean.MTB' 1.

Input the number of columns of data.
DATA> 1

Input the sample size followed by the population size.
DATA>    9 484

The sample size is
K2          9.00000

The population size is
K22         484.000
```

```
The sample mean is
K3          40.8889

The estimated variance of the sample mean is
K4          3.89003

The estimated standard deviation of the sample mean is
K5          1.97231

The bound on the error of estimation is
K6          3.94463

The following table lists the sample size, population
size, sample mean, estimated variance of the sample mean,
estimated standard deviation of the sample mean,
and the bound on the error of estimation for each column:

MATRIX M1

9.000    484.000    40.889    3.890    1.972    3.945
```

Notice that the results are essentially the same as found by hand calculation; in general, the computer version may be more accurate, however, because of rounding errors in the hand calculations.

Many sample surveys are conducted to obtain information about a population total. The federal auditor of Example 4.1 would probably be interested in verifying the computer figure for the total accounts receivable (in dollars) for the $N = 1000$ open accounts.

You recall that the mean for a population of size N is the sum of all observations in the population divided by N. The population total—that is, the sum of all observations in the population—is denoted by the symbol τ. Hence

$$N\mu = \tau$$

Intuitively, we expect the estimator of τ to be N times the estimator of μ, which it is.

Estimator of the population total τ:

$$\hat{\tau} = N\bar{y} = \frac{N\sum_{i=1}^{n} y_i}{n} \tag{4.5}$$

Estimated variance of τ:

$$\hat{V}(\hat{\tau}) = \hat{V}(N\bar{y}) = N^2\left(\frac{s^2}{n}\right)\left(\frac{N-n}{N}\right) \tag{4.6}$$

where

$$s^2 = \frac{\sum_{i=1}^{n}(y_i - \bar{y})^2}{n-1}$$

Bound on the error of estimation:

$$2\sqrt{\hat{V}(N\bar{y})} = 2\sqrt{N^2\left(\frac{s^2}{n}\right)\left(\frac{N-n}{N}\right)} \qquad (4.7)$$

Note that the estimated variance of $\hat{\tau} = N\bar{y}$ in Equation (4.6) is N^2 times the estimated variance of $\bar{y}$ given in Equation (4.3).

Example 4.4

An industrial firm is concerned about the time spent each week by scientists on certain trivial tasks. The time-log sheets of a simple random sample of $n = 50$ employees show the average amount of time spent on these tasks is 10.31 hours, with a sample variance $s^2 = 2.25$. The company employs $N = 750$ scientists. Estimate the total number of man-hours lost each week on trivial tasks and place a bound on the error of estimation.

Solution

We know the population consists of $N = 750$ employees from which a random sample of $n = 50$ time-log sheets was obtained. The average amount of time lost for the 50 employees was $\bar{y} = 10.31$ hours/week. Therefore, the estimate of τ is

$$\hat{\tau} = N\bar{y} = 750(10.31) = 7732.5 \text{ hours}$$

To place a bound on the error of estimation, we apply Equation (4.7) to obtain

$$2\sqrt{\hat{V}(\hat{\tau})} = 2\sqrt{(750)^2\left(\frac{2.25}{50}\right)\left(\frac{750-50}{750}\right)}$$

$$= 2\sqrt{23{,}625} = 307.4 \text{ hours}$$

Thus, the estimate of total time lost is $\hat{\tau} = 7732.5$ hours. We are reasonably confident that the error of estimation is less than 307.4 hours.

4.4

SELECTING THE SAMPLE SIZE FOR ESTIMATING POPULATION MEANS AND TOTALS

At some point in the design of the survey, someone must make a decision about the size of the sample to be selected from the population. So far, we have discussed a sampling procedure (simple random sampling) but have said nothing about the number of observations to be included in the sample. The implications of such a decision are obvious. Observations cost money. If the sample is too large, time and talent are wasted. Conversely, if the number of observations included in the sample is too small, we have bought inadequate information for the time and effort expended and have again been wasteful.

The number of observations needed to estimate a population mean μ with a bound on the error of estimation of magnitude B is found by setting 2 SD of the estimator, $\bar{y}$, equal to B and solving this expression for n. That is, we must solve

$$2\sqrt{V(\bar{y})} = B \tag{4.8}$$

for n.

Recall that the estimated variance of $\bar{y}$, $\hat{V}(\bar{y})$, is given by

$$\hat{V}(\bar{y}) = \frac{s^2}{n}\left(\frac{N-n}{N}\right) \tag{4.9}$$

Also,

$$V(\bar{y}) = \frac{\sigma^2}{n}\left(\frac{N-n}{N-1}\right) \tag{4.10}$$

You will recognize Equation (4.10) from an introductory course as the familiar variance of $\bar{y}$—that is, σ^2/n, multiplied by the factor

$$(N-n)/(N-1)$$

The required sample size can now be found by solving the following equation for n:

$$2\sqrt{V(\bar{y})} = 2\sqrt{\frac{\sigma^2}{n}\left(\frac{N-n}{N-1}\right)} = B \tag{4.11}$$

The solution is given in Equation (4.12).

Sample size required to estimate μ with a bound on the error of estimation B:

$$n = \frac{N\sigma^2}{(N-1)D + \sigma^2} \tag{4.12}$$

where

$$D = \frac{B^2}{4}$$

Solving for n in a practical situation presents a problem because the population variance σ^2 is unknown. Since a sample variance s^2 is frequently available from prior experimentation, we can obtain an approximate sample size by replacing σ^2 with s^2 in Equation (4.12). We will illustrate a method for guessing a value of σ^2 when very little prior information is available. If N is large, as it usually is, the $(N-1)$ can be replaced by N in the denominator of Equation (4.12).

Example 4.5 The average amount of money μ for a hospital's accounts receivable must be estimated. Although no prior data are available to estimate the population variance σ^2, that most accounts lie within a \$100 range is known. There are $N = 1000$ open accounts. Find the sample size needed to estimate μ with a bound on the error of estimation $B = \$3$.

Solution

We need an estimate of σ^2, the population variance. Since the range is often approximately equal to 4 SD (4σ), one-fourth of the range will provide an approximate value of σ. Hence,

$$\sigma \approx \frac{\text{range}}{4} = \frac{100}{4} = 25$$

and

$$\sigma^2 \approx (25)^2 = 625$$

Using Equation (4.12), we obtain

$$n = \frac{N\sigma^2}{(N-1)D + \sigma^2}$$

where

$$D = \frac{B^2}{4} = \frac{(3)^2}{4} = 2.25$$

So

$$n = \frac{1000(625)}{999(2.25) + 625} = 217.56$$

That is, we need approximately 218 observations to estimate μ, the mean accounts receivable, with a bound on the error of estimation of \$3.00.

In like manner, we can determine the number of observations needed to estimate a population total τ, with a bound on the error of estimation of magnitude B. The required sample size is found by setting 2 SD of the estimator equal to B and solving this expression for n. That is, we must solve

$$2\sqrt{V(N\bar{y})} = B$$

or, equivalently,

$$2N\sqrt{V(\bar{y})} = B \qquad (4.13)$$

[The reason for this equivalence is given directly after Equation (4.7).]

Sample size required to estimate τ with a bound on error B:

$$n = \frac{N\sigma^2}{(N-1)D + \sigma^2} \tag{4.14}$$

where

$$D = \frac{B^2}{4N^2}$$

Example 4.6 An investigator is interested in estimating the total weight gain in 0 to 4 weeks for $N = 1000$ chicks fed on a new ration. Obviously, to weigh each bird would be time-consuming and tedious. Therefore, determine the number of chicks to be sampled in this study in order to estimate τ with a bound on the error of estimation equal to 1000 grams. Many similar studies on chick nutrition have been run in the past. Using data from these studies, the investigator found that σ^2, the population variance, was approximately equal to 36.00 $(\text{grams})^2$. Determine the required sample size.

Solution

We can obtain an approximate sample size using Equation (4.14) with σ^2 equal to 36.00 and

$$D = \frac{B^2}{4N^2} = \frac{(1000)^2}{4(1000)^2} = .25$$

That is,

$$n = \frac{N\sigma^2}{(N-1)D + \sigma^2} = \frac{1000(36.00)}{999(0.25) + 36.00} = 125.98$$

The investigator therefore needs to weigh $n = 126$ chicks to estimate τ, the total weight gain for $N = 1000$ chickens in 0 to 4 weeks, with a bound on the error of estimation equal to 1000 grams.

4.5

ESTIMATION OF A POPULATION PROPORTION

The investigator conducting a sample survey is frequently interested in estimating the proportion of the population that possesses a specified characteristic. For example, a congressional leader investigating the merits of an 18-year-old voting age may want to estimate the proportion of the potential voters in the district between the ages of 18 and 21. A marketing research group may be interested in the proportion of the total sales market in diet preparations that is attributable to a particular product. That is, what percentage of sales is accounted for by a particular product? A forest manager may be interested in the proportion of trees with a diameter of 12 inches or more. Television ratings are often determined by estimating the proportion of the viewing public that watches a particular program.

You will recognize that all these examples exhibit a characteristic of the binomial experiment—that is, an observation either does belong or does not belong to the category of interest. For example, one can estimate the proportion of eligible voters in a particular district by examining population census data for several of the precincts within the district. An estimate of the proportion of voters between 18 and 21 years of age for the entire district will be the fraction of potential voters from the precincts sampled that fell into this age range.

In a subsequent discussion, we denote the population proportion and its estimator by the symbols p and $\hat{p}$, respectively. The properties of $\hat{p}$ for simple random sampling parallel those of the sample mean $\bar{y}$ if the response measurements are defined as follows: Let $y_i = 0$ if the ith element sampled does not possess the specified characteristic and $y_i = 1$ if it does. Then the total number of elements in a sample of size n possessing a specified characteristic is

$$\sum_{i=1}^{n} y_i$$

If we draw a simple random sample of size n, the sample proportion $\hat{p}$ is the fraction of the elements in the sample that possess the characteristic of interest. For example, the estimate $\hat{p}$ of the proportion of eligible voters between the ages of 18 and 21 in a certain district is

$$\hat{p} = \frac{\text{number of voters sampled between the ages of 18 and 21}}{\text{number of voters sampled}}$$

or

$$\hat{p} = \frac{\sum_{i=1}^{n} y_i}{n} = \bar{y}$$

In other words, $\hat{p}$ is the average of the 0 and 1 values from the sample. Similarly, we can think of the population proportion as the average of the 0 and 1 values for the entire population (that is, $p = \mu$).

Estimator of the population proportion p:

$$\hat{p} = \bar{y} = \frac{\sum_{i=1}^{n} y_i}{n} \tag{4.15}$$

Estimated variance of $\hat{p}$:

$$\hat{V}(\hat{p}) = \frac{\hat{p}\hat{q}}{n-1}\left(\frac{N-n}{N}\right) \qquad (4.16)$$

where

$$\hat{q} = 1 - \hat{p}$$

Bound on the error of estimation:

$$2\sqrt{\hat{V}(\hat{p})} = 2\sqrt{\frac{\hat{p}\hat{q}}{n-1}\left(\frac{N-n}{N}\right)} \qquad (4.17)$$

Example 4.7 A simple random sample of $n = 100$ college seniors was selected to estimate (1) the fraction of $N = 300$ seniors going on to graduate school and (2) the fraction of students that have held part-time jobs during college. Let y_i and x_i ($i = 1, 2, \ldots, 100$) denote the responses of the ith student sampled. We will set $y_i = 0$ if the ith student does not plan to attend graduate school and $y_i = 1$ if he does. Similarly, let $x_i = 0$ if he has not held a part-time job sometime during college and $x_i = 1$ if he has. Using the sample data presented in the accompanying table, estimate p_1, the proportion of seniors planning to attend graduate school, and p_2, the proportion of seniors who have had a part-time job sometime during their college careers (summers included).

Student	y	x
1	1	0
2	0	1
3	0	1
4	1	1
5	0	0
6	0	0
7	0	1
⋮	⋮	⋮
96	0	1
97	1	0
98	0	1
99	0	1
100	1	1
	$\sum_{i=1}^{100} y_i = 15$	$\sum_{i=1}^{100} x_i = 65$

Solution

The sample proportions from Equation (4.15) are given by

$$\hat{p}_1 = \frac{\sum_{i=1}^{n} y_i}{n} = \frac{15}{100} = .15$$

and

$$\hat{p}_2 = \frac{\sum_{i=1}^{n} x_i}{n} = \frac{65}{100} = .65$$

The bounds on the errors of estimation of p_1 and p_2 are, respectively,

$$2\sqrt{\hat{V}(\hat{p}_1)} = 2\sqrt{\frac{\hat{p}_1\hat{q}_1}{n-1}\left(\frac{N-n}{N}\right)}$$

$$= 2\sqrt{\frac{(.15)(.85)}{99}\left(\frac{300-100}{300}\right)} = 2(0.0293) = .059$$

and

$$2\sqrt{\hat{V}(\hat{p}_2)} = 2\sqrt{\frac{\hat{p}_2\hat{q}_2}{n-1}\left(\frac{N-n}{N}\right)}$$

$$= 2\sqrt{\frac{(.65)(.35)}{99}\left(\frac{300-100}{300}\right)} = 2(.0391) = .078$$

Thus, we estimate that .15 (15%) of the seniors plan to attend graduate school, with a bound on the error of estimation equal to .059 (5.9%). We estimate that .65 (65%) of the seniors have held a part-time job during college, with a bound on the error of estimation equal to .078 (7.8%).

We have shown that the population proportion p can be regarded as the average (μ) of the 0 and 1 values for the entire population. Hence, the problem of determining the sample size required to estimate p to within B units should be analogous to determining a sample size for estimating μ with a bound on the error of estimation B. Recall that the required sample size for estimating μ is given by

$$n = \frac{N\sigma^2}{(N-1)D + \sigma^2} \tag{4.18}$$

where $D = B^2/4$ [see Equation (4.12)]. The corresponding sample size needed to estimate p can be found by replacing σ^2 in Equation (4.18) with the quantity pq.

Sample size required to estimate p with a bound on the error of estimation B:

$$n = \frac{Npq}{(N-1)D + pq} \tag{4.19}$$

where

$$q = 1 - p \qquad \text{and} \qquad D = \frac{B^2}{4}$$

In a practical situation, we do not know p. An approximate sample size can be found by replacing p with an estimated value. Frequently, such an estimate can be obtained from similar past surveys. However, if no such prior information is available, we can substitute $p = .5$ into Equation (4.19) to obtain a conservative sample size (one that is likely to be larger than required).

Example 4.8

Student government leaders at a college want to conduct a survey to determine the proportion of students who favor a proposed honor code. Since interviewing $N = 2000$ students in a reasonable length of time is almost impossible, determine the sample size (number of students to be interviewed) needed to estimate p with a bound on the error of estimation of magnitude $B = .05$. Assume that no prior information is available to estimate p.

Solution

We can approximate the required sample sizes when no prior information is available by setting $p = .5$ in Equation (4.19). We have

$$D = \frac{B^2}{4} = \frac{(.05)^2}{4} = .000625$$

Hence,

$$\begin{aligned} n &= \frac{Npq}{(N-1)D + pq} \\ &= \frac{(2000)(.5)(.5)}{(1999)(.000625) + (.5)(.5)} = \frac{500}{1.499} \\ &= 333.56 \end{aligned}$$

That is, 334 students must be interviewed to estimate the proportion of students who favor the proposed honor code with a bound on the error of estimation of $B = .05$.

Example 4.9

Referring to Example 4.8, suppose that, in addition to estimating the proportion of students who favor the proposed honor code, student government leaders also want to estimate the number of students who feel the student union building adequately serves their needs. Determine the combined sample size required for a survey to estimate p_1, the proportion that favors the proposed honor code, and p_2, the proportion that believes the student union adequately serves its needs, with bounds on the errors of estimation of magnitude $B_1 = .05$ and $B_2 = .07$. Although no prior information is available to estimate p_1, approximately 60% of the students believed the union adequately met their needs in a similar survey run the previous year.

Solution

In this example, we must determine a sample size n that will allow us to estimate p_1 with a bound $B_1 = .05$ and p_2 with a bound $B_2 = .07$. First, determine the sample sizes that satisfy each objective separately. The larger of the two will then be the combined sample size for a survey to meet both objectives. From Example 4.8, the sample size required to estimate p_1 with a bound on the error of estimation of $B_1 = .05$ was $n = 334$ students. We can use data from the survey of the previous year to determine the sample size needed to estimate p_2. We have

$$D = \frac{B^2}{4} = \frac{(.07)^2}{4} = .001225$$

and hence, with $p_2 = .60$,

$$\begin{aligned} n &= \frac{Npq}{(N-1)D + pq} \\ &= \frac{(2000)(.6)(.4)}{(1999)(.001225) + (.6)(.4)} = \frac{480}{2.68877} \\ &= 178.52 \end{aligned}$$

That is, 179 students must be interviewed to estimate p_2, the proportion of the $N = 2000$ students who believe the student union meets its needs, with a bound on the error of estimation equal to .07.

The sample size required to achieve both objectives in one survey is 334, the larger of the two sample sizes.

4.6

COMPARING ESTIMATES

After estimating individual parameters, such as means or proportions, we often want to compare the estimates of two parameters. The mean incomes for two ethnic groups over the past year can be compared by looking at the difference between sample means for random samples of incomes from the two groups. Whether or not Republicans are gaining on Democrats in a congressional race can be assessed by looking at the difference between the proportions voting Republican (or, at least, reporting so) for two polls conducted a few weeks apart.

Whether using means or proportions, comparisons are usually made by way of differences. In this regard, a general result will be stated before looking at specific cases. For any two random variables, y_1 and y_2,

$$E(y_1 - y_2) = E(y_1) - E(y_2)$$

and

$$V(y_1 - y_2) = V(y_1) + V(y_2) - 2\,\mathrm{cov}(y_1, y_2)$$

If y_1 and y_2 are independent, then $\mathrm{cov}(y_1, y_2) = 0$.

Suppose $y_1, \ldots, y_n$ is a random sample from a population with mean μ_y and $x_1, \ldots, x_m$ is an independent random sample from a population with mean μ_x. Then $\mu_y - \mu_x$ can be estimated unbiased by $\bar{y} - \bar{x}$ because

$$E(\bar{y} - \bar{x}) = E(\bar{y}) - E(\bar{x}) = \mu_y - \mu_x$$

Further,

$$V(\bar{y} - \bar{x}) = V(\bar{y}) + V(\bar{x})$$

where the variance of each sample mean could be estimated as in Equation (4.3). If the two samples are not independent, some information on $\mathrm{cov}(\bar{y}, \bar{x})$ would have to be available.

Example 4.10

How do small, compact, and midsized cars compare with regard to price, gas mileage, and horsepower? Table 4.5 shows typical values for these three variables for samples of 10 small cars (size = 1), 5 compact cars (size = 2), and 11 midsized cars (size = 3). The prices are billed as midprices for each car. Let's concentrate on highway miles per gallon ('Highway mpg') for now. How do these three car sizes compare on the mean highway miles per gallon?

Solution

Accurate variance computation depends on knowledge of the population size as well as the sample size, so it is important to know that the 10 small cars were sampled from a list of 20 popular small cars, the 5 compacts were sampled from a list of 16 compacts, and the 11 midsized cars were sampled from a list of 21 such cars. The sampled populations, then, are the popular varieties of these sizes of cars.

It is good practice to plot the data before completing an analysis. Figure 4.2 shows box plots for the gas-mileage data. What do these plots show?

TABLE 4.5 **Price (in thousands of dollars), highway mpg, and horsepower for three sizes of cars (1993)**

Row	Make and Model	Size	Price	Highway mpg	Horsepower
1	Acura Integra	1	15.9	31	140
2	Dodge Colt	1	9.2	33	92
3	Honda Civic	1	12.1	46	102
4	Hyundai Excel	1	8.0	33	81
5	Mazda 323	1	8.3	37	82
6	Mitsubishi Mirage	1	10.3	33	92
7	Subaru Justy	1	8.4	37	73
8	Suzuki Swift	1	8.6	43	70
9	Toyota Tercel	1	9.8	37	82
10	Volkswagen Fox	1	9.1	33	81
11	Chevrolet Corsica	2	11.4	34	110
12	Chrysler LeBaron	2	15.8	28	141
13	Ford Tempo	2	11.3	27	96
14	Mazda 626	2	16.5	34	164
15	Subaru Legacy	2	19.5	30	130
16	BMW 535i	3	30.0	30	208
17	Chevrolet Lumina	3	15.9	29	110
18	Dodge Dynasty	3	15.6	27	100
19	Ford Taurus	3	20.2	30	140
20	Hyundai Sonata	3	13.9	27	128
21	Lexus SC300	3	35.2	23	225
22	Mitsubishi Diamante	3	26.1	24	202
23	Nissan Maxima	3	21.5	26	160
24	Pontiac Grand Prix	3	18.5	27	200
25	Toyota Camry	3	18.2	29	130
26	Volvo 850	3	26.7	28	168

	Size	N	Mean	Median	SD	SEMEAN
Price	1	10	9.970	9.150	2.408	0.762
	2	5	14.90	15.80	3.53	1.58
	3	11	21.98	20.20	6.72	2.03
Highway mpg	1	10	36.30	35.00	4.85	1.54
	2	5	30.60	30.00	3.29	1.47
	3	11	27.273	27.000	2.284	0.689
Horsepower	1	10	89.50	82.00	20.08	6.35
	2	5	128.2	130.0	26.5	11.9
	3	11	161.0	160.0	42.9	12.9

Without belaboring computational details the pertinent information on these three samples of highway gas mileages are the following:

Size	n_i	N_i	$\bar{y}_i$	$\hat{V}(\bar{y}_i)$
1	10	20	36.30	1.178
2	5	16	30.60	1.485
3	11	21	27.27	0.226

FIGURE 4.2 Box plots of gas mileage for three car sizes

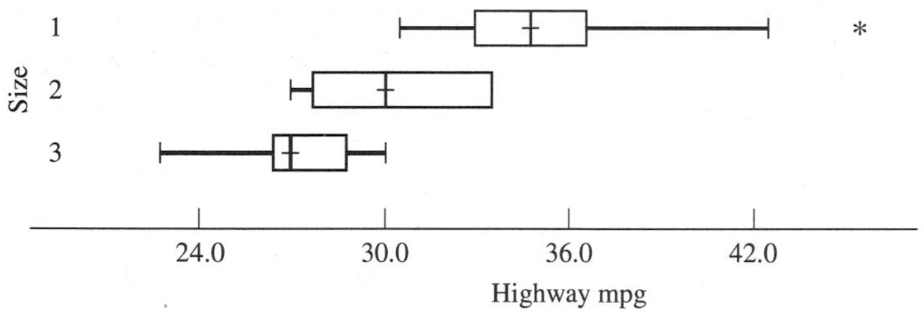

To compare the gas mileage for small and compact cars, the appropriate estimator and bound is

$$(\bar{y}_1 - \bar{y}_2) \pm 2\sqrt{\hat{V}(\bar{y}_1) + \hat{V}(\bar{y}_2)}$$
$$(36.3 - 30.6) \pm 2\sqrt{1.178 + 1.485}$$
$$5.7 \pm 3.3$$

Any true difference in expected gas mileages between 2.4 mpg and 9.0 mpg would be consistent with these samples.

To compare compact and midsized cars, the appropriate interval estimate is

$$(\bar{y}_2 - \bar{y}_3) \pm 2\sqrt{V(\bar{y}_2) + V(\bar{y}_3)}$$
$$(30.6 - 27.3) \pm 2\sqrt{1.485 + 0.226}$$
$$3.3 \pm 2.6$$

In moving from a compact to a midsize car, could one expect to realize less of a reduction in gas mileage?

When comparing means, we will consider only the independent-sample case because the dependent case becomes too complicated to handle at this level. When comparing proportions, however, a commonly occurring dependent situation does have a rather simple solution. Suppose an opinion poll asks "Do you favor the U.S. foreign policy in Latin America?" The options given are

Yes ______
No ______
No opinion ______

In the face of a sizeable number of "no opinions," you will want to compare the proportion checking "yes" with the proportion checking "no" by looking at the difference between these *dependent proportions*. To see this, suppose we are allowed to see the proportion of yes responses and it is high—say, about 80%. Then we know that the proportion of no responses must be low, below 20%. Thus, these two sample percentages are not only dependent but also negatively correlated (or have negative covariance).

This is a multinomial sampling situation because each respondent had three choices (more than two is the key). For the two sample proportions $\hat{p}_1$ and $\hat{p}_2$ arising from a multinomial sample of size n,

$$E(\hat{p}_1 - \hat{p}_2) = p_1 - p_2$$

and

$$V(\hat{p}_1 - \hat{p}_2) = V(\hat{p}_1) + V(\hat{p}_2) - 2\,\text{cov}(\hat{p}_1, \hat{p}_2)$$
$$= \frac{p_1(1-p_1)}{n} + \frac{p_2(1-p_2)}{n} + 2\frac{p_1 p_2}{n}$$

(Why do we get a + sign in front of the third term?) Because p_i is estimated by $\hat{p}_i$, it is easy to find a good estimator of this variance.

Example 4.11

Should smoking be banned from the workplace? A Time/Yankelovich poll of 800 adult Americans carried out on April 6–7, 1994, gave the following results (see *Time*, Apr. 18, 1994):

Should smoking be banned from workplaces, should there be special smoking areas, or should there be no restrictions?

	Nonsmokers	Smokers
Banned	44%	8%
Special areas	52%	80%
No restrictions	3%	11%

Based on a sample of approximately 600 nonsmokers and 200 smokers, estimate the following:

A. The true difference between the proportions choosing "banned"
B. The true difference between the proportions of nonsmokers choosing "banned" and "special areas"

Solution

A. The proportions choosing "banned" are independent of each other; a high value of one does not force a low value of the other. Thus, an appropriate estimate of this difference is

$$(.44 - .08) \pm 2\sqrt{\frac{(.44)(.56)}{600} + \frac{(.08)(.92)}{200}}$$
$$.36 \pm .06$$

Thus, a true difference anywhere between 30% and 42% would be consistent with these data.

B. The proportion of nonsmokers choosing "special areas" is dependent on the proportion choosing "banned"; if the latter is large, the former must be small. These are multinomial proportions. Thus, an appropriate estimate of the true difference is

$$(.52 - .44) \pm 2\sqrt{\frac{(.44)(.56)}{600} + \frac{(.52)(.48)}{600} + 2\frac{(.44)(.52)}{600}}$$

$$.08 \pm .08$$

Any true difference between 0 and 16% would be consistent with these data. In particular, there is no strong evidence to say that, for the population, these two proportions would differ.

It has been noted that the 2-SD bound on the error is equivalent to an approximate 95% confidence interval estimate. If two such intervals are used on the same data, however, the chance that *both* intervals contain the true parameter value is less than 95%. Thus, many interval estimates should *not* be constructed from the same sample data unless an adjustment is made to account for this multiplicity of uses. Most standard statistical methods books present a variety of such adjustments, but they will not be covered here. Instead, we will follow the philosophy of making only a few meaningful comparisons, rather than making all possible comparisons, when a situation calls for comparisons to be made.

Example 4.12

A survey of 209 licensed clinical psychologists randomly selected from all such professionals in the state of Michigan was used to study the effects of age and gender on the diagnosis of symptoms related to a feeling of depression in patients. All psychologists read the same case study, except for changes in patient age and gender. The resulting diagnoses are reported in Table 4.6.

A. Does age appear to affect the diagnosis of depression (as opposed to a diagnosis of an organic cause)?

B. Does gender appear to affect the diagnosis of depression?

Solution

A. We can make many comparisons to look at the effect of age on diagnosis. To keep these to a minimum, consider just the extremes: age 45 versus age 85. Pooling across gender, 39 of 42, or 93%, of the 45-year-olds were diagnosed as depressed, whereas 25 of 36, or 69%, of the 85-year-olds were so

TABLE 4.6 **General diagnosis by age and gender of patient**

	Age (years)											
	45		**55**		**65**		**75**		**85**		**Total**	
Gender	*n*	%	*n*	%	*n*	%	*n*	%	*n*	%	*n*	%
Male												
Depression	22	95.7	9	64.0	12	80	8	57.1	15	75	66	76.7
Organic	0	0	0	0	1	6.7	4	28.6	2	14.3	7	8.1
Adjustment	1	4.3	5	37.7	2	13.3	2	14.3	3	15.0	13	15.1
Total	23	26.7	14	16.3	15	17.4	14	16.3	20	23.3	—	—
Female												
Depression	17	89.5	18	81.8	14	87.5	18	94.7	10	62.5	77	83.7
Organic	0	0	0	0	0	0	0	0	1	6.3	1	1.1
Adjustment	2	10.5	4	18.2	2	12.5	1	6.3	5	31.3	14	15.2
Total	19	20.7	22	23.9	16	17.4	19	20.7	16	17.4	—	—

Source: *Professional Psychology: Research & Practice* 24, no. 2 (1993): 206–212.

diagnosed. The appropriate interval estimate is

$$(.93 - .69) \pm 2\sqrt{\frac{(.93)(.07)}{42} + \frac{(.69)(.31)}{36}}$$
$$.24 \pm .17$$

There is statistical evidence to say that the 85-year-olds are less likely to receive a diagnosis of depression.

B. At first glance, it looks like women may be more likely to be diagnosed as depressed than men might be, but the interval estimate of this difference in proportions is

$$(.77 - .84) \pm 2\sqrt{\frac{(.77)(.23)}{66} + \frac{(.84)(.16)}{77}}$$
$$-.07 \pm .13.$$

Women may not be more often diagnosed as depressed in the population of such patients. This observed difference could just be due to sampling error.

4.7 SUMMARY

The objective of statistics is to make inferences about a population from information contained in a sample. Two factors affect the quantity of information in a given investigation. The first is the sample size. The larger the sample size, the more information we expect to obtain about the population. The second factor is the amount of variation in the data. Variation can be controlled by the design of the sample survey, that is, the method by which observations are obtained.

In this chapter, we discussed the simplest type of sample survey design, namely, simple random sampling. This design does not attempt to reduce the effect of data variation on the error of estimation. A simple random sample of size n occurs if each sample of n elements from the population has the same chance of being selected. Random number tables are quite useful in determining the elements that are to be included in a simple random sample.

In estimating a population mean μ and total τ, we use the sample mean $\bar{y}$ and sample total $N\bar{y}$, respectively. Both estimators are unbiased; that is, $E(\bar{y}) = \mu$ and $E(N\bar{y}) = \tau$. The estimated variance and the bound on the error of estimation are given for both estimators.

Sometime during the design of an actual survey, the experimenter must decide how much information is desired, that is, how large a bound on the error of estimation can be tolerated. Sample-size requirements were presented for estimating μ and τ with a specified bound on the error of estimation.

The third parameter estimated was the population proportion p. The properties of $\hat{p}$ were presented and related to the properties of $\bar{y}$, the estimator of the population mean μ. Selecting the sample size to estimate p with a specified bound on the error of estimation was based on the same principle employed in selecting a sample size for estimating μ and τ.

Sometimes it is important to make comparisons among means or proportions by estimating differences. The variances of these differences are easily estimated for independent samples. The one dependent case considered was in comparing multinomial proportions, the type often arising in sample surveys.

CASE STUDY REVISITED

THE GALLUP POLL

If the poll discussed in the opening case study consists of 1200 randomly selected adults, then about half of the respondents should be women and half men. Thus, sample sizes of 600 will be used for responses to questions just from men or women. "Has the proportion of men who drink decreased significantly from 1992 to 1994?" is a question about two nearly independent proportions. The estimate and bound on the error is found by

$$(.72 - .70) \pm 2\sqrt{\frac{(.72)(.28)}{600} + \frac{(.70)(.30)}{600}}$$
$$.02 \pm .05$$

and, hence, there is no significant decrease.

A similar estimate and bound for the women's proportions yields an interval of $-.04 + .06$, and so there is no significant increase in the women's proportions either.

Comparison of the proportions who have decreased drinking prior to 1994 with those who decreased drinking prior to 1984 involves a difference of independent proportions again, but here the sample size is 1200 in both groups. The estimate and bound is given by

$$(.41 - .29) \pm 2\sqrt{\frac{(.41)(.59)}{1200} + \frac{(.29)(.71)}{1200}}$$
$$.12 \pm .04$$

This does show a significant increase in the proportions who have decreased drinking and, in fact, justifies the headline to the article in which these data appeared: "Number of Drinkers Holding Steady, But Drinking Less."

Comparison of the proportions who have decreased drinking over the past five years with those who have stayed the same (both for 1994) involves a difference of dependent proportions. The estimate and bound on the error is given by

$$(.41 - .51) \pm 2\sqrt{\frac{(.41)(.59)}{1200} + \frac{(.51)(.49)}{1200} + \frac{2(.41)(.51)}{1200}}$$
$$-.10 \pm .06$$

Because the interval does not overlap zero, the difference between the sample proportions appears to reflect a real difference between the population proportions.

EXERCISES

4.1 List all possible simple random samples of size $n = 2$ that can be selected from the population $\{0, 1, 2, 3, 4\}$. Calculate σ^2 for the population and $V(\bar{y})$ for the sample.

4.2 For the simple random samples generated in Exercise 4.1, calculate s^2 for each sample. Show numerically that

$$E(s^2) = \frac{N}{N-1}\sigma^2$$

4.3 Suppose you were to estimate the number of weed clusters of a certain type in a field. What is the population, and what would you use for sampling units? How would you construct a frame? How would you select a simple random sample? If a sampling unit is an area such as a square yard, does the size chosen for a sampling unit affect the accuracy of the results? What considerations would go into our choice of size of sampling unit?

4.4 The data set in Appendix Table 3 lists resident population figures per state from the 1990 census. Select a simple random sample of five states. Use the 1990 population figures for the sampled states to estimate the total U.S. population and place a bound on the error of estimation. Does your interval answer include the total population figure given in the table? Do you think every possible 95% confidence interval based on samples of size 5 would include the true total? Why?

4.5 State park officials were interested in the proportion of campers who consider the campsite spacing adequate in a particular campground. They decided to take a simple random sample of $n = 30$ from the first $N = 300$ camping parties that visit the campground. Let $y_i = 0$ if the head of the ith party sampled does not think the campsite spacing is adequate and $y_i = 1$ if he does $(i = 1, 2, \ldots, 30)$. Use the data in the accompanying table to estimate p, the proportion of campers who consider the campsite spacing adequate. Place a bound on the error of estimation.

Camper sampled	Response, y_i
1	1
2	0
3	1
⋮	⋮
29	1
30	1
	$\sum_{i=1}^{30} y_i = 25$

4.6 Use the data in Exercise 4.5 to determine the sample size required to estimate p with a bound on the error of estimation of magnitude $B = .05$.

4.7 A simple random sample of 100 water meters within a community is monitored to estimate the average daily water consumption per household over a specified dry spell. The sample mean and sample variance are found to be $\bar{y} = 12.5$ and $s^2 = 1252$. If we assume that there are $N = 10{,}000$ households within the community, estimate μ, the true average daily consumption, and place a bound on the error of estimation.

4.8 Using data in Exercise 4.7, estimate the total number of gallons of water, τ, used daily during the dry spell. Place a bound on the error of estimation.

4.9 Resources managers of forest game lands are concerned about the size of the deer and rabbit populations during the winter months in a particular forest. As an estimate of population size, they propose using the average number of pellet groups for rabbits and deer per 30-foot-square plots. From an aerial photograph, the forest was divided into $N = 10{,}000$ 30-foot-square grids. A simple random sample of $n = 500$ plots was taken, and the number of pellet groups was observed for rabbits and for deer. The results of this study are summarized in the accompanying table. Estimate μ_1 and μ_2, the average number of pellet groups for deer and rabbits, respectively, per 30-square-foot plots. Place bounds on the errors of estimation.

Deer	Rabbits
Sample mean = 2.30	Sample mean = 4.52
Sample variance = .65	Sample variance = .97

4.10 A simple random sample of $n = 40$ college students was interviewed to determine the proportion of students in favor of converting from the semester to the quarter system. Twenty-five of the students answered affirmatively. Estimate the proportion of students on campus in favor of the change. (Assume $N = 2000$.) Place a bound on the error of estimation.

4.11 A dentist was interested in the effectiveness of a new toothpaste. A group of $N = 1000$ schoolchildren participated in a study. Prestudy records showed there was an average of 2.2 cavities every six months for the group. After three months on the study, the dentist sampled $n = 10$ children to determine how they were progressing on the new toothpaste. Using the data in the accompanying table, estimate the mean number of cavities for the entire group and place a bound on the error of estimation.

Child	Number of cavities in the three-month period
1	0
2	4
3	2
4	3
5	2
6	0
7	3
8	4
9	1
10	1

4.12 The Fish and Game Department of a particular state was concerned about the direction of its future hunting programs. To provide for a greater potential for future hunting, the department wanted to determine the proportion of hunters seeking any type of game bird. A simple random sample of $n = 1000$ of the $N = 99{,}000$ licensed hunters was obtained. Suppose 430 indicated they hunted game birds. Estimate p, the proportion of licensed hunters seeking game birds. Place a bound on the error of estimation.

4.13 Using the data in Exercise 4.12, determine the sample size the department must obtain to estimate the proportion of game bird hunters, given a bound on the error of estimation of magnitude $B = .02$.

4.14 A company auditor was interested in estimating the total number of travel vouchers that were incorrectly filed. In a simple random sample of $n = 50$ vouchers taken from a group of $N = 250$, 20 were filed incorrectly. Estimate the total number of vouchers from the $N = 250$ that have been filed incorrectly and place a bound on the error of estimation. [*Hint*: If p is the population proportion of incorrect vouchers, then Np is the total number of incorrect vouchers. An estimator of Np is $N\hat{p}$, which has an estimated variance given by $N^2\hat{V}(\hat{p})$.]

4.15 A psychologist wishes to estimate the average reaction time to a stimulus among 200 patients in a hospital specializing in nervous disorders. A simple random sample of $n = 20$ patients was selected, and their reaction times were measured, with the following results:

$$\bar{y} = 2.1 \text{ seconds} \qquad s = .4 \text{ second}$$

Estimate the population mean μ and place a bound on the error of estimation.

4.16 In Exercise 4.15, how large a sample should be taken in order to estimate μ with a bound of 1 second on the error of estimation? Use 1.0 second as an approximation of the population standard deviation.

4.17 The manager of a machine shop wishes to estimate the average time an operator needs to complete a simple task. The shop has 98 operators. Eight operators are selected at random and timed. The observed results are shown in the accompanying table. Estimate the average time for completion of the task among all operators and place a bound on the error of estimation.

Time (in minutes)	
4.2	5.3
5.1	4.6
7.9	5.1
3.8	4.1

4.18 A sociological study conducted in a small town calls for the estimation of the proportion of households that contain at least one member over 65 years of age. The city has 621 households according to the most recent city directory. A simple random sample of $n = 60$ households was selected from the directory. At the completion of the fieldwork, out of the 60 households sampled, 11 contained at least one member over 65 years of age. Estimate the true population proportion p and place a bound on the error of estimation.

4.19 In Exercise 4.18, how large a sample should be taken in order to estimate p with a bound of .08 on the error of estimation? Assume the true proportion p is approximately .2.

4.20 An investigator is interested in estimating the total number of "count trees" (trees larger than a specified size) on a plantation of $N = 1500$ acres. This information is used to determine the total volume of lumber for trees on the plantation. A simple random sample of $n = 100$ 1-acre plots was selected, and each plot was examined for the number of count trees. The sample average for the $n = 100$ 1-acre plots was $\bar{y} = 25.2$ with a sample variance of $s^2 = 136$. Estimate the total number of count trees on the plantation. Place a bound on the error of estimation.

4.21 Using the results of the survey described in Exercise 4.20, determine the sample size required to estimate τ, the total number of trees on the plantation, with a bound on the error of estimation of magnitude $B = 1500$.

4.22 A large construction firm has 120 houses in various stages of completion. For estimation of the total dollar amount to be listed as inventory of construction in progress, a simple random sample of 12 of these houses is selected and accumulated costs determined on each. Assume the following costs were obtained for the 12 sample houses:

35,500	30,200	28,900
36,400	29,800	34,100
32,600	26,400	38,000
38,200	32,200	27,500

Estimate the total accumulated costs for the 120 houses and place a bound on the error of estimation.

4.23 A survey of 811 registered voters in Florida, taken at the end of September 1994, reported that 57% of the respondents are against legalized casino gambling in Florida. The article reporting the survey states that the sampling error is 3.5%.

(a) Is this the correct sampling error to report?

(b) Can a valid statistical argument be made that "most Floridians disapprove of legalizing casinos in the state?" Explain.

(c) Proponents of the casinos say that their poll shows only a little over one-third of the people in the state oppose casinos. What could cause this difference between the two poll results?

4.24 The definition of "being successful" appears to be changing. A Roper poll of 1027 adults, reported in *Working Woman* (Nov. 1993), revealed that 53% of those interviewed had changed their definition of success over the past five years to depend less on money and job and more on family and friends.

(a) Do these results justify saying that a majority of Americans have changed their definition of success? Justify your answer statistically.

(b) The poll also reports that 70% of the women and 58% of the men now want to spend more time with family. Assuming that the 1027 adults constitute a random sample, what interval would you use to estimate the true proportion of women who want to spend more time with family?

(c) In the poll, 81% of the Hispanics and 73% of the blacks said they felt more successful at the time of the poll than they did five years previously. Given that about 10% of the U.S. population is Hispanic and about 13% is black, what sampling error would you use for each of these sample percentages?

4.25 The Gallup poll explains sampling error in the following way:

> The chance that sampling error will affect a percentage based on survey results is mainly dependent upon the number of interviews on which the percentage is based. In ninety-five out of 100 cases, results based on national samples of 1000 interviews can be expected to vary by no more than 4 percentage points (plus or minus the figure obtained) from the results that would be obtained if all qualified adults were interviewed in the same way. For results based on smaller national samples or on sub-samples (such as men or persons over the age of fifty), the chance of sampling error is greater and therefore larger margins of sampling error are necessary in order to be equally confident of survey conclusions.

Do you agree with this explanation? Can you suggest any wording changes that might improve it?

4.26 The A. C. Nielsen company has electronic monitors hooked up to about 4000 of the 95 million American homes. The data from the monitors provide estimates of the proportion of homes tuned to a particular TV program. Nielsen offers the following defense of this sample size (D. Cody, "Polls and Pollsters," *Sky*, Oct. 1982, p. 116):

> Mix together 70,000 white beans and 30,000 red beans and then scoop out a sample of 1000. The mathematical odds are that the number of red beans will be between 270 and 330, or 27 to 33% of the sample, which translated to a "rating" of 30, plus or minus three, with a 20 to 1 assurance of statistical reliability. The basic statistical law wouldn't change even if the sampling came from 80 million beans rather than 100,000.

Interpret and justify this statement in terms of the results of this chapter.

4.27 Reggie Jackson, one of the great baseball batters of modern times, was called Mr. October by his fans because he seemed to excel in the World Series. During the regular season, Reggie had 2584 hits in 9864 official at bats for a batting average (proportion of hits) of .262. During League Championship Series, he had 37 hits in 163 official at bats for a .227 batting average. During World Series play, he had 35 hits in 98 official at bats for a batting average of .357. Do you think the nickname is justified, based on the data? Explain your reasoning.

4.28 Do people view Congress differently from the way they view their representatives to Congress? A poll of 1161 adults produced the results shown in Figure 4.3, as reported in the *New York Times* (Sept. 13, 1994). Provide an answer to the question posed above, justifying your answer from the data in the results of the poll.

4.29 An auditor detects that a certain firm is regularly overstating the dollar amounts of inventories because of delays in recording withdrawals. The auditor wants to estimate the total *overstated* amount on 1000 listed items by obtaining exact (audited) inventory amounts on a random sample of 15 items and comparing these exact figures with the recorded amounts. The data for the sampled items are as shown in the following table (all data in dollars). Estimate the total overstated amount on the 1000 types of items and place a bound on the error of estimation. (Ignore the fpc.)

Item number	Audited amount	Recorded amount	Overstatement (difference)
1	175	210	35
2	295	305	10
3	68	91	23
4	74	82	8

(continued)

(*continued*)

Item number	Audited amount	Recorded amount	Overstatement (difference)
5	128	140	12
6	241	250	9
7	362	384	22
8	72	80	8
9	59	82	23
10	112	140	28
11	118	124	6
12	210	230	20
13	240	260	20
14	123	147	24
15	96	108	12

FIGURE 4.3

How the Public Views Congress as a Whole...

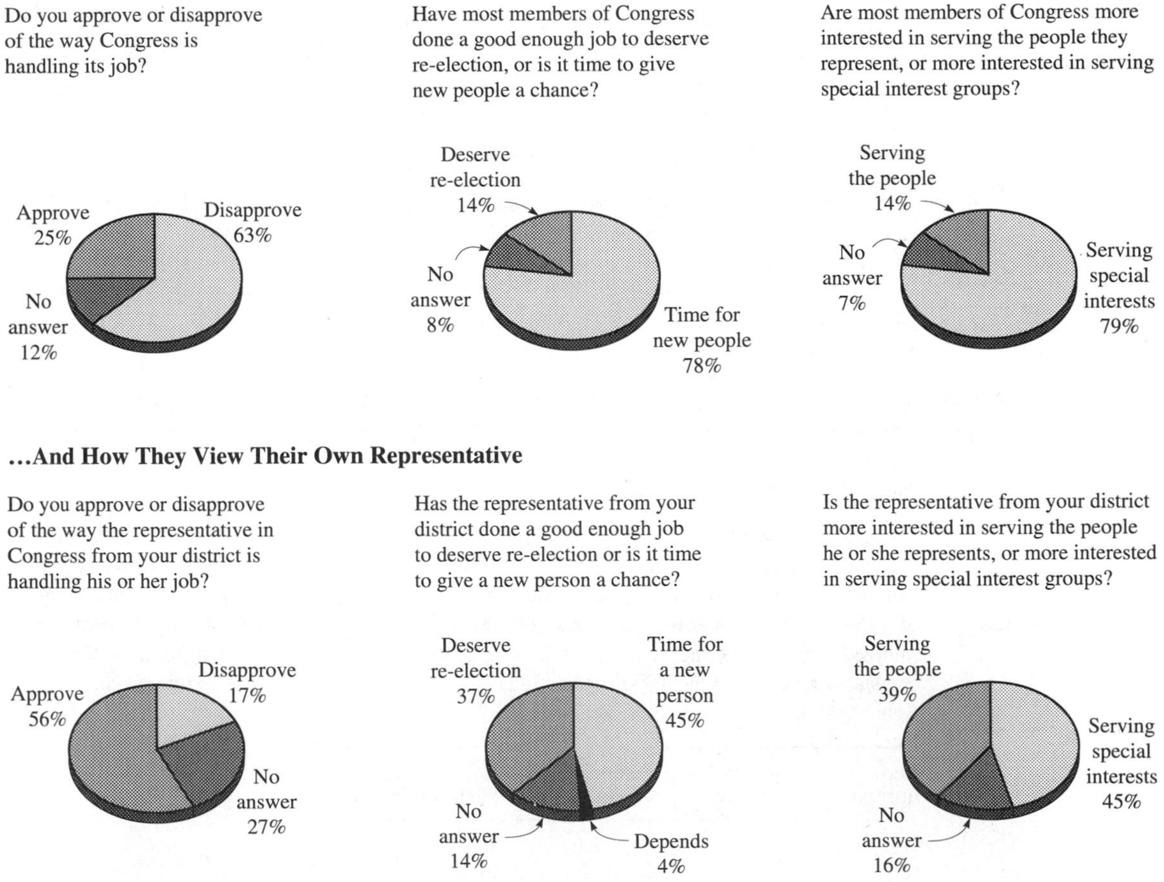

Based on nationwide telephone interviews conducted Sept. 8 to 11 with 1,161 adults.

Source: © 1994 The New York Times Company, Reprinted by permission

4.30 An auditor randomly samples 20 accounts receivable from the 500 accounts of a certain firm. The auditor lists the amount of each account and checks to see whether the underlying documents are in compliance with stated procedures. The data are as follows (amounts in dollars, Y = yes, N = no):

Account	Amount	Compliance
1	278	Y
2	192	Y
3	310	Y
4	94	N
5	86	Y
6	335	Y
7	310	N
8	290	Y
9	221	Y
10	168	Y
11	188	N
12	212	N
13	92	Y
14	56	Y
15	142	Y
16	37	Y
17	186	N
18	221	Y
19	219	N
20	305	Y

Estimate the total accounts receivable for the 500 accounts of the firm and place a bound on the error of estimation. Do you think that the *average* account receiv-able for the firm exceeds $250? Why?

4.31 Refer to Exercise 4.30. From the data given on the compliance checks, estimate the proportion of the firm's accounts that fail to comply with stated procedures. Place a bound on the error of estimation. Do you think the proportion of accounts that comply with stated procedures exceeds 80%? Why?

4.32 Refer to the data on cars of three sizes as seen in Table 4.5 and used in Example 4.10.
(a) Construct box plots to compare the price data for the three groups.
(b) Construct an estimate, with error bounds, for the difference in mean price between small and compact cars.
(c) Construct an estimate, with error bounds, for the difference in mean price between compact and midsized cars.
(d) Construct box plots to compare the horsepower data for the three groups.
(e) Choose one meaningful comparison to make with regard to mean horsepower and construct an appropriate estimate with a measure of error.

4.33 The number of days per year of poor air quality (PSI over 100) is shown below for a sample of 15 sites in each of three years. (These data were studied in Exercise 2.10.)

Sites	1981	1985	1990
Atlanta	9	9	16
Boston	0	2	1
Chicago	3	6	0
Dallas	12	11	5
Denver	4	0	0
Detroit	12	1	3
Houston	33	30	35
Kansas City	4	3	2
Los Angeles	152	153	116
New York	21	13	8
Philadelphia	22	25	11
Pittsburgh	9	2	0
San Francisco	1	1	0
Seattle	1	0	2
Washington	12	12	5
Total	295	268	204

Source: U.S. Environmental Protection Agency, Office of Air Quality and Standards.

(a) Does the mean air quality for the United States (as measured by PSI days) appear to improve between 1981 and 1985?
(b) Does the mean air quality for the United States appear to improve between 1985 and 1990?
(c) Recalculate your answers to parts (a) and (b) with Los Angeles removed from the data set.
(d) Recalculate your answers to parts (a) and (b) with both Los Angeles and Houston removed from the data set.
(e) What do you think is the fairest way to report the trend from 1981 to 1985 to 1990?

4.34 The major league baseball season came to an abrupt end in the middle of 1994. In a poll of 600 adult Americans (*Time*, Aug. 22, 1994), 29% blamed the players for this strike, 34% blamed the owners, and the rest held various other opinions. Does evidence suggest that the true proportions who blame players and owners, respectively, are really different?

4.35 "If the election for Congress were being held today, would you vote for the Democratic candidate or the Republican candidate?" This question was asked of samples of 800 adult Americans in two separate polls, with results as shown below:

	Aug. 17–18, 1994	Aug. 31–Sept. 1, 1994
Democrat	42%	38%
Republican	34%	40%

Source: *Time*, Sept. 12, 1994.

(a) Is there statistical evidence that Republicans had made a significant gain in this time period?
(b) Can one say with conviction that Republican votes would outnumber Democratic votes as of September 1, 1994?

4.36 The final polls before the presidential election of 1994 for four polling agencies showed the following results:

Poll	Size	Clinton	Bush	Perot	Undecided
CNN	1562	44	36	14	6
Gallup	1579	43	36	15	6
Harris	1675	44	39	17	0
ABC	1369	42	37	17	4
Actual		43	38	19	

All declared Clinton the winner. Were they justified in this claim?

4.37 On the other hand, based on a survey of 773 registered voters in Florida, the Florida poll said the race between Clinton and Bush was "too close to call." At that time, 39% of those polled favored Bush, and 37% favored Clinton. Was the Florida poll correct in its interpretation of the data?

4.38 Do you agree with the conclusions reached in the following article? Justify your answer statistically.

> NEW YORK—A poll comparing men's and women's attitudes on public issues suggests women want stricter law enforcement against drunken driving, guns and drug dealing.
>
> The poll was commissioned by Life magazine for a story in the June issue headlined "If Women Ran America." Life said its poll found women interested in "safety first. But fairness also, especially fairness for women at work."
>
> Two in three women polled considered unequal pay for the same work to be a very serious problem for women in the workplace. Just half the mean responded similarly.
>
> Half the women but only a third of the men think discrimination in promotions is a very serious problem for women at work.
>
> The poll said 78 percent of women, compared with 64 percent of men, think businesses should be required to provide paid maternity leave.
>
> The poll was taken by the Gallup Organization, which surveyed a national sample of 614 women and 608 men by phone March 30–April 5.
>
> The margin of error ranges from plus or minus 3 percentage points for the whole sample, up to 6 points for comparisons of results of men and women.
>
> In other words, the poll indicates a gender gap, rather than chance variation, accounts for differences of opinion such as this: 55 percent of women but only 46 percent of men said the government should make fighting crime and violence an extremely important priority.
>
> Seventy-six percent of women and only 58 percent of men said the justice system

wasn't tough enough on drunken drivers. On drug dealing, 88 percent of women and 77 percent of men wanted the system to be tougher. Seventy percent of women and 63 percent of men wanted to be tougher on illegal gun possession.

Women were more compassionate than men on some issues: 85 percent would approve of a law requiring businesses to allow employees an unpaid 12-week family medical leave. Women were more likely than men to approve of such a leave for homosexual couples, and to say they would vote for a gay candidate. (*Source*: *Gainesville Sun*, May 5, 1992.)

4.39 A study to assess the attitudes of accountants toward advertising their services involved sending questionnaires to 200 accountants selected from a list of 1400 names. A total of 82 usable questionnaires were returned. The data summary for one question is as follows:

Likelihood of advertising in the future (%)

	All respondents (82)	Those having advertised in the past (46)
Virtual certainty	22	35
Very likely	4	5
Somewhat likely	19	35
About 50–50	18	15
Somewhat unlikely	6	10
Very unlikely	12	0
Absolutely not	15	0
No response	4	0

Source: K. Traynor, "Accountant Advertising: Perceptions, Attitudes and Behaviors," *Journal of Advertising Research* 23, no. 6 (1984). Copyright © 1984 by the Advertising Research Foundation.

(a) Estimate the population proportion virtually certain to advertise in the future.
(b) Estimate the population proportion having *at least* a 50–50 chance of advertising in the future.
(c) Among those who advertised in the past, estimate the population proportion somewhat unlikely to advertise again.
(d) Among those who advertised in the past, estimate the population proportion having *at least* a 50–50 chance of advertising again.

Place bounds on the errors of estimation in all cases. Do parts (c) and (d) require further assumptions over those made for parts (a) and (b)?

4.40 A marketing research firm estimates the proportion of potential customers preferring a certain brand of lipstick by "randomly" selecting 100 women who come by their booth in a shopping mall. Of the 100 sampled, 65 women stated a preference for brand A.
(a) How would you estimate the true proportion of women preferring brand A, with a bound on the error of estimation?
(b) What is the target population in this study?

(c) Did the marketing research firm select a simple random sample?
(d) What additional problems do you see with this type of sampling?

4.41 A legal case is being formulated by a union of secretaries who claim that they are being paid unfairly low wages by their employer. The 64 secretaries in the firm have an average annual salary of $18,300, with a standard deviation of $400. The average salary for all secretaries in the city in which this firm is located is $20,100. Can you support the claim of the secretaries by statistical arguments? If so, carefully state these arguments and the assumptions underlying them.

4.42 The Equal Opportunity Employment Commission accuses a firm of violating minority-hiring standards, since of its 120 employees only 30 are nonwhite. In the labor market area for that firm, 36% of available employees are reported to be nonwhite. Can you support the accusation of the EOEC on statistical grounds? State your arguments for or against the accusation with careful attention to assumptions.

4.43 Observational studies suggest that there is a relationship between the issuing of an earnings report by a firm and market activity for that firm. A study of the lengths of these activity periods (called *mean return reaction periods*) for both large and small firms and for both interim and annual reports provided the data below. For this study, the return reaction period was divided into a preannouncement period, a postannouncement period, and an announcement period fixed at two

Descriptive statistics for estimated duration[a] of mean return reactions (RP_M)

		Total interval			*Preannouncement*			*Postannouncement*		
	n	Mean	Median	Standard deviation	Mean	Median	Standard deviation	Mean	Median	Standard deviation
Panel A										
All announcements	3823	19.0	13.0	16.4	5.1	5.7	3.6	12.0	6.1	14.6
Panel B										
By firm size										
Large	2001	19.9	12.9	17.3	5.3	6.1	3.6	12.7	5.9	15.5
Small	1822	18.0	13.0	15.3	4.9	5.3	3.7	11.2	6.2	13.5
Panel C										
By report type										
Interim	2799	18.2	12.5	15.9	4.9	4.9	3.7	11.5	6.2	14.0
Annual	1024	21.0	13.4	17.5	5.8	7.2	3.5	13.2	5.9	16.1
Panel D										
By reporting lag										
0–14	307	19.6	12.6	18.0	5.1	5.5	3.7	12.6	6.1	16.1
15–28	1930	19.4	13.2	17.1	5.1	5.7	3.6	12.6	6.0	15.2
29–42	901	18.3	12.5	15.9	5.2	6.1	3.7	11.2	5.5	14.0
43–56	426	18.7	12.9	15.3	5.0	5.3	3.7	11.8	6.3	13.8
≥ 57	259	17.8	14.3	11.9	4.9	5.2	3.5	11.0	8.3	10.7

[a] Measured in trading days.

Source: V. J. Defoe, "An Empirical Investigation of the Speed of the Market Reaction to Earnings Announcements," *Journal of Accounting Research* 24, no. 2 (1986): 349–363.

days' duration. *Reporting lag* is the number of calendar days between the end of the fiscal period and the day of the announcement.

All estimates called for below should be with approximate 95% confidence intervals.

(a) Estimate the true mean interval length for large firms.

(b) Estimate the true mean interval length for small firms. Do you think this differs significantly from the estimate for large firms?

(c) Estimate both pre- and postannouncement means for large firms. Do the same for small firms. How do you think the pre- and postannouncement means for the population compare?

(d) Estimate the proportion of firms with reporting lag under 29 days.

4.44 *Audit delay,* defined as the length of time from a company's fiscal year-end to the date of the auditor's report, can affect the market's reaction to the information released. (Late announcements seem to be more often associated with lower returns than are early announcements.) A study of audit delay produced the data on page 121. All estimates for this exercise should be with approximate 95% confidence intervals.

(a) Estimate the true mean audit delay (Audelay) for industrial companies. Do the same for financial companies. Compare the two answers.

(b) Estimate the true mean audit delay for public companies. Do the same for nonpublic. Compare the two answers.

(c) Does it make sense to compare the mean for public companies with the mean for industrial companies, given these data? Explain.

EXPERIENCES WITH REAL DATA

4.1 Data, collected by random sampling, from a class survey of introductory statistics students is shown in Appendix Data Set 1 and is on the accompanying disk (CLASSSUR). This data set was used in Chapter 2 to illustrate some graphical techniques for summarizing data. Use these data to answer the following.

(a) Estimate a population mean for one of the measurement variables, such as age, GPA, or study hours.

(b) Estimate a population proportion for one of the categorical variables, such as gender, class, or job status.

(c) Compare means on one variable for at least two different groups, such as men and women.

(d) Compare proportions on one categorical variable for at least two different groups.

4.2 Data on 1993 automobiles, some of which was used in Section 4.6, is provided on the disk as CARS93, with a portion of the table shown in Appendix Data Set 2. Even though all data are

Descriptive statistics for the audit delay industrial/financial and public/nonpublic subsamples

Variable	*Industrial and financial subsamples*						*Public and nonpublic subsamples*					
	Industrial			*Financial*			*Public*			*Nonpublic*		
	Mean	Standard deviation	Percentage	Mean	Standard deviation	Percentage	Mean	Standard deviation	Percentage	Mean	Standard deviation	Percentage
Audelay	68.04	35.72		56.74	34.87		48.79	20.08		66.39	38.14	
Totrev ($000)	$230,871	$983,537		$178,844	$1,284,796		$682,290	$2,301,197		71,595	$324,182	
Indus			—			—			55.1%			50.1%
Public			23.6%			20.2%			—			—
			39.6%			58.8%			51.4%			48.3%
Sample size		250			238			107			381	

Source: R. H. Ashton et al., "An Empirical Analysis of Audit Delay," *Journal of Accounting Research* 25, no. 2 (1987): 275–292.

available for all popular brands of cars, samples from this set will be used to study how well certain estimation procedures work.

(a) Select a random sample of cars and estimate the average city gas mileage for this population of cars. Choose an appropriate sample size.

(b) Select a random sample of large cars and a random sample of midsized cars. Compare the city gas mileage for these two sizes of cars. How much would you expect your gas mileage to increase if you buy a large car rather than a midsized one?

(c) By random sampling, estimate the proportion of these cars that have at least one air bag.

(d) By random sampling, compare the proportions of cars with at least one air bag for compact, midsized, and large cars.

4.3 Characteristics of the U.S. population are provided in Appendix Data Set 3 and in USPOP on the disk. Use these data to answer the following.

(a) Select a random sample of states and use the sample data to estimate the projected total population of the United States in 2010.

(b) From a random sample of states, estimate the total number of people over the age of 65 years in the United States in 1991. Then, turn this estimate into an estimate of the proportion of people over the age of 65 years in 1991.

4.4 Characteristics of the world population are provided on the disk in WORLDPOP, with a small portion of this table displayed in Appendix Data Set 4. Use these data to answer the following.

(a) Select a random sample of European countries and use it to estimate the population and the per-capita gross national product (GNP) of Europe.

(b) Select a random sample of African countries and use it to estimate the population and per-capita GNP of Africa.

(c) Select a random sample of Latin American countries and use it to estimate the population and per-capita GNP of Latin America.

(d) Compare the estimates in parts (a)–(c). Write a short report on the similarities and differences seen among these three regions.

4.5 Identify a problem in your own area of interest for which you can actually draw a simple random sample to estimate a population mean, total, or proportion. Clearly define the population and the sampling units and construct a frame. Select a simple random sample from the frame by using the random number table in the appendix. Then collect the data and make the necessary calculations.

Some suggested projects are as follows.

Business: Estimate the average gross income for firms of a certain type in your area or the average amount spent for entertainment among college men.

Social sciences: Estimate the proportion of registered voters favoring some current political proposal or the average number of persons per household for a certain section of your city.

Physical sciences: Consider a laboratory experiment such as measuring the tensile strength of wire or the diameter of a machined rod. Take n independent observations on such an experiment and treat them as a simple random sample. Construct an interval estimate of the "population" mean. Here the population is merely conceptual (one could take many measurements of the phenomenon in question), and its mean represents the average strength of wire of this type or the average diameter of the rod.

Biological sciences: Estimate the average weight of animals fed on a certain diet for a specified time period or the average height of trees in a certain plot. As an example of working with totals instead of means, estimate the total number of insect colonies (of a certain type) infesting a plot. Be careful here in selecting the sampling units and constructing the frame.

C A S E   S T U D Y

CAN WE ESTIMATE THE TOTAL COST OF HEALTH CARE?

An important problem of national concern involves the estimation of the cost of health care. These costs are studied by various agencies, in both government and private sectors, in order to establish government policies and to assess business decisions, such as rates for insurance policies.

A method of estimating hospital costs for one disease is considered in the article "Economic Impact of Kidney Stones in White Adult Males," by J. Shuster and R. L. Scheaffer (*Urology*, vol. 24, no. 4, 1984). In this work, two regions of the United States, the Carolinas and the Rocky Mountain states, were singled out for special study. A sample of $n_1 = 363$ stone patients in the Carolinas had an average cost for first hospitalization of \$1350; a sample of $n_2 = 258$ stone patients in the Rockies had an average cost for first hospitalization of \$1150. Can we estimate the total annual hospitalization costs for this disease for both regions combined? The methods of Chapter 5 will show us how to do so if some additional information is available. The methods can then be used to find an estimate for the entire United States if sample information is available for other regions.

CHAPTER 5

STRATIFIED RANDOM SAMPLING

5.1

INTRODUCTION

The purpose of sample survey design is to maximize the amount of information for a given cost. Simple random sampling, the basic sampling design, often provides good estimates of population quantities at low cost. In this chapter we define a second sampling procedure, stratified random sampling, which in many instances increases the quantity of information for a given cost.

> **Definition 5.1**
>
> A *stratified random sample* is one obtained by separating the population elements into nonoverlapping groups, called *strata*, and then selecting a simple random sample from each stratum.

Suppose a public opinion poll designed to estimate the proportion of voters who favor spending more tax revenue on an improved ambulance service is to be conducted in a certain county. The county contains two cities and a rural area. The population elements of interest for the poll are all men and women of voting age who reside in the county. A *stratified random sample* of adults residing in the county can be obtained by selecting a simple random sample of adults from each city and another simple random sample of adults from the rural area. That is, the two cities and the rural area represent three *strata* from which we obtain simple random samples.

In the county poll, why should we choose a stratified random sample rather than a simple random sample? First, keep in mind that our goal in designing surveys is to maximize the information obtained (or to minimize the bound on the error of estimation) for a fixed expenditure. Samples displaying small variability among the measurements will produce small bounds on the errors of estimation. Thus, if all the adults in one city (say, city A) tend to think alike on the ambulance service issue, we can obtain a very accurate estimate of the proportion in question with a relatively small sample. Similarly, if all the adults in the second city (city B) tend to think alike on this issue, though they may differ in opinion from those in city A, then we can again obtain an accurate estimate with a small sample. This situation may arise if city A has a hospital and hence has no great need for improved ambulance service, whereas city B does not have a hospital and hence has great need for an improved ambulance service. The opinions in the rural area may be more varied, but a smaller number of adults may reside here, and enough resources may be available for careful study of this area. When results of the stratified random sample are combined, the final estimate of the proportion of voters favoring more expenditures for an ambulance service may have a much smaller bound on the error of estimation than would an estimate from a simple random sample of comparable size.

Second, the cost of obtaining observations varies with the design of the survey. The cost of selecting the adults to be sampled, the cost of interviewer time and travel, and the cost of administering the overall sampling procedure may all be minimized by a carefully planned stratified random sample in compact, well-defined geographic areas. Such cost savings may allow the investigators to use a larger sample size than they could use for a simple random sample of the same total cost.

Third, estimates of a population parameter may be desired for certain subsets of the population. In the county poll each city commission may want to see an estimate of the proportion of voters favoring an expanded ambulance service for its own city. Stratified random sampling allows for separate estimates of population parameters within each stratum.

In summary, the principal reasons for using stratified random sampling rather than simple random sampling are as follows:

1. Stratification may produce a smaller bound on the error of estimation than would be produced by a simple random sample of the same size. This result is particularly true if measurements within strata are homogeneous.
2. The cost per observation in the survey may be reduced by stratification of the population elements into convenient groupings.

3. Estimates of population parameters may be desired for subgroups of the population. These subgroups should then be identifiable strata.

These three reasons for stratification should be kept in mind when one is deciding whether or not to stratify a population or deciding how to define strata. Sampling hospital patients on a certain diet to assess weight gain may be more efficient if the patients are stratified by gender, since men tend to weigh more than women. A poll of college students at a large university may be more coveniently administered and carried out if students are stratified into on-campus and off-campus residents. A quality control–sampling plan in a manufacturing plant may be stratified by production lines because estimates of proportions of defective products may be required by the manager of each line.

Most major surveys have some degree of stratification incorporated into the design. As examples, we will look at three important groups of surveys conducted by the U.S. Bureau of Labor Statistics.

The consumer price index (CPI) is a measure of the average change in prices for a fixed collection of goods and services for urban consumers. The CPI is actually calculated from at least four different types of surveys: surveys of cities, surveys of urban families, surveys of outlets providing goods and services, and surveys of specific goods and services. In the design of most CPI surveys, 1166 sampling units (counties or groups of contiguous counties) are identified in the population and then grouped into 85 strata. Strata are chosen on the basis of geography, population size, percentage population increase from 1960 to 1970, major industry, percentage nonwhite, and percentage urban. The sampling units within a stratum are chosen to be as much alike as possible with regard to these characteristics.

The Current Population Survey (CPS) measures aspects of employment, unemployment, and persons not in the labor force. It stratifies 1931 sampling units across the United States into strata similar to those used in the CPI surveys, except rural sampling units are used and number of farms becomes an important quantity for stratification.

The Establishment Survey (ES) collects data on work hours and earnings for nonagricultural establishments in the United States. Establishments are stratified according to industry type and size, primarily for homogeneity of measurements but also for provision of estimates for various types of industries. For example, information is provided for such industrial categories as mining, construction, manufacturing, transportation, and finance, insurance, and real estate.

In this chapter, stratification will always be used with simple random sampling in each stratum, as stated in Definition 5.1. However, stratification can be used with other types of sampling within strata. We will see some examples in later chapters.

5.2 HOW TO DRAW A STRATIFIED RANDOM SAMPLE

The first step in the selection of a stratified random sample is to clearly specify the strata; then each sampling unit of the population is placed into its appropriate stratum.

This step may be more difficult than it sounds. For example, suppose you plan to stratify the sampling units—say, households—into rural and urban units. What should be done with households in a town of 1000 inhabitants? Are these households rural or urban? They may be rural if the town is isolated in the country, or they may be urban if the town is adjacent to a large city. Hence to specify what is meant by *urban* and *rural* is essential so that each sampling unit clearly falls into only one stratum.

After the sampling units are divided into strata, we select a simple random sample from each stratum by using the techniques given in Chapter 4. We discuss the problem of choosing appropriate sample sizes for the strata later in this chapter. We must be certain that the samples selected from the strata are independent. That is, different random sampling schemes should be used within each stratum so that the observations chosen in one stratum do not depend on those chosen in another.

Some additional notation is required for stratified random sampling. Let

$$\begin{aligned} L &= \text{number of strata} \\ N_i &= \text{number of sampling units in stratum } i \\ N &= \text{number of sampling units in the population} \\ &= N_1 + N_2 + \cdots + N_L \end{aligned}$$

The following example illustrates a situation in which stratified random sampling may be appropriate.

Example 5.1

An advertising firm, interested in determining how much to emphasize television advertising in a certain county, decides to conduct a sample survey to estimate the average number of hours each week that households within the county watch television. The county contains two towns, A and B, and a rural area. Town A is built around a factory, and most households contain factory workers with school-aged children. Town B is an exclusive suburb of a city in a neighboring county and contains older residents with few children at home. There are 155 households in town A, 62 in town B, and 93 in the rural area. Discuss the merits of using stratified random sampling in this situation.

Solution

The population of households falls into three natural groupings, two towns and a rural area, according to geographic location. Thus, to use these divisions as three strata is quite natural simply for administrative convenience in selecting the samples and carrying out the fieldwork. In addition, each of the three groups of households should have similar behavior patterns among residents within the group. We expect to see relatively small variability in number of hours of television viewing among households within a group, and this is precisely the situation in which stratification produces a reduction in a bound on the error of estimation.

The advertising firm may wish to produce estimates on average television-viewing hours for each town separately. Stratified random sampling allows for these estimates.

For the stratified random sample, we have $N_1 = 155$, $N_2 = 62$, and $N_3 = 93$, with $N = 310$.

5.3

ESTIMATION OF A POPULATION MEAN AND TOTAL

How can we use the data from a stratified random sample to estimate the population mean? Let $\bar{y}_i$ denote the sample mean for the simple random sample selected from stratum i, n_i the sample size for stratum i, μ_i the population mean for stratum i, and τ_i the population total for stratum i. Then the population total τ is equal to $\tau_1 + \tau_2 + \cdots + \tau_L$. We have a simple random sample within each stratum. Therefore, we know from Chapter 4 that $\bar{y}_i$ is an unbiased estimator of μ_i and $N_i\bar{y}_i$ is an unbiased estimator of the stratum total $\tau_i = N_i\mu_i$. It seems reasonable to form an estimator of τ, which is the sum of the τ_i's, by summing the estimators of the τ_i's. Similarly, since the population mean μ equals the population total τ divided by N, an unbiased estimator of μ is obtained by summing the estimators of the τ_i's over all strata and then dividing by N. We denote this estimator by $\bar{y}_{st}$, where the subscript st indicates that stratified random sampling is used.

Estimator of the population mean μ:

$$\bar{y}_{st} = \frac{1}{N}[N_1\bar{y}_1 + N_2\bar{y}_2 + \cdots + N_L\bar{y}_L] = \frac{1}{N}\sum_{i=1}^{L} N_i\bar{y}_i \tag{5.1}$$

Estimated variance of $\bar{y}_{st}$:

$$\begin{aligned}\hat{V}(\bar{y}_{st}) &= \frac{1}{N^2}[N_1^2\hat{V}(\bar{y}_1) + N_2^2\hat{V}(\bar{y}_2) + \cdots + N_L^2\hat{V}(\bar{y}_L)] \\ &= \frac{1}{N^2}\left[N_1^2\left(\frac{N_1 - n_1}{N_1}\right)\left(\frac{s_1^2}{n_1}\right) + \cdots + N_L^2\left(\frac{N_L - n_L}{N_L}\right)\left(\frac{s_L^2}{n_L}\right)\right] \\ &= \frac{1}{N^2}\sum_{i=1}^{L} N_i^2\left(\frac{N_i - n_i}{N_i}\right)\left(\frac{s_i^2}{n_i}\right)\end{aligned} \tag{5.2}$$

As seen in Chapter 4, the bound on the error as used in this book is always two times the estimated standard deviation of the estimator. The result is an approximate 95% confidence interval estimate of the parameter in question. Thus, we will not keep writing general formulas for bounds on the error, since they can be derived easily from the estimated variance.

Example 5.2

Suppose the survey planned in Example 5.1 is carried out. The advertising firm has enough time and money to interview $n = 40$ households and decides to select random samples of size $n_1 = 20$ from town A, $n_2 = 8$ from town B, and $n_3 = 12$ from the rural area. (We will discuss the choice of sample sizes later.) The simple random samples are selected and the interviews conducted. The results, with measurements of television-viewing time in hours per week, are shown on Table 5.1.

Estimate the average television-viewing time, in hours per week, for (a) all households in the county and (b) all households in town B. In both cases, place a bound on the error of estimation.

TABLE 5.1 **Television-viewing time, in hours per week**

Town A	Town B	Rural
35	27	8
43	15	14
36	4	12
39	41	15
28	49	30
28	25	32
29	10	21
25	30	20
38		34
27		7
26		11
32		24
29		
40		
35		
41		
37		
31		
45		
34		

Solution

A good way to view the key features of these three samples of hours is through parallel box plots, as shown in Figure 5.1. The medians are decreasing as we go from town A to town B to the rural area. Town B has the largest amount of variability in the sample data, but there are no outliers or other unusual features to be concerned about.

FIGURE 5.1 Box plots of television-viewing time

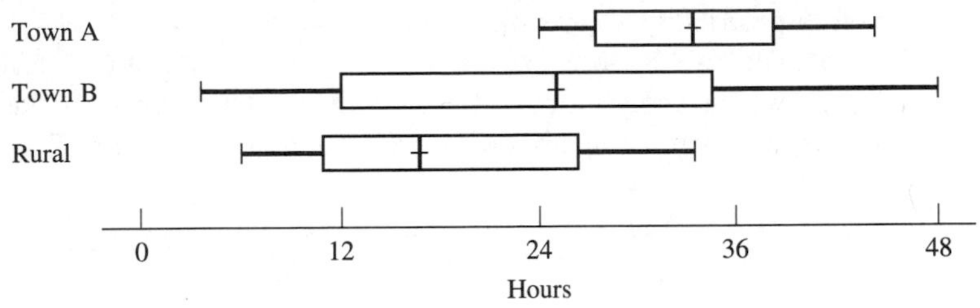

A numerical descriptive summary of the data is shown in Table 5.2.

(a) From Table 5.1 and Equation (5.1),

$$\begin{aligned}\bar{y}_{st} &= \frac{1}{N}[N_1\bar{y}_1 + N_2\bar{y}_2 + N_3\bar{y}_3] \\ &= \frac{1}{310}[(155)(33.900) + (62)(25.125) + (93)(19.00)] \\ &= 27.7\end{aligned}$$

is the best estimate of the average number of hours per week that all households in the county spend watching television. Also,

$$\begin{aligned}\hat{V}(\bar{y}_{st}) &= \frac{1}{N^2}\sum_{i=1}^{3} N_i^2\left(\frac{N_i - n_i}{N_i}\right)\left(\frac{s_i^2}{n_i}\right) \\ &= \frac{1}{(310)^2}\left[\frac{(155)^2(.871)(5.95)^2}{20} + \frac{(62)^2(.871)(15.25)^2}{8}\right. \\ &\quad \left. + \frac{(93)^2(.871)(9.36)^2}{12}\right] \\ &= 1.97\end{aligned}$$

The estimate of the population mean with an approximate 2-SD bound on the error of estimation is given by

$$\bar{y}_{st} \pm 2\sqrt{\hat{V}(\bar{y}_{st})}, \qquad 27.675 \pm 2\sqrt{1.97}, \qquad 27.7 \pm 2.8$$

TABLE 5.2 **Summary of the data from Table 5.1**

	N	Mean	Median	SD
Town A	20	33.90	34.50	5.95
Town B	8	25.12	26.00	15.25
Rural	12	19.00	17.50	9.36

Thus, we estimate the average number of hours per week that households in the county view television to be 27.7 hours. The error of estimation should be less than 2.8 hours with a probability approximately equal to .95.

(b) The $n_2 = 8$ observations from stratum 2 constitute a simple random sample; hence, we can apply formulas from Chapter 4. The estimate of the average viewing time for town B with an approximate 2-SD bound on the error of estimation is given by

$$\bar{y}_2 \pm 2\sqrt{\left(\frac{N_2 - n_2}{N_2}\right)\left(\frac{s_2^2}{n_2}\right)} \qquad \text{or} \qquad 25.1 \pm 2\sqrt{\left(\frac{62 - 8}{62}\right)\left(\frac{232.56}{8}\right)}$$

or

$$25.1 \pm 10.0$$

This estimate has a large bound on the error of estimation because s_2^2 is large and the sample size n_2 is small. Thus, the estimate $\bar{y}_{st}$ of the population mean is quite good, but the estimate $\bar{y}_2$ of the mean of stratum 2 is poor. If an estimate is desired for a particular stratum, the sample from that stratum must be large enough to provide a reasonable bound on the error of estimation.

As mentioned previously, the best way to do the extensive computations that the analysis of sampling data will require is with the aid of modern technology, either a computer or a graphing calculator. The computations within this textbook are done through MINITAB, but any good statistical software program will do essentially the same work. For purposes of simplifying the rather complex calculations related to stratified random sampling, some programs (called macros) written for MINITAB will be illustrated here. Copies of these macros, and similar ones written for SAS, are provided in the appendix. These programs could be adapted to other computer software and to programmable graphing calculators.

MINITAB MACRO FOR ESTIMATING THE MEAN; STRATIFIED RANDOM SAMPLING

The macro is called 'stmean.MTB' and can be copied from the disk that accompanies this text. The program will ask for key pieces of data that are to be typed after the DATA> prompt. The data may appear in any set of columns on the worksheet, but the columns must be consecutive. For this illustration, the data, as in Table 5.1, is arrayed in columns 1 through 3. Further notes on the operation of the macros are provided in the appendix.

The following is a printout of the macro in operation for Example 5.2:

```
MTB > Execute 'stmean.MTB' 1.
MTB > #
MTB > Exec 'mvarcolm'b.
MTB > #
```

```
Input the column number of the first column of data followed
by the number of the last column of data.
DATA >  1 3
Input the sample size followed by the population size.
DATA > 20 155
The sample size is
K2       20.0000

The population size is
K22      155.000

The sample mean is
K3      33.9000

The estimated variance of the sample mean is
K4      1.53978

The estimated standard deviation of the sample mean is
K5      1.24088

The bound on the error of estimation is
K6      2.48175

Input the sample size followed by the population size.
DATA > 8 62
The sample size is
K2      8.00000

The population size is
K22      62.0000

The sample mean is
K3      25.1250

The estimated variance of the sample mean is
K4      25.3028

The estimated standard deviation of the sample mean is
K5      5.03019

The bound on the error of estimation is
K6      10.0604
```

```
Input the sample size followed by the population size.
DATA > 12 93
The sample size is
K2     12.0000

The population size is
K22     93.0000

The sample mean is
K3     19.0000

The estimated variance of the sample mean is
K4     6.36070

The estimated standard deviation of the sample mean is
K5     2.52204

The bound on the error of estimation is
K6     5.04409

The following table lists the sample size, population size,
sample mean, estimated variance of the sample mean,
estimated standard deviation of the sample mean, and the
bound on the error of estimation for each column:
```

ROW	samsize	popsize	mean	variance	std dev	bound
1	20	155	33.900	1.5398	1.24088	2.4818
2	8	62	25.125	25.3028	5.03019	10.0604
3	12	93	19.000	6.3607	2.52204	5.0441

```
MTB >
MTB >
The estimator of the mean, where stratified sampling is used,
is
K9     27.6750

The estimated variance of the mean, using stratified
sampling, is
K13     1.96952

The estimated standard deviation of the mean, using
stratified sampling, is
K14     1.40340
```

```
The bound on the error of estimation, using stratified
sampling, is
K15      2.80679
```

Looking through the printout, observe that the computations for estimating a mean are made separately for each stratum and summarized in a table. Then the results are combined to form the stratified estimator of the population mean and an estimator of its variance. The results from the computer program are slightly more accurate than those from the hand calculations because of rounding error, but in this case, they agree quite closely. Note that the result for just stratum 2 of Example 5.2(b) is given on the second line of the enclosed table.

Procedures for the estimation of a population total τ follow directly from the procedures presented for estimating μ. Since τ is equal to $N\mu$, an unbiased estimator of τ is given by $N\bar{y}_{st}$.

Estimator of the population total τ:

$$N\bar{y}_{st} = N_1\bar{y}_1 + N_2\bar{y}_2 + \cdots + N_L\bar{y}_L = \sum_{i=1}^{L} N_i\bar{y}_i \tag{5.3}$$

Estimated variance of $N\bar{y}_{st}$:

$$\hat{V}(N\bar{y}_{st}) = N^2\hat{V}(\bar{y}_{st}) = \sum_{i=1}^{L} N_i^2\left(\frac{N_i - n_i}{N_i}\right)\left(\frac{s_i^2}{n_i}\right) \tag{5.4}$$

Example 5.3

Refer to Example 5.2 and estimate the total number of hours each week that households in the county view television. Place a bound on the error of estimation.

Solution

For the data in Table 5.1,

$$N\bar{y}_{st} = 310(27.7) = 8587 \text{ hours}$$

The estimated variance of $N\bar{y}_{st}$ is given by

$$\hat{V}(N\bar{y}_{st}) = N^2\hat{V}(\bar{y}_{st}) = (310)^2(1.97) = 189{,}278.560$$

The estimate of the population total with a bound on the error of estimation is given by

$$N\bar{y}_{st} \pm 2\sqrt{\hat{V}(N\bar{y}_{st})} \qquad \text{or} \qquad 8587 \pm 2\sqrt{189{,}278.560}$$

or

$$8587 \pm 870$$

Thus, we estimate the total weekly viewing time for households in the county to be 8587 hours. The error of estimation should be less than 870 hours.

Since the simple random samples chosen within each stratum are independent of one another, differences between stratum means can be estimated in a straightforward fashion, as introduced in Chapter 4. Recall that the variance of the difference between two independent random variables is the sum of their respective variances.

Example 5.4 In the study of television-viewing time per family, the families of town A tended to be younger and have more children than those of town B. To see if there is a significant difference between average television-viewing time for families of these towns, estimate the difference between their population means.

Solution

The basic computations necessary for this estimate are produced in the table inside the MINITAB macro. There, we see that town A has a sample mean of 33.9 hours/week, and this mean has an estimated variance of 1.5. Town B has a sample mean of 25.1 hours/week, and this mean has an estimated variance of 25.3. It follows that the estimate of the difference between the population means for these two strata is given by

$$(33.9 - 25.1) \pm 2\sqrt{1.5 + 25.3}$$
$$8.8 \pm 10.4$$

Since this interval estimate overlaps zero, we cannot rule out the fact that there may be no real difference between the two population means, even though town A has a much larger sample mean. This sample difference does not show up as statistically significant because of the large variation in the measurements from town B.

5.4

SELECTING THE SAMPLE SIZE FOR ESTIMATING POPULATION MEANS AND TOTALS

The amount of information in a sample depends on the sample size n, since $V(\bar{y}_{st})$ decreases as n increases. Let us examine a method of choosing the sample size to obtain a fixed amount of information for estimating a population parameter. Suppose we specify that the estimate $\bar{y}_{st}$ should lie within B units of the population mean, with probability approximately equal to .95. Symbolically, we want

$$2\sqrt{V(\bar{y}_{st})} = B \qquad \text{or} \qquad V(\bar{y}_{st}) = \frac{B^2}{4}$$

This equation contains the actual population variance of $\bar{y}_{st}$ rather than the estimated variance. For large N, the actual variance, $V(\bar{y}_{st})$, looks very similar to Equation (5.2), with $s_1^2, s_2^2, \ldots, s_L^2$ replaced by $\sigma_1^2, \sigma_2^2, \ldots, \sigma_L^2$.

Although we set $V(\bar{y}_{st})$ equal to $B^2/4$, we cannot solve for n unless we know something about the relationships among $n_1, n_2, \ldots, n_L$ and n. There are many ways of allocating a sample of size n among the various strata. In each case, however, the number of observations n_i allocated to the ith stratum is some fraction of the total sample size n. We denote this fraction by w_i. Hence, we can write

$$n_i = nw_i \qquad i = 1, 2, \ldots, L \tag{5.5}$$

Using Equation (5.5), we can then set $V(\bar{y}_{st})$ equal to $B^2/4$ and solve for n.

Similarly, estimation of the population total τ with a bound of B units on the error of estimation leads to the equation

$$2\sqrt{V(N\bar{y}_{st})} = B$$

or, using Equation (5.4),

$$V(\bar{y}_{st}) = \frac{B^2}{4N^2}$$

Approximate sample size required to estimate μ or τ with a bound B on the error of estimation:

$$n = \frac{\sum_{i=1}^{L} N_i^2 \sigma_i^2 / w_i}{N^2 D + \sum_{i=1}^{L} N_i \sigma_i^2} \tag{5.6}$$

where w_i is the fraction of observations allocated to stratum i, σ_i^2 is the population variance for stratum i, and

$$D = \frac{B^2}{4} \qquad \text{when estimating } \mu$$

$$D = \frac{B^2}{4N^2} \qquad \text{when estimating } \tau$$

We must obtain approximations of the population variances $\sigma_1^2, \sigma_2^2, \ldots, \sigma_L^2$ before we can use Equation (5.6). One method of obtaining these approximations is to use the sample variances $s_1^2, s_2^2, \ldots, s_L^2$ from a previous experiment to estimate $\sigma_1^2, \sigma_2^2, \ldots, \sigma_L^2$. A second method requires knowledge of the range of the observations within each stratum. From Tchebysheff's theorem and the normal distribution, the range should be roughly 4 to 6 SD.

Methods of choosing the fractions $w_1, w_2, \ldots, w_L$ are given in Section 5.5.

Example 5.5

A prior survey suggests that the stratum variances for Example 5.1 are approximately $\sigma_1^2 \approx 25, \sigma_2^2 \approx 225$, and $\sigma_3^2 \approx 100$. We wish to estimate the population mean by using $\bar{y}_{st}$. Choose the sample size to obtain a bound on the error of estimation equal to 2 hours if the allocation fractions are given by $w_1 = \frac{1}{3}, w_2 = \frac{1}{3}$, and $w_3 = \frac{1}{3}$. In other words, you are to take an equal number of observations from each stratum.

Solution

A bound on the error of 2 hours means that

$$2\sqrt{V(\bar{y}_{st})} = 2 \qquad \text{or} \qquad V(\bar{y}_{st}) = 1$$

Therefore, $D = 1$.

In Example 5.1, $N_1 = 155$, $N_2 = 62$, and $N_3 = 93$. Therefore,

$$\begin{aligned}\sum_{i=1}^{3} \frac{N_i^2\sigma_i^2}{w_i} &= \frac{N_1^2\sigma_1^2}{w_1} + \frac{N_2^2\sigma_2^2}{w_2} + \frac{N_3^2\sigma_3^2}{w_3} \\ &= \frac{(155)^2(25)}{(1/3)} = \frac{(62)^2(225)}{(1/3)} + \frac{(93)^2(100)}{(1/3)} \\ &= (24{,}025)(75) + (3844)(675) + (8649)(300) \\ &= 6{,}991{,}275\end{aligned}$$

$$\begin{aligned}\sum_{i=1}^{3} N_i\sigma_i^2 &= N_1\sigma_1^2 + N_2\sigma_2^2 + N_3\sigma_3^2 \\ &= (155)(25) + (62)(225) + (93)(100) = 27{,}125\end{aligned}$$

$$N^2D = (310)^2(1) = 96{,}100$$

From Equation (5.6), we then have

$$n = \frac{\sum_{i=1}^{3} N_i^2\sigma_i^2/w_i}{N^2D + \sum_{i=1}^{3} N_i\sigma_i^2} = \frac{6{,}991{,}275}{96{,}100 + 27{,}125} = \frac{6{,}991{,}275}{123{,}225} = 56.7$$

Thus, the experimenter should take $n = 57$ observations with

$$n_1 = n(w_1) = 57\left(\frac{1}{3}\right) = 19$$
$$n_2 = 19$$
$$n_3 = 19$$

MINITAB MACRO FOR CALCULATING THE SAMPLE SIZE

As in the case of estimating a mean, a computer macro is available for making the approximate sample-size calculations. The MINITAB macro, called 'stsampsm.MTB', is described in the appendix and can be found on the disk. Certain information must be fed into the macro at the DATA> prompts, as in the following illustration based on Example 5.4. The macro is less prone to rounding error than are hand calculations, so it may produce slightly different results.

```
MTB > Execute 'stsampsm.MTB' 1.
MTB > #

Input the number of strata.
DATA > 3

Input the weight for each stratum.
DATA > .33 .33 .33

Input the population VARIANCE for each stratum.
DATA > 25 225 100

Input the population size for each stratum.
DATA > 155 62 93

Input the actual number desired for D.
DATA > 1

The approximate sample size required to estimate mu or tau
with a bound B on the error of estimation is
K67     57.3089

The approximate individual sample sizes for each stratum are
given below:

C70
 18.9119 18.9119 18.9119
```

Example 5.6 As in Example 5.5, suppose the variances of Example 5.1 are approximated by $\sigma_1^2 \approx 25$, $\sigma_2^2 \approx 225$, and $\sigma_3^2 \approx 100$. We wish to estimate the population total τ with a bound of 400 hours on the error of estimation. Choose the appropriate sample size if an equal number of observations is to be taken from each stratum.

Solution

The bound on the error of estimation is to be 400 hours and, therefore,

$$D = \frac{B^2}{4N^2} = \frac{(400)^2}{4N^2} = \frac{40{,}000}{N^2}$$

To calculate n from Equation (5.6), we need the following quantities:

$$\sum_{i=1}^{3} \frac{N_i^2 \sigma_i^2}{w_i} = 6{,}991{,}275 \qquad \text{(from Example 5.5)}$$

$$\sum_{i=1}^{3} N_i \sigma_i^2 = 27{,}125 \qquad \text{(from Example 5.5)}$$

$$N^2 D = N^2 \left(\frac{40{,}000}{N^2} \right) = 40{,}000$$

Using Equation (5.6) yields

$$n = \frac{\sum_{i=1}^{3} N_i^2 \sigma_i^2 / w_i}{N^2 D + \sum_{i=1}^{3} N_i \sigma_i^2} = \frac{6{,}991{,}275}{40{,}000 + 27{,}125} = 104.2 \text{ or } 105$$

Then $n_1 = n_2 = n_3 = 35$.

5.5

ALLOCATION OF THE SAMPLE

Recall that the objective of a sample survey design is to provide estimators with small variances at the lowest possible cost. After the sample size n is chosen, there are many ways to divide n into the individual stratum sample sizes, $n_1, n_2, \ldots, n_L$. Each division may result in a different variance for the sample mean. Hence, our objective is to use an allocation that gives a specified amount of information at minimum cost.

In terms of our objective, the best allocation scheme is affected by three factors:

1. The total number of elements in each stratum.

2. The variability of observations within each stratum.
3. The cost of obtaining an observation from each stratum.

The number of elements in each stratum affects the quantity of information in the sample. A sample size 20 from a population of 200 elements should contain more information than a sample of 20 from 20,000 elements. Thus, large-sample sizes should be assigned to strata containing large numbers of elements.

Variability must be considered because a larger sample is needed to obtain a good estimate of a population parameter when the observations are less homogeneous.

If the cost of obtaining an observation varies from stratum to stratum, we will take small samples from strata with high costs. We will do so because our objective is to keep the cost of sampling at a minimum.

Approximate allocation that minimizes cost for a fixed value of $V(\bar{y}_{st})$ or minimizes $V(\bar{y}_{st})$ for a fixed cost:

$$n_i = n\left(\frac{N_i\sigma_i/\sqrt{c_i}}{N_1\sigma_1/\sqrt{c_1} + N_2\sigma_2/\sqrt{c_2} + \cdots + N_L\sigma_L/\sqrt{c_L}}\right) \quad (5.7)$$

$$= n\left(\frac{N_i\sigma_i/\sqrt{c_i}}{\sum_{k=1}^{L} N_k\sigma_k/\sqrt{c_k}}\right)$$

where N_i denotes the size of the ith stratum, σ_i^2 denotes the population variance for the ith stratum, and c_i denotes the cost of obtaining a single observation from the ith stratum. Note that n_i is directly proportional to N_i and σ_i and inversely proportional to $\sqrt{c_i}$.

One must approximate the variance of each stratum before sampling in order to use the allocation formula (5.7). The approximations can be obtained from earlier surveys or from knowledge of the range of the measurements within each stratum.

Substituting the n_i/n given by (5.7) for w_i in Equation (5.6) gives

$$n = \frac{\left(\sum_{k=1}^{L} N_k\sigma_k/\sqrt{c_k}\right)\left(\sum_{i=1}^{L} N_i\sigma_i\sqrt{c_i}\right)}{N^2D + \sum_{i=1}^{L} N_i\sigma_i^2} \quad (5.8)$$

for optimal allocation with the variance of $\bar{y}_{st}$ fixed at D.

Example 5.7 The advertising firm in Example 5.1 finds that obtaining an observation from a rural household costs more than obtaining a response in town A or B. The increase is due to costs of traveling from one rural household to another. The cost per observation in each town is estimated to be \$9 (that is, $c_1 = c_2 = 9$), and the costs per observation in the rural area to be \$16 (that is, $c_3 = 16$). The stratum standard deviations (approximated by the strata sample variances from a prior survey) are $\sigma_1 \approx 5, \sigma_2 \approx 15$, and $\sigma_3 \approx 10$. Find the overall sample size n and the stratum sample sizes, n_1, n_2, and n_3, that allow the firm to estimate, at minimum cost, the average television-viewing time with a bound on the error of estimation equal to 2 hours.

Solution

We have

$$\sum_{k=1}^{3} \frac{N_k \sigma_k}{\sqrt{c_k}} = \frac{N_1 \sigma_1}{\sqrt{c_1}} + \frac{N_2 \sigma_2}{\sqrt{c_2}} + \frac{N_3 \sigma_3}{\sqrt{c_3}}$$
$$= \frac{155(5)}{\sqrt{9}} + \frac{62(15)}{\sqrt{9}} + \frac{93(10)}{\sqrt{16}} = 800.83$$

and

$$\sum_{i=1}^{3} N_i \sigma_i \sqrt{c_i} = N_1 \sigma_1 \sqrt{c_1} + N_2 \sigma_2 \sqrt{c_2} + N_3 \sigma_3 \sqrt{c_3}$$
$$= 155(5)\sqrt{9} + 62(15)\sqrt{9} + 93(10)\sqrt{16} = 8835$$

Thus

$$n = \frac{\left(\sum_{k=1}^{3} N_k \sigma_k / \sqrt{c_k}\right)\left(\sum_{i=1}^{3} N_i \sigma_i \sqrt{c_i}\right)}{N^2 D + \sum_{i=1}^{3} N_i \sigma_i^2}$$
$$= \frac{(800.83)(8835)}{(310)^2(1) + 27{,}125} = 57.42 \text{ or } 58$$

Then

$$n_1 = n\left(\frac{N_1 \sigma_1 / \sqrt{c_1}}{\sum_{k=1}^{3} N_k \sigma_k / \sqrt{c_k}}\right) = n\left[\frac{155(5)/3}{800.83}\right] = .32n = 18.5 \text{ or } 18$$

Similarly,

$$n_2 = n\left[\frac{62(15)/3}{800.83}\right] = .39n = 22.6 \text{ or } 23$$
$$n_3 = n\left[\frac{93(10)/4}{800.83}\right] = .29n = 16.8 \text{ or } 17$$

Hence, the experimenter should select 18 households at random from town A, 23 from town B, and 17 from the rural area. He can then estimate the average number of hours spent watching television at minimum cost with a bound of 2 hours on the error of estimation.

MINITAB MACRO FOR ALLOCATING THE SAMPLE

For this cumbersome calculation, use the MINITAB macro called 'stallocm.MTB', which reproduces the calculations done by hand in the previous example. The results are as shown below:

```
MTB >
MTB > Execute 'stallocm.MTB' 1.
MTB > #
Input the number of strata.
DATA > 3
Input the population size for each stratum.
DATA > 155 62 93
Input the population VARIANCE for each stratum.
DATA > 25 225 100
Input the cost of obtaining an observation from each
stratum.
DATA > 9 9 16
Input the actual number desired for D.
DATA > 1
The weights for each stratum are given below:

C36
 0.322581 0.387097 0.290323

The approximate sample size required to estimate mu or tau
with a bound B on the error of estimation is
K43 57.4182

The approximate individual sample sizes for each stratum are
given below:

C37
 18.5220 22.2264 16.6698
```

In some stratified sampling problems, the cost of obtaining an observation is the same for all strata. If the costs are unknown, we may be willing to assume that the costs per observation are equal. If $c_1 = c_2 = \cdots = c_L$, then the cost terms cancel in Equations (5.7) and

$$n_i = n\left(\frac{N_i\sigma_i}{\sum_{i=1}^{L} N_i\sigma_i}\right) \tag{5.9}$$

This method of selecting $n_1, n_2, \ldots, n_L$ is called *Neyman allocation*. Under Neyman allocation, Equation (5.8) for the total sample size n becomes

$$n = \frac{\left(\sum_{i=1}^{L} N_i\sigma_i\right)^2}{N^2D + \sum_{i=1}^{L} N_i\sigma_i^2} \tag{5.10}$$

Example 5.8

The advertising firm of Example 5.1 decides to use telephone interviews rather than personal interviews because all households in the country have telephones, and this method reduces costs. The cost of obtaining an observation is then the same in all three strata. The stratum standard deviations are again approximated by $\sigma_1 \approx 5, \sigma_2 \approx 15,$ *and* $\sigma_3 \approx 10$. The firm desires to estimate the population mean μ with a bound on the error of estimation equal to 2 hours. Find the appropriate sample size n and stratum sample sizes, n_1, n_2, and n_3.

Solution

We will now use Equations (5.9) and (5.10), since the costs are the same in all strata. Therefore to find the allocation fractions, w_1, w_2, and w_3, we use Equation (5.9). Then

$$\sum_{i=1}^{3} N_i\sigma_i = N_1\sigma_1 + N_2\sigma_2 + N_3\sigma_3$$
$$= (155)(5) + (62)(15) + (93)(10) = 2635$$

And from Equation (5.9)

$$n_1 = n\left(\frac{N_1\sigma_1}{\sum_{i=1}^{3} N_i\sigma_i}\right) = n\left[\frac{(155)(5)}{2635}\right] = n(.30)$$

Similarly,

$$n_2 = n\left[\frac{(62)(15)}{2635}\right] = n(.35)$$

$$n_3 = n\left[\frac{(93)(10)}{2635}\right] = n(.35)$$

Thus, $w_1 = .30$, $w_2 = .35$, and $w_3 = .35$.

Now let us use Equation (5.10) to find n. A bound of 2 hours on the error of estimation means that

$$2\sqrt{V(\bar{y}_{st})} = 2 \qquad \text{or} \qquad V(\bar{y}_{st}) = 1$$

Therefore,

$$D = \frac{B^2}{4} = 1 \qquad \text{and} \qquad N^2D = (310)^2(1) = 96{,}100$$

Also,

$$\sum_{i=1}^{3} N_i\sigma_i^2 = 27{,}125$$

from Example 5.6, and Equation (5.10) gives

$$n = \frac{\left(\sum_{i=1}^{3} N_i\sigma_i\right)^2}{N^2D + \sum_{i=1}^{3} N_i\sigma_i^2}$$

$$= \frac{(2635)^2}{96{,}100 + 27{,}125} = 56.34 \text{ or } 57$$

Then

$$n_1 = nw_1 = (57)(.30) = 17$$
$$n_2 = nw_2 = (57)(.35) = 20$$
$$n_3 = nw_3 = (57)(.35) = 20$$

The sample size n in Example 5.8 is nearly the same as in Example 5.7, but the allocation has changed. More observations are taken from the rural area because these observations no longer have a higher cost.

Example 5.9 An experimenter wants to estimate the average weight of 90 rats (50 male and 40 female) being fed a certain diet. The rats were separated by sex; hence, using stratified random sampling with two strata seems appropriate. To approximate the variability within each stratum, the experimenter selects the smallest and largest rats in each stratum and weighs them. She finds that the range is 10 grams for the males and 8 grams for the females. How large a sample should be taken in order to estimate the population average with a bound of 1 gram on the error of estimation? Assume the cost of sampling is the same for both strata.

Solution

Let us denote males as stratum 1 and females as stratum 2. To use Equation (5.9), we must first approximate σ_1 and σ_2. The standard deviation should be about one-fourth of the range, assuming that the weights have a bell-shaped distribution. Thus,

$$\sigma_1 \approx \frac{10}{4} = 2.5 \qquad \text{and} \qquad \sigma_2 \approx \frac{8}{4} = 2.0$$

From Equation (5.9),

$$n_i = n\left(\frac{N_i\sigma_i}{\sum_{i=1}^{2} N_i\sigma_i}\right)$$

where

$$\sum_{i=1}^{2} N_i\sigma_i = (50)(2.5) + (40)(2.0) = 125 + 80 = 205$$

Then

$$n_1 = n\left(\frac{N_1\sigma_1}{\sum_{i=1}^{2} N_i\sigma_i}\right) = n\left(\frac{125}{205}\right) = .61n$$

and

$$n_2 = n\left(\frac{80}{205}\right) = .39n$$

Thus, $w_1 = .61$ and $w_2 = .39$.

We must calculate the following quantities in order to find n:

$$\sum_{i=1}^{2} N_i\sigma_i^2 = (50)(2.5)^2 + (40)(2.0)^2 = 472.50$$

$$D = \frac{B^2}{4} = \frac{(1)^2}{4} = .25$$

Using Equation (5.10), we have

$$n = \frac{\left(\sum_{i=1}^{2} N_i\sigma_i\right)^2}{N^2D + \sum_{i=1}^{2} N_i\sigma_i^2}$$

$$= \frac{(205)^2}{(90)^2(.25) + 472.50} = 16.83$$

The sample size n should have been 17 with

$$n_1 = nw_1 = (17)(.61) = 10$$

and

$$n_2 = nw_2 = (17)(.39) = 7$$

In addition to encountering equal costs, we sometimes encounter approximately equal variances, $\sigma_1^2, \sigma_2^2, \ldots, \sigma_L^2$. In that case, the σ_i's cancel in Equation (5.9) and

$$n_i = n\left(\frac{N_i}{\sum_{i=1}^{L} N_i}\right) = n\left(\frac{N_i}{N}\right) \tag{5.11}$$

This method of assigning sample sizes to the strata is called *proportional allocation* because sample sizes $n_1, n_2, \ldots, n_L$ are proportional to stratum sizes $N_1, N_2, \ldots, N_L$. Of course, proportional allocation can be, and often is, used when stratum variances and costs are not equal. One advantage to using this allocation is that the estimator $\bar{y}_{st}$ becomes simply the sample mean for the entire sample. This feature can be an important timesaving feature in some surveys.

Under proportional allocation Equation (5.6) for the value of n, which yields $V(\bar{y}_{st}) = D$, becomes

$$n = \frac{\sum_{i=1}^{L} N_i\sigma_i^2}{ND + \frac{1}{N}\sum_{i=1}^{L} N_i\sigma_i^2} \tag{5.12}$$

Example 5.10 The advertising firm in Example 5.1 thinks that the approximate variances used in previous examples are in error and that the stratum variances are approximately equal. The common value of σ_i was approximated by 10 in a preliminary study. Telephone interviews are to be used, and hence costs will be equal in all strata. The firm desires to estimate the average number of hours per week that households in the county watch television, with a bound on the error of estimation equal to 2 hours. Find the sample size and stratum sample sizes necessary to achieve this accuracy.

Solution

We have

$$\begin{aligned}\sum_{i=1}^{3} N_i\sigma_i^2 &= N_1\sigma_1^2 + N_2\sigma_2^2 + N_3\sigma_3^2 \\ &= (155)(100) + (62)(100) + (93)(100) \\ &= 310(100) = 31{,}000\end{aligned}$$

Thus, since $D = 1$, Equation (5.12) gives

$$n = \frac{31{,}000}{310(1) + (1/310)(31{,}000)} = 75.6 \text{ or } 76$$

Therefore,

$$n_1 = n\left(\frac{N_1}{\sum_{i=1}^{3} N_i}\right) = n\left(\frac{N_1}{N}\right) = n\left(\frac{155}{310}\right) = n(.5) = 38$$

$$n_2 = n\left(\frac{N_2}{N}\right) = n\left(\frac{62}{310}\right) = n(.2) = 15$$

$$n_3 = n\left(\frac{N_3}{N}\right) = n\left(\frac{93}{310}\right) = n(.3) = 23$$

These results differ from those of Example 5.8 because here the variances are assumed to be equal in all strata and are approximated by a common value.

The macro for allocation introduced earlier will work with either Neyman or proportional allocation. In the case of Neyman allocation, set the costs equal to 1 (or any other constant amount) for each stratum, and the cost terms will cancel out of the allocation expression. For proportional allocation, set the costs all equal to 1 and set the stratum variances all equal to the same constant, which could also be 1. Then, both the costs and the variances will cancel out of the allocation expression.

The amount of money to be spent on sampling is sometimes fixed before the experiment is started. Then the experimenter must find a sample size and allocation scheme that minimizes the variance of the estimator for a fixed expenditure.

Example 5.11 In the television-viewing example, suppose the costs are as specified in Example 5.7. That is, $c_1 = c_2 = 9$ and $c_3 = 16$. Let the stratum standard deviations be approximated by

$\sigma_1 \approx 5, \sigma_2 \approx 15$, and $\sigma_3 \approx 10$. Given that the advertising firm has only \$500 to spend on sampling, choose the sample size and the allocation that minimize $V(\bar{y}_{st})$.

Solution

The allocation scheme is still given by Equation (5.7). In Example 5.7, we found $w_1 = .32$, $w_2 = .39$, and $w_3 = .29$.

Since the total cost must equal \$500, we have

$$c_1 n_1 + c_2 n_2 + c_3 n_3 = 500$$
$$9n_1 + 9n_2 + 16n_3 = 500$$

Since $n_i = nw_i$, we can substitute as follows:

$$9nw_1 + 9nw_2 + 16nw_3 = 500$$
$$9n(.32) + 9n(.39) + 16n(.29) = 500$$

Solving for n, we obtain

$$11.03n = 500$$
$$n = \frac{500}{11.03} = 45.33$$

Therefore, we must take $n = 45$ to ensure that the cost remains below \$500. The corresponding allocation is given by

$$n_1 = nw_1 = (45)(.32) = 14$$
$$n_2 = nw_2 = (45)(.39) = 18$$
$$n_3 = nw_3 = (45)(.29) = 13$$

We can make the following summary statement on stratified random sampling: In general, stratified random sampling with proportional allocation will produce an estimator with smaller variance than that produced by simple random sampling (with the same sample size) if there is considerable variability among the stratum means. If sampling costs are nearly equal from stratum to stratum, stratified random sampling with optimal allocation [Equation (5.6)] will yield estimators with smaller variance than will proportional allocation when there is variability among the stratum variances.

5.6

ESTIMATION OF A POPULATION PROPORTION

In our numerical examples, we have been interested in estimating the average or the total number of hours per week spent watching television. In contrast, suppose that the advertising firm wants to estimate the proportion (fraction) of households that watches a particular show. The population is divided into strata, just as before, and a simple random sample is taken from each stratum. Interviews are then conducted to determine the proportion $\hat{p}_i$ of households in stratum i that view the show. This $\hat{p}_i$ is an unbiased estimator of p_i, the population proportion in stratum i (as described in Chapter 4). Reasoning as we did in Section 5.3, we conclude that $N_i\hat{p}_i$ is an unbiased estimator of the total number of households in stratum i that view this particular show. Hence $N_1\hat{p}_1 + N_2\hat{p}_2 + \cdots + N_L\hat{p}_L$ is a good estimator of the total number of viewing households in the population. Dividing this quantity by N, we obtain an unbiased estimator of the population proportion p of households viewing the show.

Estimator of the population proportion p:

$$\hat{p}_{st} = \frac{1}{N}(N_1\hat{p}_1 + N_2\hat{p}_2 + \cdots + N_L\hat{p}_L) = \frac{1}{N}\sum_{i=1}^{L} N_i\hat{p}_i \quad (5.13)$$

Estimated variance of $\hat{p}_{st}$:

$$\begin{aligned}\hat{V}(\hat{p}_{st}) &= \frac{1}{N^2}[N_1^2\hat{V}(\hat{p}_1) + N_2^2\hat{V}(\hat{p}_2) + \cdots + N_L^2\hat{V}(\hat{p}_L)] \\ &= \frac{1}{N^2}\sum_{i=1}^{L} N_i^2\hat{V}(\hat{p}_i) \\ &= \frac{1}{N^2}\sum_{i=1}^{L} N_i^2\left(\frac{N_i - n_i}{N_i}\right)\left(\frac{\hat{p}_i\hat{q}_i}{n_i - 1}\right) \quad (5.14)\end{aligned}$$

Example 5.12

The advertising firm wants to estimate the proportion of households in the county of Example 5.1 that view show X. The county is divided into three strata, town A, town B, and the rural area. The strata contain $N_1 = 155$, $N_2 = 62$, and $N_3 = 93$ households, respectively. A stratified random sample of $n = 40$ households is chosen with proportional allocation. In other words, a simple random sample is taken from each stratum; the sizes of the samples are $n_1 = 20$, $n_2 = 8$, and $n_3 = 12$. Interviews are conducted in the 40 sampled households; results are shown in Table 5.3. Estimate the proportion of households viewing show X, and place a bound on the error of estimation.

Solution

The estimate of the proportion of households viewing show X is given by $\hat{p}_{st}$. Using Equation (5.13), we calculate

$$\hat{p}_{st} = \frac{1}{310}[(155)(.80) + 62(.25) + 93(.50)] = .60$$

TABLE 5.3 Data for Example 5.12

Stratum	Sample size	Number of households viewing show X	$\hat{p}_i$
1	$n_1 = 20$	16	.80
2	$n_2 = 8$	2	.25
3	$n_3 = 12$	6	.50

The variance of $\hat{p}_{st}$ can be estimated by using Equation (5.14). First, let us calculate the $\hat{V}(\hat{p}_i)$ terms. We have

$$\hat{V}(\hat{p}_1) = \left(\frac{N_1 - n_1}{N_1}\right)\left(\frac{\hat{p}_1\hat{q}_1}{n_1 - 1}\right) = \left(\frac{155 - 20}{155}\right)\left[\frac{(.8)(.2)}{19}\right]$$
$$= (.871)(.008) = .007$$
$$\hat{V}(\hat{p}_2) = \left(\frac{N_2 - n_2}{N_2}\right)\left(\frac{\hat{p}_2\hat{q}_2}{n_2 - 1}\right) = \left(\frac{62 - 8}{62}\right)\left[\frac{(.25)(.75)}{7}\right]$$
$$= (.871)(.027) = .024$$
$$\hat{V}(\hat{p}_3) = \left(\frac{N_3 - n_3}{N_3}\right)\left(\frac{\hat{p}_3\hat{q}_3}{n_3 - 1}\right) = \left(\frac{93 - 12}{93}\right)\left[\frac{(.5)(.5)}{11}\right]$$
$$= (.871)(.023) = .020$$

From Equation (5.14)

$$\hat{V}(\hat{p}_{st}) = \frac{1}{N^2}\sum_{i=1}^{3} N_i^2 \hat{V}(\hat{p}_i)$$
$$= \frac{1}{(310)^2}\,[(155)^2(.007) + (62)^2(.024) + (93)^2(.020)]$$
$$= .0045$$

Then the estimate of proportion of households in the county that view show X, with a bound on the error of estimation, is given by

$$\hat{p}_{st} \pm 2\sqrt{\hat{V}(\hat{p}_{st})}, \qquad .60 \pm 2\sqrt{.0045}$$
$$.60 \pm 2(.07), \qquad .60 \pm .14$$

The bound on the error in Example 5.12 is quite large. We could reduce this bound and make the estimator more precise by increasing the sample size. The problem of choosing a sample size is considered in the next section.

5.7

SELECTING THE SAMPLE SIZE AND ALLOCATING THE SAMPLE TO ESTIMATE PROPORTIONS

To estimate a population proportion, we first indicate how much information we desire by specifying the size of the bound; the sample size is chosen accordingly.

The formula for the sample size n (for a given bound B on the error of estimation) is the same as Equation (5.6) except that σ_i^2 becomes $p_i q_i$.

Approximate sample size required to estimate p with a bound B on the error of estimation:

$$n = \frac{\sum_{i=1}^{L} N_i^2 p_i q_i / w_i}{N^2 D + \sum_{i=1}^{L} N_i p_i q_i} \tag{5.15}$$

where w_i is the fraction of observations allocated to stratum i, p_i is the population proportion for stratum i, and $D = \frac{B^2}{4}$.

The allocation formula that gives the variance of $\hat{p}_{st}$ equal to some fixed constant at minimum cost is the same as Equation (5.7) with σ_i replaced by $\sqrt{p_i q_i}$.

Approximate allocation that minimizes cost for a fixed value of $V(\hat{p}_{st})$ or minimizes $V(\hat{p}_{st})$ for a fixed cost:

$$\begin{aligned} n_i &= n \frac{N_i \sqrt{p_i q_i / c_i}}{N_1 \sqrt{p_1 q_1 / c_1} + N_2 \sqrt{p_2 q_2 / c_2} + \cdots + N_L \sqrt{p_L q_L / c_L}} \\ &= n \frac{N_i \sqrt{p_i q_i / c_i}}{\sum_{k=1}^{L} N_k \sqrt{p_k q_k / c_k}} \end{aligned} \tag{5.16}$$

where N_i denotes the size of the ith stratum, p_i denotes the population proportion for the ith stratum, and c_i denotes the cost of obtaining a single observation from the ith stratum.

Example 5.13 The data of Table 5.3 were obtained from a sample conducted last year. The advertising firm now wants to conduct a new survey in the same county to estimate the proportion of households viewing show X. Although the fractions p_1, p_2, and p_3 that appear in Equations (5.15) and (5.16) are unknown, they can be approximated by the estimates from the earlier study—that is, $\hat{p}_1 = .80$, $\hat{p}_2 = .25$, and $\hat{p}_3 = .50$. The cost of obtaining an observation is \$9 for either town and \$16 for the rural area—that is, $c_1 = c_2 = 9$ and $c_3 = 16$. The number of households within the strata are $N_1 = 155$, $N_2 = 62$, and $N_3 = 93$. The firm wants to estimate the population proportion p with a bound on the error of estimation equal to .1. Find the sample size n and the strata sample sizes, n_1, n_2, and n_3, that will give the desired bound at minimum cost.

Solution

We first use Equation (5.16) to find the allocation fractions w_i. Using $\hat{p}_i$ to approximate p_i, we have

$$\begin{aligned}\sum_{i=1}^{3} N_i\sqrt{\frac{\hat{p}_i\hat{q}_i}{c_i}} &= N_1\sqrt{\frac{\hat{p}_1\hat{q}_1}{c_1}} + N_2\sqrt{\frac{\hat{p}_2\hat{q}_2}{c_2}} + N_3\sqrt{\frac{\hat{p}_3\hat{q}_3}{c_3}}\\ &= 155\sqrt{\frac{(.8)(.2)}{9}} + 62\sqrt{\frac{(.25)(.75)}{9}} + 93\sqrt{\frac{(.5)(.5)}{16}}\\ &= \frac{62.000}{3} + \frac{26.846}{3} + \frac{46.500}{4}\\ &= 20.667 + 8.949 + 11.625 = 41.241\end{aligned}$$

and

$$n_1 = n\frac{N_1\sqrt{\hat{p}_1\hat{q}_1/c_1}}{\sum_{i=1}^{3} N_i\sqrt{\hat{p}_i\hat{q}_i/c_i}} = n\left(\frac{20.667}{41.241}\right) = n(.50)$$

Similarly,

$$n_2 = n\left(\frac{8.949}{41.241}\right) = n(.22)$$

$$n_3 = n\left(\frac{11.625}{41.241}\right) = n(.28)$$

Thus, $w_1 = .50$, $w_2 = .22$, and $w_3 = .28$.

The next step is to use Equation (5.15) to find n. First, the following quantities must be calculated:

$$\sum_{i=1}^{3} \frac{N_i^2 \hat{p}_i \hat{q}_i}{w_i} = \frac{N_1^2 \hat{p}_1 \hat{q}_1}{w_1} + \frac{N_2^2 \hat{p}_1 \hat{q}_2}{w_2} + \frac{N_3^2 \hat{p}_3 \hat{q}_3}{w_3}$$

$$= \frac{(155)^2(.8)(.2)}{.50} + \frac{(62)^2(.25)(.75)}{.22} + \frac{(93)^2(.5)(.5)}{.28}$$

$$= 18,686.46$$

$$\sum_{i=1}^{3} N_i \hat{p}_i \hat{q}_i = N_1 \hat{p}_1 \hat{q}_1 + N_2 \hat{p}_2 \hat{q}_2 + N_3 \hat{p}_3 \hat{q}_3$$

$$= (155)(.8)(.2) + (62)(.25)(.75) + (93)(.5)(.5)$$

$$= 59.675$$

To find D, we let $2\sqrt{V(\hat{p}_{st})} = .1$ (the bound on the error of estimation). Then

$$V(\hat{p}_{st}) = \frac{(.1)^2}{4} = .0025 = D$$

and

$$N^2 D = (310)^2(.0025) = 240.25$$

Finally, from Equation (5.15), n is given approximately by

$$n = \frac{\sum_{i=1}^{3} N_i^2 \hat{p}_i \hat{q}_i / w_i}{N^2 D + \sum_{i=1}^{3} N_i \hat{p}_i \hat{q}_i} = \frac{18,686.46}{240.25 + 59.675} = 62.3 \text{ or } 63$$

Hence

$$n_1 = nw_1 = (63)(.50) = 31$$
$$n_2 = nw_2 = (63)(.22) = 14$$
$$n_3 = nw_3 = (63)(.28) = 18$$

MINITAB MACROS FOR ESTIMATING A PROPORTION AND ALLOCATING THE SAMPLE; STRATIFIED RANDOM SAMPLING

Based on Examples 5.12 and 5.13, the MINITAB macros 'stprop.MTB' and 'stallocp.MTB' produce results for estimating a proportion and allocating a sample as shown below. Again, the computer results are slightly more accurate and may slightly disagree with the hand calculations.

```
MTB > Execute 'stprop.MTB' 1.
MTB > #
```

```
MTB > Exec 'mvarcolp'
MTB > #

Input the sample size followed by the population size.

Input the sample proportion.

The sample size is
K11     20.0000

The population size is
K22     155.000

The sample proportion is
K1      0.800000

The estimated variance of the sample proportion is
K2     0.00733446

The estimated standard deviation of the sample proportion is
K3     0.0856415

The bound on the error of estimation is
K4     0.171283

Input the sample size followed by the population size.

Input the sample proportion.

The sample size is
K11    8.00000

The population size is
K22    62.0000

The sample proportion is
K1     0.250000

The estimated variance of the sample proportion is
K2     0.0233295

The estimated standard deviation of the sample proportion is
K3     0.152740
```

```
The bound on the error of estimation is
K4      0.305480

Input the sample size followed by the population size.

Input the sample proportion.

The sample size is
K11   12.0000

The population size is
K22   93.0000

The sample proportion is
K1     0.500000

The estimated variance of the sample proportion is
K2      0.0197947

The estimated standard deviation of the sample proportion is
K3      0.140694

The bound on the error of estimation is
K4      0.281387

The following table lists the sample size, population size,
sample proportion, estimated variance of the sample
proportion, estimated standard deviation of the sample
proportion, and the bound on the error of estimation for
each column:

 ROW  samsize  popsize  samprop   variance   std dev     bound

   1       20      155     0.80  0.0073345  0.085641  0.171283
   2        8       62     0.25  0.0233295  0.152740  0.305480
   3       12       93     0.50  0.0197947  0.140694  0.281387

The estimator of the proportion, where stratified sampling
is used, is
K9      0.600000

The estimated variance of the proportion, using stratified
sampling, is
K13     0.00454832
```

```
The estimated standard deviation of the proportion, using
stratified sampling, is
K14    0.0674412

The bound on the error of estimation, using stratified
sampling, is
K15    0.134882

MTB > Execute 'stallop.MTB' 1.
MTB > #
Input the number of strata.

Input the population size for each stratum.

Input the population proportion for each stratum.

Input the cost of obtaining an observation from each
stratum.

Input the actual number desired for D.

The weights for each stratum are given below:

C36
 0.501124 0.216993 0.281882

The approximate sample size required to estimate mu or tau
with a bound B on the error of estimation is
K43    62.2257

The approximate individual sample sizes for each stratum are
given below:

C37
 31.1828 13.5025 17.5403
```

Example 5.14 Recalling that town A has more families with children than does town B, the firm wants to see if there is a significant difference between the proportions of families watching show X. Estimate the true difference between these proportions.

Solution

The output of the macro shows the computations of sample proportions and variances for each stratum, as does the solution to Example 5.12. Thus, the estimate of the difference becomes

$$(.80 - .25) \pm 2\sqrt{.007 + .023}$$
$$.55 \pm .35$$

It does, indeed, look like the proportion of families watching the show in question is larger for town A than it is for town B.

If the cost of sampling does not vary from stratum to stratum, then the cost factors c_i cancel from Equation (5.16).

Example 5.15 Suppose that in Example 5.13 telephone interviews are to be conducted, and hence the cost of sampling is the same in all strata. The fraction p_i will be approximated by $\hat{p}_i$, $i = 1, 2, 3$. We desire to estimate the population proportion p with a bound of .1 on the error of estimation. Find the appropriate sample size to achieve this bound at minimum cost.

Solution

Equation (5.16) is used to find the fractions w_1, w_2, and w_3, but now all c_i terms can be replaced by 1. Hence,

$$\sum_{i=1}^{3} N_i\sqrt{\hat{p}_i\hat{q}_i} = 155\sqrt{(.8)(.2)} + 62\sqrt{(.25)(.75)} + 93\sqrt{(.5)(.5)}$$
$$= 62.000 + 26.846 + 46.500 = 135.346$$

and

$$n_1 = n\left(\frac{N_1\sqrt{\hat{p}_1\hat{q}_1}}{\sum_{i=1}^{3} N_i\sqrt{\hat{p}_i\hat{q}_i}}\right) = n\left(\frac{62.000}{135.346}\right) = n(.46)$$

Similarly,

$$n_2 = n\left(\frac{26.846}{135.346}\right) = n(.20)$$

$$n_3 = n\left(\frac{46.500}{135.346}\right) = n(.34)$$

Thus, $w_1 = .46$, $w_2 = .20$, and $w_3 = .34$.

Equation (5.15) or Equation (5.10) with $\sigma_i = \sqrt{p_i q_i}$ can be used to find n. Using (5.10), we have

$$\sum_{i=1}^{3} N_i \hat{p}_i \hat{q}_i = 59.675 \qquad \text{(from Example 5.13)}$$

$$N^2 D = 240.25 \qquad \text{(from Example 5.13)}$$

and

$$n = \frac{\left(\sum_{i=1}^{3} N_i \sqrt{\hat{p}_i \hat{q}_i}\right)^2}{N^2 D + \sum_{i=1}^{3} N_i \hat{p}_i \hat{q}_i}$$

$$= \frac{(135.346)^2}{240.25 + 59.675} = 61.08 \text{ or } 62$$

Hence, we take a sample of 62 observations to estimate p with a bound on the error of magnitude $B = .1$. The corresponding allocation is given by

$$n_1 = nw_1 = 62(.46) = 29$$
$$n_2 = nw_2 = 62(.20) = 12$$
$$n_3 = nw_3 = 62(.34) = 21$$

These answers are close to those of Example 5.13. The changes in allocation result because costs do not vary in Example 5.15.

Recall that the allocation formula (5.7) assumes a very simple form when the variances as well as costs are equal for all strata. Equation (5.16) simplifies in the same way, provided all stratum proportions p_i are equal and all costs c_i are equal. Then Equation (5.16) becomes

$$n_i = n\left(\frac{N_i}{N}\right) \qquad i = 1, 2, \ldots, L \tag{5.17}$$

As previously noted, this method for assignment of sample sizes to the strata is called *proportional allocation.*

Example 5.16 In the television survey of Example 5.13, the advertising firm plans to use telephone interviews; therefore, the cost of sampling will not vary from stratum to stratum. The stratum sizes are $N_1 = 155$, $N_2 = 62$, and $N_3 = 93$. The results of last year's survey (see Table 5.3) do not appear to hold for this year. The firm believes that the proportion of households viewing show X is close to .4 in each of the three strata. The firm desires to estimate the population proportion p with a bound of .1 on the error of estimation. Find the sample size n and the allocation that gives this bound at minimum cost.

Solution

The allocation fractions are found by using Equation (5.16) with $p_1, \ldots, p_L$ and $c_1, \ldots, c_l$ replaced by 1. Thus,

$$n_1 = n\left(\frac{N_1}{\sum_{i=1}^{3} N_i}\right) = n\left(\frac{N_1}{N}\right) = n\left(\frac{155}{310}\right) = n(.5)$$

$$n_2 = n\left(\frac{N_2}{N}\right) = n\left(\frac{62}{310}\right) = n(.2)$$

$$n_3 = n\left(\frac{N_3}{N}\right) = n\left(\frac{93}{310}\right) = n(.3)$$

or

$$w_1 = .5, \qquad w_2 = .2, \qquad w_3 = .3$$

The sample size n is found from Equation (5.15) using .4 as an approximation to p_1, p_2, and p_3, or it can be found by setting $\sigma_i^2 = p_i q_i$ in Equation (5.11). Using the latter approach, with $p_i = .4$, yields.

$$\sum_{i=1}^{3} N_i p_i q_i = 155(.4)(.6) + 62(.4)(.6) + 93(.4)(.6)$$
$$= 74.4$$
$$ND = (310)(.0025) = .775$$

and

$$n = \frac{\sum_{i=1}^{3} N_i p_i q_i}{ND + \frac{1}{N}\sum_{i=1}^{3} N_i p_i q_i}$$
$$= \frac{74.4}{0.775 + (1/310)(74.4)} = 73.3 \text{ or } 74$$

Then

$$n_1 = nw_1 = (74)(.5) = 37$$
$$n_2 = nw_2 = (74)(.2) = 15$$
$$n_3 = nw_3 = (74)(.3) = 22$$

5.8

ADDITIONAL COMMENTS ON STRATIFIED SAMPLING

Stratified random sampling does not always produce an estimator with a smaller variance than that of the corresponding estimator in simple random sampling. The following example illustrates this point.

Example 5.17 A wholesale food distributor in a large city wants to know whether demand is great enough to justify adding a new product to his stock. To aid in making his decision, he plans to add this product to a sample of the stores he services in order to estimate average monthly sales. He only services four large chains in the city. Hence, for administrative convenience, he decides to use stratified random sampling with each chain as a stratum. There are 24 stores in stratum 1, 36 in stratum 2, 30 in stratum 3, and 30 in stratum 4. Thus $N_1 = 24$, $N_2 = 36$, $N_3 = 30$, $N_4 = 30$, and $N = 120$. The distributor has enough time and money to obtain data on monthly sales in $n = 20$ stores. Because he has no prior information on the stratum variances, and because the cost of sampling is the same in each stratum, he decides to use proportional allocation, which gives

$$n_1 = n\left(\frac{N_1}{N}\right) = 20\left(\frac{24}{120}\right) = 4$$

Similarly,

$$n_2 = 20\left(\frac{36}{120}\right) = 6$$

$$n_3 = 20\left(\frac{30}{120}\right) = 5$$

$$n_4 = 5$$

The new product is introduced in four stores chosen at random from chain 1, six stores from chain 2, and five stores each from chains 3 and 4. The sales figures after a month show the results given in the accompanying table. Estimate the average sales for the month, and place a bound on the error of estimation.

Stratum 1	Stratum 2	Stratum 3	Stratum 4
94	91	108	92
90	99	96	110
102	93	100	94
110	105	93	91
	111	93	113
	101		
$\bar{y}_1 = 99$	$\bar{y}_2 = 100$	$\bar{y}_3 = 98$	$\bar{y}_4 = 100$
$s_1^2 = 78.67$	$s_2^2 = 55.60$	$s_3^2 = 39.50$	$s_4^2 = 112.50$

Solution

From Equation (5.1)

$$\bar{y}_{\text{st}} = \frac{1}{N}\sum_{i=1}^{4} N_i\bar{y}_i = 99.3$$

Note that the estimate $\bar{y}_{\text{st}}$ of the population mean is the average of all sample observations when proportional allocation is used.

The estimated variance of $\bar{y}_{\text{st}}$, from Equation (5.2), is

$$\hat{V}(\bar{y}_{\text{st}}) = \frac{1}{N^2}\sum_{i=1}^{4} N_i^2\left(\frac{N_i - n_i}{N_i}\right)\left(\frac{s_i^2}{n_i}\right)$$

where for this example

$$\left(\frac{N_i - n_i}{N_i}\right) = \frac{5}{6} \qquad i = 1, 2, 3, 4$$

Then

$$\hat{V}(\bar{y}_{\text{st}}) = \frac{1}{(120)^2}\left(\frac{5}{6}\right)\left[(24)^2\left(\frac{78.67}{4}\right) + (36)^2\left(\frac{55.60}{6}\right) + (30)^2\left(\frac{39.50}{5}\right) + (30)^2\left(\frac{112.50}{5}\right)\right]$$
$$= 2.93$$

and the estimate of average monthly sales with a bound on the error of estimation is

$$\bar{y}_{\text{st}} \pm 2\sqrt{\hat{V}(\bar{y}_{\text{st}})}, \qquad 99.3 \pm 2\sqrt{2.93}, \qquad 99.3 \pm 3.4$$

Suppose the distributor had decided to take a simple random sample of $n = 20$ stores and the same 20 stores as in Example 5.17 were selected. In other words, suppose the 20 stores constitute a simple random sample rather than a stratified random sample. Then the estimator of the population mean has the same value as that calculated in the example, that is,

$$\bar{y} = \bar{y}_{\text{st}} = 99.3$$

But the estimated variance becomes

$$\hat{V}(\bar{y}) = \left(\frac{N - n}{N}\right)\left(\frac{s^2}{n}\right) = \left(\frac{5}{6}\right)\left(\frac{59.8}{20}\right) = 2.49$$

We see that the estimated variance is *smaller* for simple random sampling. Thus, we conclude simple random sampling may have been better than stratified random sampling for this problem. The experimenter did not consider the fact that sales vary greatly among stores within a chain when he stratified on chains. He could have obtained a smaller variance for his estimator by stratifying on amount of sales, that is, by putting stores with low monthly sales in one stratum, stores with high sales in another, and so forth.

In many sample survey problems, more than one measurement is taken on each sampling unit in order to estimate more than one population parameter. This situation causes complications in selecting the appropriate sample size and allocation, as illustrated in the following example.

Example 5.18

A state forest service is conducting a study of the people who use state-operated camping facilities. The state has two camping areas, one located in the mountains and one located along the coast. The forest service wishes to estimate the average number of people per campsite and the proportion of campsites occupied by out-of-state campers during a particular weekend when all sites are expected to be used. The average number of people is to be estimated with a bound of 1 on the error of estimation, and the proportion of out-of-state users is to be estimated with a bound of .1. The two camping areas conveniently form two strata, the mountain location forming stratum 1 and the coastal location stratum 2. It is known that $N_1 = 120$ campsites and $N_2 = 80$ campsites. Find the sample size and allocation necessary to achieve both of the bounds.

Solution

Assuming that the costs of sampling are the same in each stratum, we can achieve the smallest sample size by using Neyman allocation. However, this allocation depends on the stratum variances and gives different allocations for the two different types of measurements involved in the problem. Instead, we use proportional allocation because it is usually close to optimum and it gives the same allocation for any desired measurement. Thus,

$$w_1 = \frac{N_1}{N} = \frac{120}{200} = .6$$

$$w_2 = \frac{N_2}{N} = \frac{80}{200} = .4$$

Now the sample size must be determined separately for each of the desired estimates. First, consider estimating the average number of persons per campsite. We must have an approximation of the stratum variances in order to use Equation (5.6) for the sample size. The forest service knows from experience that most sites contain from 1 to 9 persons. Therefore, we can use the approximation

$$\sigma_i \approx \frac{9-1}{4} = 2 \qquad i = 1, 2$$

Hence,

$$\sum_{i=1}^{2} \frac{N_i^2\sigma_i^2}{w_i} = \frac{(120)^2(4)}{.6} + \frac{(80)^2(4)}{.4} = 160{,}000$$

$$\sum_{i=1}^{2} N_i\sigma_i^2 = (120)(4) + (80)(4) = 800$$

$$N^2D = N^2\left(\frac{B^2}{4}\right) = (200)^2\left(\frac{1}{4}\right) = 10{,}000$$

From Equation (5.6),

$$n = \frac{\sum_{i=1}^{2} N_i^2\sigma_i^2/w_i}{N^2D + \sum_{i=1}^{2} N_i\sigma_i^2} = \frac{160{,}000}{10{,}000 + 800} = 14.8 \text{ or } 15$$

is the required sample size.

Now let us consider estimating the proportion of out-of-state users. No prior estimates of the stratum proportions p_i are available, so we let $p_1 = p_2 = .5$ to obtain a maximum sample size. We use Equation (5.15) to find n, and hence we must find

$$\sum_{i=1}^{2} \frac{N_i^2 p_i q_i}{w_i} = \frac{(120)^5(.5)(.5)}{.6} + \frac{(80)^2(.5)(.5)}{.4} = 10{,}000$$

$$N^2D = N^2\left(\frac{B^2}{4}\right) = (200)^2\left(\frac{.01}{4}\right) = 100$$

$$\sum_{i=1}^{2} N_i p_i q_i = 120(.5)(.5) + 80(.5)(.5) = 50$$

From Equation (5.15),

$$n = \frac{\sum_{i=1}^{2} N_i^2 p_i q_i/w_i}{N^2D + \sum_{i=1}^{2} N_i p_i q_i} = \frac{10{,}000}{100 + 50} = 67$$

Thus,

$$n_1 = nw_1 = (67)(.6) = 40$$
$$n_2 = nw_2 = (67)(.4) = 27$$

are the sample sizes required in order to achieve both bounds. Note that these sample sizes give an estimate of the average number of persons per campsite with a much smaller bound than required.

5.9

AN OPTIMAL RULE FOR CHOOSING STRATA

If our only objective of stratification is to produce estimators with small variance, then the best criterion by which to define strata is the set of values that the response can take on. For example, suppose we wish to estimate the average income per household in a community. We could estimate this average quite accurately if we could put all low-income households in one stratum and all high-income households in another before actually sampling. Of course, this allocation is often impossible because detailed knowledge of incomes before sampling might make the statistical problem unnecessary in the first place. However, we sometimes have some relating frequency data on broad categories of the variable of interest or on some highly correlated variable. In these cases, the "cumulative square root of the frequency method" works well for delineating strata. Rather than attempt to explain this method in theory, we will simply show how it works in practice. Generally, we would not choose more than five or six strata when using this method.

Example 5.19 An investigator wishes to estimate the average yearly sales for 56 firms, using a sample of $n = 15$ firms. Frequency data on these firms is available in the form of classification by \$50,000 increments and appears in the accompanying table. How can we best allocate the firms to $L = 3$ strata?

Income (in thousands)	Frequency	$\sqrt{\text{Frequency}}$	Cumulative $\sqrt{\text{frequency}}$
100–150	11	3.32	3.32
150–200	14	3.74	7.06
200–250	9	3.00	10.06
250–300	4	2.00	12.06
300–350	5	2.24	14.30
350–400	8	2.83	17.13
400–450	3	1.73	18.86
450–500	2	1.41	20.27
	56		

Solution

Note that we have added two columns to the frequency data for the population, namely, the square root of the frequencies and the cumulative square root. The approximately optimal method for stratification is to mark off equal intervals on the cumulative square root scale. (*Note*: On this scale 7.06 is $3.32 + 3.74$, and so on.) Thus, $(20.27)/3 = 6.76$, and our stratum boundaries should be as close as possible to 6.76 and $2(6.76) = 13.52$.

On the actual scale, 7.06 is closest to 6.76, and 14.30 is closest to 13.52. Thus, the following three strata result:

Stratum 1 : Firms with scales from 100,000 to 200,000

Stratum 2 : Firms with scales from 200,001 to 350,000

Stratum 3 : Firms with scales from 350,001 to 500,000

Assuming that firms in these strata can be identified before sampling, the sample of $n = 15$ can be allocated five to each stratum. (Equal stratum sample sizes are nearly optimal with this technique.)

5.10

STRATIFICATION AFTER SELECTION OF THE SAMPLE

Occasionally, sampling problems arise in which we would like to stratify on a key variable, but we cannot place the sampling units into their correct strata until after the sample is selected. For example, we may wish to stratify a public opinion poll by gender of the respondent. If the poll is conducted by sampling telephone numbers, then respondents cannot be placed into the male or female stratum until after they are contacted. Similarly, an auditor may want to stratify accounts according to whether they are wholesale or retail, but she may not have this information until after an account is actually pulled for the sample.

Suppose a simple random sample of n people is selected for a poll. The sample can be divided into n_1 men and n_2 women after the sample is interviewed. Then instead of using $\bar{y}$ to estimate μ, we can use $\bar{y}_{st}$ *provided* that N_i/N is known for both men and women. Note that in this situation n_1 and n_2 are *random* since they can change from sample to sample even though n is fixed. Thus, this sample is not exactly a stratified random sample according to Definition 5.1. However, if N_i/N is known and if $n_i \geq 20$ for each stratum, then this method of stratification after selection of the sample is nearly as accurate as stratified random sampling with proportional allocation.

Stratification after the selection of a sample (or *poststratification*) is often appropriate when a simple random sample is not properly balanced according to major groupings of the population. Suppose, for example, that a simple random sample of $n = 100$ people is selected from a population that should be equally divided between men and women. The sample measurement of interest is the weight of the respondent, and the goal is to estimate the average weight of people in the population. The sample gives the following information:

Men	Women
$n_1 = 20$	$n_2 = 80$
$\bar{y}_1 = 180$ pounds	$\bar{y}_2 = 110$ pounds
$\bar{y} = 124$	

With men underrepresented in the sample, the estimate $\bar{y} = 124$ seems unduly low. We can adjust this estimate by calculating

$$\bar{y}_{\text{st}} = \left(\frac{N_1}{N}\right)\bar{y}_1 + \left(\frac{N_2}{N}\right)\bar{y}_2 = .5(180) + .5(110) = 145$$

This estimate seems to be more realistic, since men and women are now equally weighted. Note that N_i/N is known, to a good degree of approximation, even though neither N_1 nor N_2 is given.

The poststratification estimator $\bar{y}_{\text{st}}$ will not have the same variance as the stratified sampling mean of Chapter 5, because the stratification was not designed into the sampling plan. However, an approximate variance can be worked out along the following lines.

To simplify the writing, we let $N_i/N = W_i$, $i = 1, \ldots, L$. In poststratification, the n_i are random variables with

$$E(n_i) = nW_i, \qquad i = 1, \ldots, L$$

Thus, the poststratification estimator should behave something like a stratified sampling estimator under proportional allocation. We will see that this is indeed the case.

If the n_i were fixed, Equation (5.2) gives

$$\begin{aligned} \hat{V}(\bar{y}_{\text{st}}) &= \sum_{i=1}^{L} W_i^2 \frac{s_i^2}{n_i}\left(\frac{N_i - n_i}{N_i}\right) \\ &= \sum_{i=1}^{L} W_i^2 \frac{s_i^2}{n_i} - \frac{1}{N}\sum_{i=1}^{L} W_i s_i^2 \end{aligned} \tag{5.18}$$

Note, however, that the n_i are random. In this situation, a general expression for $\hat{V}(\bar{y}_{\text{st}})$ can be approximated by replacing $1/n_i$ by its expected value. Unfortunately, it is difficult to find the expected value of the reciprocal of a random variable, but a good approximation here is given by

$$E\left(\frac{1}{n_i}\right) \approx \frac{1}{nW_i} + \frac{1 - W_i}{n^2 W_i^2} \tag{5.19}$$

Substitution of the expression in (5.19) for $1/n_i$ in Equation (5.18) yields

$$\begin{aligned} \hat{V}_p(\bar{y}_{\text{st}}) &= \frac{1}{n}\sum_{i=1}^{n} W_i s_i^2 + \frac{1}{n^2}\sum_{i=1}^{L}(1 - W_i)s_i^2 - \frac{1}{N}\sum W_i s_i^2 \\ &= \left(\frac{1}{n} - \frac{1}{N}\right)\sum_{i=1}^{L} W_i s_i^2 + \frac{1}{n^2}\sum_{i=1}^{L}(1 - W_i)s_i^2 \\ &= \frac{N - n}{Nn}\sum_{i=1}^{L} W_i s_i^2 + \frac{1}{n^2}\sum_{i=1}^{L}(1 - W_i)s_i^2, \end{aligned} \tag{5.20}$$

where the subscript p refers to poststratification. The first term in $\hat{V}_p(\bar{y}_{\text{st}})$ is the variance one would get from a stratified sample mean under proportional allocation. The second term is always nonnegative and shows the amount of increase in variance one can expect from post- rather than prestratification. Note that the divisor of the increase term is n^2 and consequently that term is usually quite small.

To summarize, the approximation (5.19) only works well when n is large and n_i is guaranteed to be positive. Also, the increase in variance is small when n is large. Thus, poststratification produces good results only when n is large and all n_i terms are relatively large as well. A practical consequence of this is that we cannot poststratify too finely. For example, we could poststratify a sample of people into age groups, but we would be forced to stick with only two or three such age groups, not eight or ten.

Example 5.20 A large firm knows that 40% of its accounts receivable are wholesale and 60% are retail. However, to identify individual accounts without pulling a file and looking at it is difficult. An auditor wishes to sample $n = 100$ of these accounts in order to estimate the average amount of accounts receivable for the firm. A simple random sample turns out to contain 70% wholesale accounts and 30% retail accounts. The data are separated into wholesale and retail accounts after sampling, with the following results (in dollars):

Wholesale	Retail
$n_1 = 70$	$n_2 = 30$
$\bar{y}_1 = 520$	$\bar{y}_2 = 280$
$s_1 = 210$	$s_2 = 90$

Estimate μ, the average amount of accounts receivable for the firm and place a bound on the error of estimation.

Solution

Since the observed proportion of wholesale accounts (.7) is far from the true population proportion (.4), stratifying *after* a simple random sample is selected seems appropriate. This procedure is justified since n_1 and n_2 both exceed 20.

Now

$$\bar{y}_{st} = \left(\frac{N_1}{N}\right)\bar{y}_1 + \left(\frac{N_2}{N}\right)\bar{y}_2 = (.4)(520) + (.6)(280) = 376$$

and ignoring the finite population correction, we have from (5.20):

$$\begin{aligned}\hat{V}_p(\bar{y}_{st}) &= \frac{1}{n}\sum_{i=1}^{2} W_i s_i^2 + \frac{1}{n^2}\sum_{i=1}^{2}(1 - W_i)s_i^2 \\ &= \frac{1}{100}[.4(210)^2 + .6(90)^2] + \frac{1}{(100)^2}[.6(210)^2 + .4(90)^2] \\ &= 225 + 2.97 \\ &= 227.97\end{aligned}$$

and

$$2\sqrt{\hat{V}_p(\bar{y}_{st})} = 30.$$

Hence, we are quite confident that the interval \$346 to \$406 contains the true value of μ.

In the expression for $\hat{V}_p(\bar{y}_{st})$, the first term is what we would have obtained if we had stratified in advance of sampling (and obtained these sample results). The second term is the penalty we pay for not stratifying in advance.

Two notes of caution are in order. If N_i/N is not known or if it cannot be closely approximated, this method of stratification should *not* be used since errors in the weights, N_i/N, can cause this stratified estimator to be very poor. Somctimcs, this method of stratification is used to adjust for nonresponse. For example, if many nonrespondents to a simple random sample were men, then the sample proportion of men would be low, and an adjusted estimate could be produced by stratification after sampling. This method may still lead to serious bias in the results if the nonresponse biases the simple random sample. The point to remember is that the original sample must still be a simple random sample from the population.

5.11 DOUBLE SAMPLING FOR STRATIFICATION

Up to this point, it has been assumed that $W_i = N_i/N$, $i = 1, \ldots, L$, are known constants before sampling begins. A little thought will suggest that this often is not the case, even though stratified sampling seems to be appropriate. For example, we may want to stratify a population of voters according to gender, income level, or education level, but information to perform the stratification is not available from voter registration rolls. We may want to stratify automobiles into classes of miles driven (for purposes of obtaining detailed information on owner satisfaction), but we will not be able to do that until we talk to the owners.

The basic idea of double sampling (two-phase sampling) is rather simple, but it does complicate the estimation of variances. Suppose that preliminary information (such as gender of a voter or mileage on a car) on which to base stratification is easy to obtain, whereas detailed information on the variables under study (such as opinions on political issues or on the quality of a car) is not. We then can take a large sample for identifying strata and a much smaller sample for collecting detailed data. We could, for example, call many voters to identify gender or income level (phase 1 sample), when only a few could be interviewed (phase 2 sample) for purposes of completing a detailed questionnaire.

Suppose the phase 1 sample, of size n', is used to determine which elements fall into the various strata. Let

$$w_i' = \frac{n_i'}{n'}, \qquad i = 1, \ldots, L \tag{5.21}$$

denote the proportion of the first sampling falling into stratum i. Then w_i' is an unbiased estimator of W_i, assuming randomness of the phase 1 sample.

At the second phase of sampling, n_i elements are randomly sampled from the n_i' elements identified as belonging to stratum i. Measurements are obtained from these n_i elements, and $\bar{y}_i$ and s_i can be calculated for each stratum. Following Equation (5.1), we can then make up an estimator of the population mean μ as

$$\bar{y}_{\text{st}}' = \sum_{i=1}^{L} w_i' \bar{y}_i \tag{5.22}$$

If the phase 2 sampling fractions for each stratum, n_i/N_i, are all small and N is large, an approximate variance for $\bar{y}_{\text{st}}'$ is given by

$$\hat{V}(\bar{y}_{\text{st}}') = \frac{n'}{n'-1} \sum_{i=1}^{L} \left[\left(w_i'^2 - \frac{w_i'}{n'} \right) \frac{s_i^2}{n_i} + \frac{w_i'(\bar{y}_i - \bar{y}_{\text{st}}')^2}{n'} \right] \tag{5.23}$$

If n' is so large that w_i'/n' is negligible, this variance estimate reduces to

$$\hat{V}(\bar{y}_{\text{st}}') = \sum_{i=1}^{L} \left[\frac{w_i'^2 s_i^2}{n_i} + \frac{w_i'(\bar{y}_i - \bar{y}_{\text{st}}')^2}{n'} \right] \tag{5.24}$$

Ignoring finite population correction terms, the first part of Equation (5.24) looks like (5.2), with w_i' replacing N_i/N. The second part is the additional component of variance that comes about because we did not know the correct weights W_i and had to estimate them from a sample. Note that we do not necessarily want to make the second term small, because that would necessitate making the $\bar{y}_i$'s about equal. Recall that stratification pays dividends over simple random sampling when the $\bar{y}_i$ are quite different. Thus, choosing strata that produce different $\bar{y}_i$ still may be better than simple random sampling, even though double sampling might have to be employed to estimate the W_i.

We illustrate the computations in the following example.

Example 5.21 From a list of enrollments and faculty sizes for American four-year colleges and universities, it is desired to estimate the average enrollment (for the 1986–1987 academic year). Private institutions tend to be smaller than public ones, so stratification is in order. However, the list is not broken up this way, even though the data is coded to indicate the type of college or university. Thus, the type of college (public or private) can be obtained quickly, while the enrollment data is more cumbersome to handle.

A one-in-ten systematic sample was done to obtain information on type of college.

This resulted in the following:

Private	Public	Total
$n'_1 = 84$	$n'_2 = 57$	$n' = 141$

Subsamples of 11 private and 12 public colleges gave the following data on enrollments and faculty size. (The faculty data will be used later.) Estimate the average enrollment for American colleges and universities in 1986–1987.

Private $n_1 = 11$		***Public*** $n_2 = 12$	
Enrollment	**Faculty**	**Enrollment**	**Faculty**
1618	122	7332	452
1140	88	2356	131
1000	65	21879	996
1225	55	935	50
791	79	1293	106
1600	79	5894	326
746	40	8500	506
1701	75	6491	371
701	32	781	108
6918	428	7255	298
1050	110	2136	128
		5380	280

Source: The World Almanac & Book of Facts, 1988 edition. Copyright © 1987, Newspaper Enterprise Association, Inc., New York, pp. 234–250.

Solution

From the data given above,

$$\begin{aligned}\bar{y}'_{st} &= w'_1\bar{y}_1 + w'_2\bar{y}_2 \\ &= \left(\frac{84}{141}\right)(1681) + \left(\frac{57}{141}\right)(5853) \\ &= (.60)(1681) + (.40)(5853) = 3349.8\end{aligned}$$

Using Equation (5.24),

$$\begin{aligned}
\hat{V}(\bar{y}'_{st}) &= \frac{1}{n_1}(w'_1 s_1)^2 + \frac{1}{n_2}(w'_2 s_2)^2 \\
&\quad + \frac{1}{n'}[w'_1(\bar{y}_1 - \bar{y}'_{st})^2 + w'_2(\bar{y}_2 - \bar{y}'_{st})^2] \\
&= \frac{1}{11}[(.60)(1773)]^2 + \frac{1}{12}[(.40)(5763)]^2 \\
&\quad + \frac{1}{141}[(.60)(1681 - 3349.8)^2 + (.40)(5853 - 3349.8)^2] \\
&= 545{,}708.05 + 29{,}626.52 \\
&= 575{,}334.57
\end{aligned}$$

and

$$\sqrt{\hat{V}(\bar{y}'_{st})} = 758.5.$$

The second part of the variance (the one due to estimating the true stratum weights) may look large, but it constitutes only 5% of the final variance. The resulting estimate of the standard deviation of $\bar{y}'_{st}$ is still quite large due to the small sample sizes and large variation among college enrollments, but it is much smaller than the error associated with a single random sample of 23 colleges from the list.

5.12

SUMMARY

A stratified random sample is obtained by separating the population elements into groups, or strata, such that each element belongs to one and only one stratum, and then independently selecting a simple random sample from each stratum. This sample survey design has three major advantages over simple random sampling. First, the variance of the estimator of the population mean is usually reduced because the variance of observations within each stratum is usually smaller than the overall population variance. Second, the cost of collecting and analyzing the data is often reduced by the separation of a large population into smaller strata. Third, separate estimates can be obtained for individual strata without selecting another sample and, hence, without additional cost.

An unbiased estimator, $\bar{y}_{st}$, of the population mean is a weighted average of the sample means for the strata; it is given by Equation (5.1). An unbiased estimator of the variance of $\bar{y}_{st}$ is given by Equation (5.2); this estimator is used in placing bounds on the error of estimation. An unbiased estimator of the population total is also given, along with its estimated variance.

Before conducting a survey, experimenters should consider how large an error of estimation they will tolerate and then should select the sample size accordingly. The sample size n is given by Equation (5.6) for a fixed bound B on the error of estimation.

The sample must then be allocated among the various strata. The allocation that gives a fixed amount of information at minimum cost is given by Equation (5.7); it is affected by the stratum sizes, the stratum variances, and the costs of obtaining observations.

The estimator $\hat{p}_{st}$ of a population proportion has the form as $\bar{y}_{st}$ and is given by Equation (5.13). An unbiased estimator of $V(\hat{p}_{st})$ is given by Equation (5.14). The related allocation and sample size problems have the same solutions as before, except that σ^2 is replaced by $p_i q_i$.

CASE STUDY REVISITED

THE ESTIMATION OF HEALTH CARE COSTS

In the problem of estimating total first hospitalization costs for kidney stone patients, the Carolinas and the Rockies were selected as strata because they have very different incident rates for the disease, and information was desired for each region separately. Also, this stratification into geographic regions simplified the sampling procedures.

The sample data is summarized as follows:

Carolinas	Rockies
$n_1 = 363$	$n_2 = 258$
$\bar{y}_1 = 1350$	$\bar{y}_2 = 1150$
$\frac{s_1^2}{n_1} = 3600$	$\frac{s_2^2}{n_2} = 3600$

To estimate the total annual cost for the regions, we must first find N_1 and N_2, the numbers of stone patients expected to be found in the respective regions in a typical year. We can approximate these figures if we can find the incident rates for the disease and if we know the total population of the regions.

A companion study showed the number of stone incidents in the Carolinas to be 454 out of 100,000 population and the number in the Rockies to be 263 out of 100,000. The population of the Carolinas is 8,993,000, and the population of the Rocky Mountain states is 7,351,000, according to the 1980 census. Thus,

$$N_1 = 8{,}993{,}000\left(\frac{454}{100{,}000}\right) = 40{,}828$$

and

$$N_2 = 7{,}351{,}000\left(\frac{263}{100{,}000}\right) = 19{,}333$$

We can now estimate the total annual first hospitalization cost for stone patients in the two regions combined as

$$N_1\bar{y}_1 + N_2\bar{y}_2$$

or

$$(40{,}828)(1350) + (19{,}333)(1150) = 77{,}350{,}750$$

The bound on the error of estimation is (since population sizes are large compared with sample sizes)

$$2\sqrt{\frac{N_1^2(s_1^2)}{n_1} + \frac{N_2^2(s_2^2)}{n_2}} = 2\sqrt{(40{,}828)^2(3600) + (19{,}333)^2(3600)}$$

$$= 5{,}420{,}880$$

Thus, we estimate the total annual cost for the two regions to be between, roughly, \$72 million and \$82 million.

This method can be used to estimate the total cost for the entire United States, but sample data would be required for the remaining geographic regions.

EXERCISES

5.1 A chain of department stores is interested in estimating the proportion of accounts receivable that are delinquent. The chain consists of four stores. So that the cost of sampling is reduced, stratified random sampling is used, with each store as a stratum. Since no information on population proportions is available before sampling, proportional allocation is used. From the accompanying table, estimate p, the proportion of delinquent accounts for the chain, and place a bound on the error of estimation.

	Stratum I	Stratum II	Stratum III	Stratum IV
Number of accounts receivable	$N_1 = 65$	$N_2 = 42$	$N_3 = 93$	$N_4 = 25$
Sample size	$n_1 = 14$	$n_2 = 9$	$n_3 = 21$	$n_4 = 6$
Sample number of delinquent accounts	4	2	8	1

5.2 A corporation desires to estimate the total number of man-hours lost, for a given month, because of accidents among all employees. Since laborers, technicians, and administrators have different accident rates, the researcher decides to use stratified random sampling, with each group forming a separate stratum. Data from previous years suggest the variances shown in the accompanying table for the number of man-hours lost per employee in the three groups, and current data give the stratum sizes. Determine the Neyman allocation for a sample of $n = 30$ employees.

I (Laborers)	II (Technicians)	III (Administrators)
$\sigma_1^2 = 36$	$\sigma_2^2 = 25$	$\sigma_3^2 = 9$
$N_1 = 132$	$N_2 = 92$	$N_3 = 27$

5.3 For Exercise 5.2, estimate the total number of man-hours lost during the given month and place a bound on the error of estimation. Use the data in the accompanying table, obtained from sampling 18 laborers, 10 technicians, and 2 administrators. Make a plot of the data to check for unusual features.

I (Laborers)			II (Technicians)		III (Administrators)
8	24	0	4	5	1
0	16	32	0	24	8
6	0	16	8	12	
7	4	4	3	2	
9	5	8	1	8	
18	2	0			

5.4 A report from the Census Bureau in October 1994 provided data on new one-family houses for a sample of 28 metropolitan statistical areas (MSAs) and consolidated metropolitan statistical areas (CMSAs) from around the country. (CMSAs tend to be larger than MSAs and can be subdivided into other metropolitan areas for purposes of census data summaries.) Data on total housing units sold, median sales price, and median floor area per house are reported, as shown below. The median sales price can be thought of as a typical price for that area. Similarly, the median floor area can be thought of as a typical floor area for houses in that area. There were 250 MSAs and 18 CMSAs in the United States for the year in which these data were reported.

(a) Plot the sales prices in parallel box plots, one for MSAs and one for CMSAs, and comment on any unusual features you see. Do you see any reason to make adjustments to the data before proceeding to estimate the mean typical selling price for the country?

(b) Treating these data as a stratified random sample, with the MSAs and CMSAs being the two strata, estimate the mean typical sales price per house for all metropolitan areas of the United States. Calculate a bound for the error of estimation.

(c) Plot the total number of units sold in parallel box plots, one for each stratum. Do you see any unusual features here?

(d) Estimate the total number of houses sold in all metropolitan areas of the United States in 1993 and calculate a bound for the error of estimation.

(e) Suppose you are to estimate the population mean or total for each of the three variables in the data set. For which of the three outcome variables—total units sold, price, or square footage—will stratification produce the *least* gain in precision over simple random sampling? Explain.

(f) Estimate the difference in average typical selling price between the two strata. Can we say that houses in the CMSAs are, on the average, higher priced than those in the MSAs?

Metropolitan area	1993 total sold (thous.)	Median Sales price (in dollars)	Median Finished floor area (in sq ft)
Atlanta, GA MSA	27.8	$118,200	2120
Charlotte–Gastonia–Rock Hill, NC–SC MSA	7.8	115,100	1945
Chicago–Gary–Kenosha, IL–IN–WI CMSA	18.9	159,500	2020
Colorado Springs, CO MSA	3.0	138,600	2210
Dallas–Fort Worth, TX CMSA	19.2	123,000	2325
Denver–Boulder–Greeley, CO CMSA	11.9	174,600	2225
Houston–Galveston–Brazoria, TX CMSA	10.2	114,200	2680
Jacksonville, FL MSA	5.0	95,000	1995
Kansas City, MO–KS MSA	6.5	99,300	1720
Las Vegas, NV–AZ MSA	18.5	121,700	1770
Los Angeles–Riverside–Orange County, CA CMSA	23.4	139,800	1820
Miami–Fort Lauderdale, FL CMSA	11.2	131,500	2185
Minneapolis–St. Paul, MN-WI MSA	10.6	155,600	2030
New Orleans, LA MSA	1.6	99,200	2045
New York–Northern NJ–Long Island, NY–NJ–CT–PA CMSA	17.9	191,400	2140
Norfolk–Virginia Beach–Newport News, VA–NC MSA	6.0	120,200	2245
Orlando, FL MSA	10.5	107,500	1725
Phoenix–Mesa, AZ MSA	21.9	113,900	2180
Sacramento–Yolo, CA CMSA	6.5	144,000	1540
St. Louis, MO–IL MSA	6.9	144,500	1995
Salt Lake City–Ogden, UT MSA	6.5	100,300	1545
San Antonio, TX MSA	3.0	117,900	2375
San Diego, CA MSA	3.8	225,000	2375
Seattle–Tacoma–Bremerton, WA CMSA	10.8	159,700	1885
Tampa–St. Petersburg–Clearwater, FL MSA	8.4	113,700	2240
Tucson, AZ MSA	3.8	106,600	1810
Washington-Baltimore, DC–MD–VA–WV CMSA	31.1	184,400	2305
West Palm Beach–Boca Raton, FL MSA	5.6	158,400	2150

5.5 A corporation wishes to obtain information on the effectiveness of a business machine. A number of division heads will be interviewed by telephone and asked to rate the equipment on a numerical scale. The divisions are located in North America, Europe, and Asia. Hence, stratified sampling is used. The costs are larger for interviewing division heads located outside North America. The accompanying table gives the costs per interview, approximate variances of the ratings, and N_i's that have been established. The corporation wants to estimate the average rating with $V(\bar{y}_{st}) = .1$. Choose the sample size n that achieves this bound, and find the appropriate allocation.

Stratum I (North America)	Stratum II (Europe)	Stratum III (Asia)
$c_1 = \$9$	$c_2 = \$25$	$c_3 = \$36$
$\sigma_1^2 = 2.25$	$\sigma_2^2 = 3.24$	$\sigma_3^2 = 3.24$
$N_1 = 112$	$N_2 = 68$	$N_3 = 39$

5.6 A school desires to estimate the average score that may be obtained on a reading comprehension exam for students in the sixth grade. The school's students are grouped into three tracks, with the fast learners in track I and the slow learners in track III. The school decides to stratify on tracks since this method should reduce variability of test scores. The sixth grade contains 55 students in track I, 80 in track II, and 65 in track III. A stratified random sample of 50 students is proportionally allocated and yields simple random samples of $n_1 = 14$, $n_2 = 20$, and $n_3 = 16$ from tracks I, II, and III. The test is administered to the sample of students, with the results as shown in the table.
(a) Estimate the average score for the sixth grade, and place a bound on the error of estimation.

Track I		Track II		Track III	
80	92	85	82	42	32
68	85	48	75	36	31
72	87	53	73	65	29
85	91	65	78	43	19
90	81	49	69	53	14
62	79	72	81	61	31
61	83	53	59	42	30
		68	52	39	32
		71	61		
		59	42		

(b) Construct parallel box plots for these data and comment upon the patterns you see. Do you think there could be a problem in placing students in tracks?
(c) Estimate the difference in average scores between track I and track II students. Are track I students significantly better, on the average, than track II students?

5.7 Suppose the average test score for the class in Exercise 5.6 is to be estimated again at the end of the school year. The costs of sampling are equal in all strata, but the variances differ. Find the optimum (Neyman) allocation of a sample of size 50, using the data of Exercise 5.6 to approximate the variances.

5.8 Using the data of Exercise 5.6, find the sample size required to estimate the average score, with a bound of 4 points on the error of estimation. Use proportional allocation.

5.9 Repeat Exercise 5.8, by using Neyman allocation. Compare the results with the answer to Exercise 5.8.

5.10 A forester wants to estimate the total number of farm acres planted in trees for a state. Since the number of acres of trees varies considerably with the size of the farm, he decides to stratify on farm sizes. The 240 farms in the state are placed in one of four categories according to size. A stratified random sample of 40 farms, selected by using proportional allocation, yields the results shown in the accompanying table on number of acres planted in trees. Estimate the total number of acres of trees on farms in the state, and place a bound on the error of estimation. Graph the data on an appropriate plot and comment on the variation as one moves from I to IV.

Stratum I 0–200 acres		Stratum II 200–400 acres		Stratum III 400–600 acres		Stratum IV Over 600 acres	
$N_1 = 86$		$N_2 = 72$		$N_3 = 52$		$N_4 = 30$	
$n_1 = 14$		$n_2 = 12$		$n_3 = 9$		$n_4 = 5$	
97	67	125	155	142	256	167	655
42	125	67	96	310	440	220	540
25	92	256	47	495	510	780	
105	86	310	236	320	396		
27	43	220	352	196			
45	59	142	190				
53	21						

5.11 The study of Exercise 5.10 is to be made yearly, with the bound on the error of estimation of 5000 acres. Find an approximate sample size to achieve this bound if Neyman allocation is used. Use the data in Exercise 5.10.

5.12 A psychologist working with a group of mentally retarded adults desires to estimate their average reaction time to a certain stimulus. She feels that men and women probably will show a difference in reaction times, so she wants to stratify on gender. The group of 96 people contains 43 men. In previous studies of this type, researchers have found that the times range from 5 to 20 seconds for men and from 3 to 14 seconds for women. The costs of sampling are the same for both strata. Using optimum allocation, find the approximate sample size necessary to estimate the average reaction time for the group to within 1 second.

5.13 A county government is interested in expanding the facilities of a day-care center for mentally retarded children. The expansion will increase the cost of enrolling a child in the center. A sample survey will be conducted to estimate the proportion of families with retarded children that will make use of the expanded facilities. The families are divided into those who use the existing facilities and those who do not. Some families live in the city in which the center is located, and some live in the surrounding suburban and rural areas. Thus, stratified random sampling is used, with users in the city, users in the surrounding country, nonusers in the city, and nonusers in the country forming strata 1, 2, 3, and 4, respectively. Approximately 90% of the present users and 50% of the present nonusers will use the expanded facilities. The cost of obtaining an observation from a user is \$4 and from a nonuser is \$8. The difference in cost results because nonusers are difficult to locate.

Existing records give $N_1 = 97$, $N_2 = 43$, $N_3 = 145$, and $N_4 = 68$. Find the approximate sample size and allocation necessary to estimate the population proportion with a bound of .05 on the error of estimation.

5.14 The survey of Exercise 5.13 is conducted and yields the following proportion of families who will use the new facilities:

$$\hat{p}_1 = .87, \qquad \hat{p}_2 = .93, \qquad \hat{p}_3 = .60, \qquad \hat{p}_4 = .53$$

Estimate the population proportion p, and place a bound on the error of estimation. Was the desired bound achieved?

5.15 Suppose in Exercise 5.13 that the total cost of sampling is fixed at \$400. Choose the sample size and allocation that minimizes the variance of the estimator $\hat{p}_{st}$ for this fixed cost.

5.16 Refer to the information on 56 business firms given in Example 5.19.
(a) Suppose that the $n = 15$ observations are to constitute a stratified random sample with only two strata. Find the optimal dividing point between the strata. With $n_1 = 7$ and $n_2 = 8$, assume that the resulting sample measurements (in thousands of dollars) turns out to be 110, 142, 212, 227, 167, 130, 194 for stratum 1 and 387, 345, 465, 308, 280, 480, 355, 405 for stratum 2. Estimate μ by $\bar{y}_{st}$ and calculate the estimated variance of $\bar{y}_{st}$.
(b) Now suppose the dividing point between the two strata is shifted to 300,000. Suppose the same 15 sample measurements are drawn in a stratified random sample with $n_1 = 8$ and $n_2 = 7$. Note that this sampling shifts the 280 value from stratum 2 to stratum 1. (This result would not be likely to happen in practice and is only used here for illustrative purposes.) Find $\bar{y}_{st}$ and calculate the estimated variance of $\bar{y}_{st}$. The numerical answer should indicate the superiority of the cumulative square root of frequencies method.

5.17 If no information is available on the variable of primary interest—say y—then optimal stratification can be approximated by looking at a variable—say x—that is highly correlated with y. Suppose an investigator wishes to estimate the average number of days of sick leave granted by a certain group of firms in a given year. No information on sick leave is available, but data on the number of employees per firm can be found. Assume that for these firms total days of sick leave are highly correlated with number of employees. Use the frequency data in the accompanying table to optimally divide the 97 firms into $L = 4$ strata for which equal sample sizes can be used.

Number of employees	Frequency
0–10	2
11–20	4
21–30	6
31–40	6
41–50	5
51–60	8
61–70	10
71–80	14
81–90	19
91–100	13
101–110	3
111–120	7

5.18 Refer to Exercise 4.30. The auditor now wants to subsample some accounts from the 20 for more detailed auditing. Separate the 20 accounts into two strata by applying the cumulative square root of frequencies method to the amounts given.

5.19 A standard quality control check on automobile batteries involves simply measuring their weight. One particular shipment from the manufacturer consisted of batteries produced in two different months, with the same number of batteries from each month. The investigator decides to stratify on months in the sampling inspection in order to observe month-to-month variation.

Simple random samples of battery weights for the two months yielded the following measurements (in pounds):

Month A	Month B
61.5	64.5
63.5	63.8
63.5	63.5
64.0	66.5
63.8	63.5
64.5	64.0

Estimate the average weight of the batteries in the population (shipment), and place a bound on the error of estimation. Ignore the fpc. The manufacturing standard for this type of battery is 65 pounds. Do you think this shipment meets the standard on the average?

5.20 In Exercise 5.19, do you think stratifying on month is desirable, or would simple random sampling work just as well? Assume that taking a simple random sample is just as convenient as taking a stratified random sample.

5.21 A quality control inspector must estimate the proportion of defective microcomputer chips coming from two different assembly operations. She knows that, among the chips in the lot to be inspected, 60% are from assembly operation A and 40% are from assembly operation B. In a random sample of 100 chips, 38 turn out to be from operation A and 62 from operation B. Among the sampled chips from operation A, 6 are defective. Among the sampled chips from operation B, 10 are defective.
(a) Considering only the simple random sample of 100 chips, estimate the proportion of defectives in the lot, and place a bound on the error of estimation.
(b) Stratifying the sample, after selection, into chips from operation A and B, estimate the proportion of defectives in the population, and place a bound on the error of estimation. Ignore the fpc's in both cases. Which answers do you find more acceptable?

5.22 When does stratification produce large gains in precision over simple random sampling? (Assume costs of observations are constant under both designs.)

5.23 A market research analyst wants to estimate the proportion of people who favor his company's product over a similar product from a rival company. The test area for his research is the state of New York. He is also interested in separate estimates of this proportion for those between the ages of 18 and 25 and for those over age 25. Discuss possible designs for this survey.

5.24 A researcher wishes to estimate the average income of employees in a large firm. Records have the employees listed by seniority, and, generally speaking, salary increases with seniority. Discuss the relative merits of simple random sampling and stratified random sampling in this case. Which would you recommend, and how would you set up the sampling scheme?

5.25 In the use of $\bar{y}_{st}$ as an estimator of μ, finding an allocation and a sample size that minimizes the $V(\bar{y}_{st})$ for fixed cost c is sometimes advantageous. That is, the cost c allowed for the survey is fixed, and we want to find the best allocation of resources in terms of maximizing the information on μ. The optimum allocation in this case is still given by Equation (5.7). Show that the appropriate choice for n is

$$n = \frac{(c - c_0) \sum_{i=1}^{L} N_i\sigma_i/\sqrt{c_i}}{\sum_{i=1}^{L} N_i\sigma_i\sqrt{c_i}}$$

where c_0 is a fixed overhead cost for the survey.

5.26 A water management board wants to estimate the total amount of water used for irrigation in one growing season among all farms in a certain management district. There is available a list of farms in which the farms are listed by acreage, from smallest to largest. Would you suggest using a simple random sample, a pps sample, or a stratified random sample of farms? Why? (For a stratified random sample, state how you would stratify and which allocation procedure you would use.)

5.27 The personnel manager of a corporation wants to estimate, for one year, the total number of days used for sick leave among all 46 plants in his firm. The 46 plants are divided into 20 "small" plants and 26 "large" plants. From past experience, the manager figures that the small plants may use from 0 to 100 days of sick leave, whereas the large plants may use from 10 to 200 days of sick leave. If it is desired to estimate the total to within 100 days:
(a) find the appropriate allocation of the sample to the two strata.
(b) find the appropriate sample size.

5.28 The data in Example 5.21 lists faculty sizes for a double sample (two-phased sample) of American colleges. Use the information given there to estimate the average faculty size for American colleges in 1986–1987. Construct an estimated variance for your estimate of the average.

5.29 A question on a proposed annexation is to be asked to residents of a suburban area, but responses for registered voters could be quite different from those who are not registered. Of 1000 residents of the area who were telephoned, 80% were registered voters. Ten percent of each group (registered and nonregistered) were asked to complete a follow-up questionnaire, on which one question was "Do you favor annexation into the city?" The data are summarized as follows:

	Voters	Nonvoters
n_i	80	20
y_i	60	8

y_i = number answering "yes" to the question on annexation

Estimate the proportion of residents who will respond yes to the question of interest and estimate the variance of your estimator.

5.30 Wage earners in a large firm are stratified into management and clerical classes, the first having 300 and the second having 500 employees. To assess attitude on sick-leave policy, independent random samples of 100 workers each were selected, one sample from each of the classes. After the sample data were collected, the responses were divided according to gender. In the table of results,

$a =$ Number who **like** the policy

$b =$ Number who **dislike** the policy

$c =$ Number who have **no opinion** on the policy

	Management $N_1 = 300$	Clerical $N_2 = 500$	Total $N = 800$
Male	$a = 60$	$a = 24$	110
	$b = 15$	$b = 4$	
	$c = 5$	$c = 2$	
Female	$a = 10$	$a = 42$	90
	$b = 7$	$b = 20$	
	$c - 3$	$c - 8$	
Total	$n_1 = 100$	$n_2 = 100$	$n = 200$

Find an estimate and an estimated variance of that estimate for each parameter listed below:
(a) Proportion of managers who like the policy
(b) Proportion of wage earners who like the policy
(c) Total number of female wage earners who dislike the policy
(d) Difference between the proportion of male managers who like the policy and the proportion of female managers who like the policy
(e) Difference between the proportion of managers who like the policy and the proportion of managers who dislike the policy

5.31 Are anesthesiologists overworked and therefore putting patients at risk? This question was investigated as part of a survey carried out at the University of Florida. The population of those practicing anesthesiology was stratified into three groups: anesthesiologists (comprising about 50% of the population), anesthesiology residents (comprising about 10% of the population), and nurse anesthetists (comprising about 40% of the population). The frequencies of those in each stratum who thought they had worked without a break beyond a safe limit sometime during the last 6 months are shown in the accompanying table:

Job classification	Worked without break beyond safe limit	Frequency	Percentage
Anesthesiologist	No	417	31.4
	Yes	913	68.7
Anesthesiology resident	No	29	17.6
	Yes	136	82.4
Nurse anesthetist	No	240	21.8
	Yes	860	78.2

(a) Estimate the population proportion of those who think they have worked beyond a safe limit. Calculate a bound on the error of estimation.

(b) Do anesthesiologists differ significantly from residents in this matter?
(c) Do anesthesiologists differ significantly from nurse anesthetists in this matter?

5.32 In the same survey discussed in Exercise 5.31, the respondents were asked for the longest continuous time (in hours) of administering anesthesia without a break over the last 6 months. A summary of the results is as follows:

Job	Mean	SD*	Sample size
Anesthesiologist	7.63	0.15	1347
Anesthesiology resident	7.74	0.35	163
Nurse anesthetist	6.55	0.11	1095

* SD = standard deviation of the mean, ignoring fpc.

(a) Estimate the mean time for the population of those giving anesthesia, with an estimated bound on the error.
(b) Do residents have a significantly higher average than the other groups? Justify your answer statistically.

EXPERIENCES WITH REAL DATA

5.1 Data on the population of the United States is given in the appendix and on the data disk under USPOP. Suppose it is possible to get good projections for the population in 2010 from only a sample of states. Based on this sample, the goal is to estimate the projected U.S. population in the year 2010. The states are divided into four geographic regions. Using these regions as strata, select an appropriately sized stratified random sample of states and, from projected populations given for these states, estimate the total U.S. population in 2010. Since the total projected population is available from the data on all the states, check to see if this figure is within the error bound you established for your estimate.

5.2 Data on the world population is also given in the appendix and on the data disk under WORLD. Stratifying by continent, select a stratified random sample of countries of the world and use their projected populations for the year 2010 to estimate the total world population in 2010.

5.3 The Florida Survey Research Center has completed a telephone survey on opinions on recycling for a group of cities in Florida. The questionnaire reproduced on the following pages shows the information that was coded to keep track of the city, county, and interviewer, as well as the questions that were asked in the survey. The data is on the data disk in a file called RECYCLE.

The survey design was a stratified random sample with three strata defined by the amount of education on recycling that goes on in these cities. Stratum 1 is low, stratum 2 is moderate, and stratum 3 is high in terms of educational programs. A stratum code is given for each response in the data set. The sample sizes are equal across the three strata. Assume that the population sizes for the three strata are nearly equal as well.

Your job is to analyze the data from this questionnaire for *two* questions of your choosing by following the outline given below.

(a) Estimate the true population proportion for two of the questions on the survey. You may select any questions you like.
(b) For each of these questions, does the proportion of men responding in the category of interest differ from the proportion of women responding in the same category?
(c) For one of the questions you selected, estimate the true population proportions within each of the three strata.
(d) For the question used in part (c), compare the true proportions among the three strata. That is, do the true proportions on these questions seem to differ among the strata?
It may be convenient to display the sampled data on two-way tables in order to describe the results and make appropriate calculations. Most statistical software, including MINITAB, has options for producing such tables.

THIS QUESTIONNAIRE IS THE PROPERTY OF
FLORIDA SURVEY RESEARCH CENTER 1994 JOB 1015

D CODE: ____ ____ ____ ____ 1–4 INTERVIEWER: ____ ____ 5–6

DATE: ____/____/____ COUNTY CODE: ____ ____ 7–8

TIME BEGUN: ________ CITY CODE: ____ ____ 9–10

TIME ENDED: ________

Hello, my name is (first and last), calling from the UF Survey Research Center. We are doing a survey of people's opinions on recyling programs in Florida communities and this number was chosen at random. I assure you that your name will not be associated with the results of this study and that this is not a sales call. You do not have to answer any question that you do not wish to answer. This survey will take about 12 minutes.

REASK "A" UNTIL YOU GET A "YES" OR "NO" OR A "DON'T KNOW" FROM ACTUAL R.

A. Do you live within the city limits of ____________?

Yes....................CONTINUE
No.....................TERMINATE
Don't know.......TERMINATE

B. Who in your household takes care of the recycling? Are they at home?

* AT HOME → CONTINUE
* NOT AT HOME → CALL PERSON BACK
* DON'T RECYCLE → ASK FOR FEMALE HEAD OF HOUSEHOLD
(IF NONE ASK FOR MALE HEAD OF HOUSE)

1. Some people regularly recycle a lot of different items, while others find that separating trash and rinsing cans is too inconvenient. Would you say that the overall recycling program in your community is convenient or inconvenient? (IF CHOICE) Would you say it is very (CONVENIENT/INCONVENIENT) or only somewhat (CONVENIENT/INCONVENIENT)?

Very Convenient.................. 1 11
Somewhat Convenient 2
Somewhat Inconvenient...... 3
Very Inconvenient................ 4
Don't Know 5

2. As you may know, some materials can be recycled in your community while others cannot. Do you happen to know whether your community's recycling program allows you to recycle. . .(READ FIRST ITEM)?

	Yes	No	Don't Know	
a. Newspapers	1	2	3	12
b. Magazines	1	2	3	13
c. Aluminum soda cans	1	2	3	14
d. Glass bottles	1	2	3	15
e. Plastic soda bottles	1	2	3	16
f. Plastic milk jugs	1	2	3	17
g. Paper milk cartons	1	2	3	18
h. Vegetable cans	1	2	3	19

3. When you buy soda, do you usually buy cans, glass bottles, two liter plastic bottles, or more than one kind of container?

Cans	1	20
Glass bottles	2	
Two liter plastic bottles	3	
More than one kind	4	
(VOL) Don't buy soda	5	
Don't Know	6	

4. When you buy milk, do you usually buy plastic jugs, paper cartons, glass bottles, or more than one kind of container?

Plastic jugs	1	21
Paper cartons	2	
Glass bottles	3	
More than one kind	4	
(VOL) Don't buy milk	5	
Don't Know	6	

5. When you buy condiments like ketchup and mustard, do you usually buy glass bottles, plastic bottles, or both?

Glass bottles	1	22
Plastic bottles	2	
Both	3	
(VOL) Don't buy	4	
Don't Know	5	

6. When you buy vegetables, do you usually buy canned, frozen, fresh, or more than one kind?

Canned	1	23
Frozen	2	
Fresh	3	
More than one kind	4	
(VOL) Don't buy	5	
Don't Know	6	

7. Do you subscribe to a daily newspaper?

Yes	1	24
No	2	
Don't Know	3	

8. Do you subscribe to any magazines?

Yes	1	25
No	2	
Don't Know	3	

9. Thinking back over the past month, have you recycled any of your trash?

Yes 1 26
No 2
Don't Know 3

10. Some recycling programs require people to take some special steps before they recycle their materials. Some people don't mind these things, while others think they are a real hassle. As I read a list of steps a recycling program might require, please tell me how much trouble you think they would be. Here's the first one. . .—do you feel it would be a real hassle, a minor nuisance, or no trouble at all?

	Real Hassle	Minor Nuisance	No Trouble	Don't Know	
a. Rinsing soda cans and plastic milk jugs	1	2	3	4	27
b. Rinsing cans with sticky substances like dog food or tomato paste.	1	2	3	4	28
c. Separating recyclable materials from other trash.	1	2	3	4	29
d. Taking recyclables to a collection center at a shopping center.	1	2	3	4	30
e. Bundling newspapers.	1	2	3	4	31
f. Storing recyclables until they are picked up.	1	2	3	4	32
g. Taking a special recycling bin out to the curb for pickup.	1	2	3	4	33

11. Over the past month, how often did you recycle your soda cans—all of the time, most of the time, some of the time, none of the time, or didn't you buy this item during the past month?

All of the time 1 34
Most of the time...................2
Some of the time..................3
None of the time 4
Didn't buy..............................5
(VOL) Not recyclable..........6
Don't Know 7

12. Over the past month, how often did you recycle your glass bottles —all of the time, most of the time, some of the time, none of the time, or didn't you buy this item during the past month?

All of the time 1 35
Most of the time...................2
Some of the time..................3
None of the time 4
Didn't buy..............................5
(VOL) Not recyclable..........6
Don't Know 7

13. How often did you recycle your plastic milk jugs?

All of the time 1 36
Most of the time...................2
Some of the time..................3
None of the time 4
Didn't buy..............................5
(VOL) Not recyclable..........6
Don't Know 7

14. How about your newspapers?

All of the time	1	37
Most of the time	2	
Some of the time	3	
None of the time	4	
Didn't buy	5	
(VOL) Not recyclable	6	
Don't Know	7	

15. How about your magazines?

All of the time	1	38
Most of the time	2	
Some of the time	3	
None of the time	4	
Didn't buy	5	
(VOL) Not recyclable	6	
Don't Know	7	

16. How about your vegetable cans?

All of the time	1	39
Most of the time	2	
Some of the time	3	
None of the time	4	
Didn't buy	5	
(VOL) Not recyclable	6	
Don't Know	7	

17. How about your plastic soda bottles?

All of the time	1	40
Most of the time	2	
Some of the time	3	
None of the time	4	
Didn't buy	5	
(VOL) Not recyclable	6	
Don't Know	7	

18. How about your paper milk cartons?

All of the time	1	41
Most of the time	2	
Some of the time	3	
None of the time	4	
Didn't buy	5	
(VOL) Not recyclable	6	
Don't Know	7	

19. Now I'd like to read you a few statements with which some people agree and others disagree. I'd like your own opinion. Here's the first one. . .—do you strongly agree, somewhat agree, somewhat disagree, or strongly disagree?

	Strongly Agree	Somewhat Agree	Somewhat Disagree	Strongly Disagree	Don't Know	
a. People like me can be effective in working for good environmental quality.	1	2	3	4	5	42
b. Major decisions about the future of our environment are ultimately made by those in power.	1	2	3	4	5	43
c. What happens to the environment is basically the result of chance events.	1	2	3	4	5	44
d. Citizens' active participation in groups concerned with environmental issues can help solve ecological problems.	1	2	3	4	5	45
e. People's individual efforts to improve the quality of the environment are often undermined by those in power.	1	2	3	4	5	46
f. It doesn't pay to worry about environmental quality because only good or bad luck will affect it.	1	2	3	4	5	47
g. Although there is continual contamination of our lakes, streams and air, nature's purifying processes soon return them to normal.	1	2	3	4	5	48
h. The government will have to introduce harsh measures to halt pollution since few people will regulate themselves.	1	2	3	4	5	49
i. Industry is trying its best to develop effective anti-pollution technology.	1	2	3	4	5	50
j. If asked, I would contribute time, money or both to an organization that works to improve the quality of the environment.	1	2	3	4	5	51

20. How effective do you think recycling can be as a means of reducing trash sent to the dump or landfill—extremely effective, very effective, moderately effective, slightly effective, or not at all effective?

Extremely effective 1 52
Very effective 2
Moderately effective 3
Slightly effective...................4
Not at all effective 5
Don't Know 6

21. In disputes involving economic growth and environmental protection, do you feel the federal government should generally promote economic growth or protect the environment? (IF

Economic growth first.........1 53
Lean toward growth.............2
Lean toward environment...3

CHOICE) Should they put (ECONOMIC GROWTH/THE ENVIRONMENT) first or just lean toward (ECONOMIC GROWTH/THE ENVIRONMENT)?

Environment first 4
Don't Know 5

22. In these disputes, do you feel the federal government has generally promoted economic growth or protected the environment? (IF CHOICE) Have they put (ECONOMIC GROWTH/THE ENVIRONMENT) first or just leaned toward (ECONOMIC GROWTH/THE ENVIRONMENT)?

Economic growth first 1 54
Lean toward growth 2
Lean toward environment...3
Environment first 4
Don't Know 5

23. Now I'd like to ask you about the amount of money the federal government spends in several different areas. Do you feel the government should spend more or less than they currently do on [READ FIRST ITEM]? And should they spend much (MORE/LESS) or only somewhat (MORE/LESS)?

	Much More	Smwht More	Smwht Less	Much Less	(VOL) No Change	Don't Know	
a. Health care.	1	2	3	4	5	6	55
b. Assistance for the unemployed.	1	2	3	4	5	6	56
c. Space and scientific research.	1	2	3	4	5	6	57
d. Improving and protecting the environment.	1	2	3	4	5	6	58
e. The war on drugs.	1	2	3	4	5	6	59

Now, I'd like to ask you some questions for statistical purposes only.

24. SEX. Don't ask, just record.

Male.......................................1 60
Female 2

25. What is the highest level of education completed by either parent who raised you?

8th Grade or less..................1 61
Some high school 2
High School graduate 3
Technical/Vocational............4
Some College 5
College Graduate 6
Graduate School/Prof..........7
Don't Know/Refused...........8

26. And just to make sure we have a representative sample—could you please tell me your race? (IF NEEDED) Well, most people classify themselves as white, Hispanic, African American, or Asian American. . .

Black.......................................1 62
White 2
Hispanic 3
Asian..4
(VOL) Other..........................5
Refused 6

Thank you for participating in this survey. Have a nice evening.

5.4 The CARS93 data, used in Activity 4.2, has cars classified as to being one of six different types, small, compact, midsize, large, sporty, or van. A numerical type code is given in the data set, in addition to the actual name of the type. The goal of this activity is to see if poststratification on car type pays any dividends when estimating average city gasoline mileage or proportion of cars with air bags for the cars in this population.
(a) Select a random sample of cars from this population. As in Activity 4.2, estimate the average city mpg for these cars, with a bound on the error of estimation.
(b) Estimate the proportion of these cars that have at least one air bag, with a bound on the error of estimation.
(c) Using the data from part (a), poststratify on the car type and then estimate the average city mpg by this method.
(d) Using the data from part (b), poststratify on car type and then estimate the proportion of cars that have at least one air bag by this method.
(e) Comparing the above results, comment on when poststratification might produce big gains in terms of the error of estimation.

5.5 This activity is an extension of Example 5.21 on the use of double sampling for stratification. The *World Almanac* (and possibly others) has a list of U.S. colleges and universities, along with a designation as to whether they are public or private and data on enrollment and faculty size. Bring the results of Example 5.21 up to date by completing the following.
(a) Find a current edition of the *World Almanac* (or equivalent). Select a large sample (size of your choosing) of colleges and universities for the purpose of checking off whether they are private or public. This process is easy to do and goes quickly.
(b) Select a subsample of the institutions sampled in part (a). For the subsample, write down the enrollment and faculty size for each institution.
(c) Use the data from parts (a) and (b) to estimate the total enrollment at U.S. colleges and universities for the year of your data.
(d) Use the data from parts (a) and (b) to estimate the average faculty size at U.S. colleges and universities for the year of your data.

5.6 We now move from selecting samples from real sets of data to selecting samples from probability distributions. The probability distributions given on the following table and graphs represent the heights of adults in America. PROB-M denotes the probabilities of various heights (in inches) for males, PROB-F denotes the probabilities for females, and PROB denotes the combined probabilities for adults. The goal is to select samples from these distributions to compare estimates of the average height from stratified random sampling to estimates from simple random sampling.

Height	PROB-M	PROB-F	PROB
56	0.000000	0.000500	0.000250
57	0.000000	0.003800	0.001900
58	0.000000	0.005100	0.002550
59	0.000000	0.012800	0.006400
60	0.000000	0.020000	0.010000
61	0.001800	0.049100	0.025450
62	0.001600	0.086200	0.043900
63	0.002700	0.113100	0.057900
64	0.017600	0.127500	0.072550
65	0.014800	0.162800	0.088800
66	0.043900	0.166700	0.105300
67	0.079400	0.106100	0.092750
68	0.105000	0.069300	0.087150
69	0.122100	0.039300	0.080700
70	0.147700	0.021100	0.084400
71	0.145900	0.010400	0.078150
72	0.118900	0.000000	0.059450
73	0.084000	0.000000	0.042000
74	0.042000	0.000000	0.021000
75	0.034300	0.000000	0.017150
76	0.022300	0.000000	0.011150

Row	Male	Female	Adults
1	70	63	62
2	70	67	75
3	66	67	68
4	67	67	63
5	75	68	69
6	75	66	71
7	73	64	62
8	66	66	74
9	71	65	68
10	70	57	58
11	72	66	73
12	70	63	72
13	68	66	69
14	70	62	69
15	67	65	69
16	72	64	66
17	71	61	71
18	73	63	63
19	70	67	72
20	73	70	72
21			66
22			65
23			71
24			67
25			70
26			68
27			66
28			70
29			67
30			62
31			70
32			63
33			70
34			73
35			71
36			64
37			70
38			66
39			71
40			63

	n	Mean	Median	SD
Male	20	70.450	70.000	2.685
Female	20	64.850	65.500	2.870
Adults	40	67.975	69.000	3.932

FIGURE 5.2

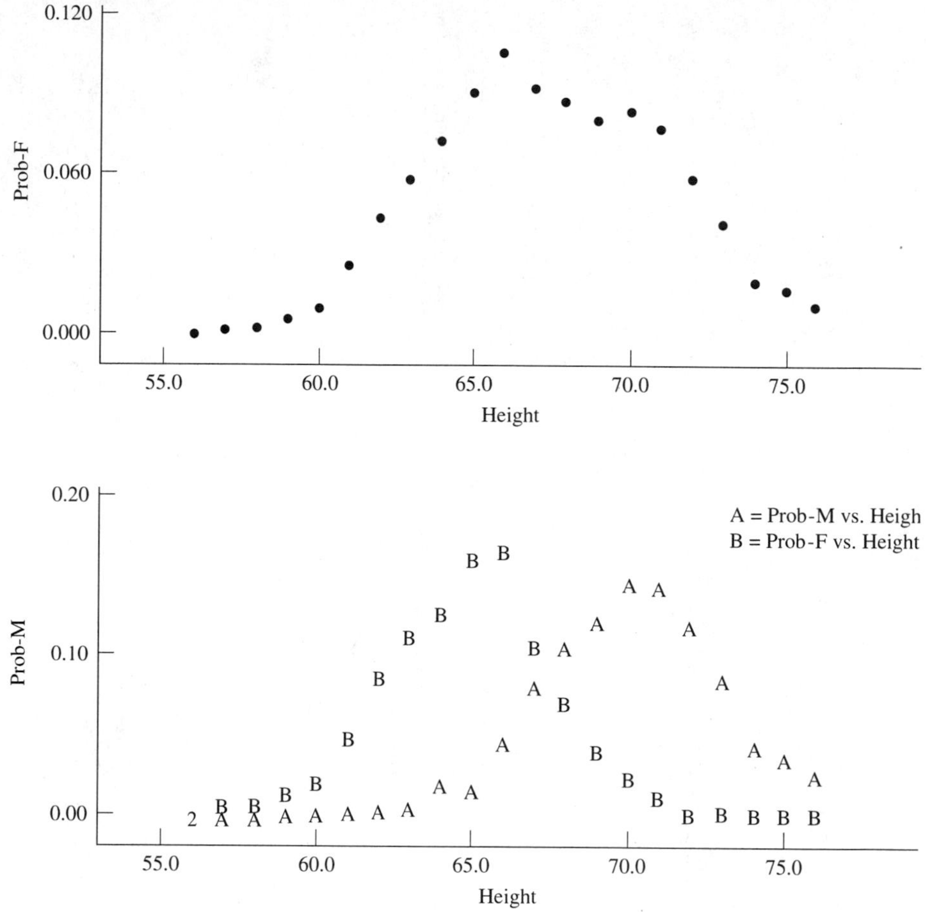
0.120
0.060
0.000
Prob-F
55.0
60.0
65.0
70.0
75.0
Height
0.20
0.10
0.00
Prob-M
A = Prob-M vs. Heigh
B = Prob-F vs. Height
55.0
60.0
65.0
70.0
75.0
Height

(a) Find a computer program that will produce random samples from specified discrete distributions. (MINITAB does this nicely.) Enter these three distributions, one for males, one for females, and one for adults without regard to gender.
(b) Select a random sample of 20 male heights and a separate random sample of 20 female heights. From these data, estimate the average height of all adults and calculate a bound on the error of estimation. (Assume that about 50% of adults are male.)
(c) Select a simple random sample of 40 heights for the height distribution of adults. From these data, estimate the average height of all adults and calculate a bound on the error of estimation.
(d) If time permits, repeat steps (b) and (c) a number of times, to as to generate a sampling distribution of estimates in each case.
(e) Compare the results of (b) and (c). Comment on when stratification seems to produce gains in precision of estimates.
(If a computer for selecting the samples is unavailable, one such set is provided.)

C A S E S T U D Y

HOW ACCURATE IS THE INVENTORY?

An important task of an auditor is to assess the accuracy of a firm's reported inventory figures. This assessment is done by sampling items from the inventory list, determining appropriate dollar values for each sampled item, and then estimating the total inventory error. If the total error is denoted by τ_e, the total reported inventory amount (book amount) by τ_x and the total audited inventory amount by τ_y, then the auditor wants to estimate

$$\tau_e = \tau_x - \tau_y$$

Since τ_x is known (it is the firm's reported figure), the only problem is to estimate τ_y.

The techniques of Chapter 6 can be used to estimate total audited amounts of inventory and total inventory error. One such analysis is presented later in the chapter.

CHAPTER 6

RATIO, REGRESSION, AND DIFFERENCE ESTIMATION

6.1

INTRODUCTION

Estimation of the population mean and total in preceding chapters was based on a sample of response measurements, $y_1, y_2, \ldots, y_n$, obtained by simple random sampling (Chapter 4) and stratified random sampling (Chapter 5). Sometimes other variables are closely related to the response y. By measuring y and one or more subsidiary variables, we can obtain additional information for estimating the population mean. You are probably familiar with the use of subsidiary variables to estimate the mean of a response y. It is basic to the concept of correlation and provides means for development of a prediction equation relating y and x by the method of least squares. This topic is ordinarily covered in introductory courses in statistics (Mendenhall and Beaver, 1994, Chapter 10).

Chapters 4 and 5 presented simple estimators of population parameters utilizing the response measurements $y_1, y_2, \ldots, y_n$; however, primary emphasis was placed on the design of the sample survey (simple and stratified random sampling). In contrast, this chapter presents three new methods of estimation based on the use of a subsidiary variable x. The methods are called ratio, regression, and difference estimation. All three require the measurement of two variables, y and x, on each element of the sample. A variety of sampling designs can be employed in conjunction with ratio, regression, or difference estimation, but we will discuss mainly simple random sampling. The basic ideas of how these techniques carry over to stratified random sampling will, however, be illustrated for ratio estimation.

6.2

SURVEYS THAT REQUIRE THE USE OF RATIO ESTIMATORS

Efficient estimation of a population total sometimes requires the use of subsidiary variables. We illustrate the use of a *ratio estimator* for one of the situations. The wholesale price paid for oranges in large shipments is based on the sugar content of the load. The exact sugar content cannot be determined prior to the purchase and extraction of the juice from the entire load; however, it can be estimated. One method of estimating this quantity is to first estimate the mean sugar content per orange, μ_y, and then to multiply by the number of oranges N in the load. Thus, we could randomly sample n oranges from the load to determine the sugar content y for each. The average of these sample measurements, $y_1, y_2, \ldots, y_n$, will estimate μ_y; $N\bar{y}$ will estimate the total sugar content for the load, τ_y. Unfortunately, this method is not feasible because it is too time-consuming and costly to determine N (that is, to count the total number of oranges in the load).

We can avoid the need to know N by noting the following two facts. First, the sugar content of an individual orange, y, is closely related to its weight x; second, the ratio of the total sugar content τ_y to the total weight of the truckload τ_x is equal to the ratio of the mean sugar content per orange, μ_y, to the mean weight μ_x. Thus,

$$\frac{\mu_y}{\mu_x} = \frac{N\mu_y}{N\mu_x} = \frac{\tau_y}{\tau_x}$$

Solving for the total sugar content of the load, we have

$$\tau_y = \frac{\mu_y}{\mu_x}(\tau_x)$$

We can estimate μ_y and μ_x by using $\bar{y}$ and $\bar{x}$, the averages of the sugar contents and weights for the sample of n oranges. Also, we can measure τ_x, the total weight of the oranges on the truck. Then a *ratio estimate* of the total sugar content τ_y is

$$\hat{\tau}_y = \frac{\bar{y}}{\bar{x}}(\tau_x)$$

or, equivalently (multiplying numerator and denominator by n),

$$\hat{\tau}_y = \frac{n\bar{y}}{n\bar{x}}(\tau_x) = \frac{\sum_{i=1}^{n} y_i}{\sum_{i=1}^{n} x_i}(\tau_x)$$

In this case, the number of elements in the population, N, is unknown, and therefore we cannot use the simple estimator $N\bar{y}$ of the population total τ_y (Section 4.3). Thus, a ratio estimator or its equivalent is necessary to accomplish the estimation objective. However, if N is known, we have the choice of using the estimator $N\bar{y}$ or the ratio estimator to estimate τ_y. If y and x are highly correlated—that is, if x contributes information for the prediction of y—the ratio estimator should be better than $N\bar{y}$, which depends solely on $\bar{y}$.

In addition to the population total τ_y, there are often other parameters of interest. We may want to estimate the population mean μ_y by using a ratio estimation procedure. For example, suppose we wish to estimate the average sugar content per orange in a large shipment. We could use the sample mean $\bar{y}$ to estimate μ_y. However, if x and y are correlated, a ratio estimator that uses information from the auxiliary variable x frequently provides a more precise estimator of μ_y.

The population ratio is another parameter that may be of interest to an investigator. For example, assume we want to estimate the ratio of total automobile sales for the first quarter of this year to the number of sales during the corresponding period of the previous year. Let τ_x be the total number of sales for the first quarter of last year and let τ_y be the total number of sales for the same period this year. We are interested in estimating the ratio

$$R = \frac{\tau_y}{\tau_x}$$

The concept of ratio estimation is used in the analysis of data from many important and practical surveys used by government, business, and academic researchers. For instance, the consumer price index (CPI) is actually a ratio of costs of purchasing a fixed set of items of constant quality and quantity for two points in time. Currently, the CPI compares today's prices with those of 1967. The CPI is based, in part, on data collected every month or every other month from 24,000 establishments (stores, hospitals, filling stations, and so on) selected from 85 urban areas around the country. The CPI is used mainly as a measure of inflation (see Chapter 1).

The Current Population Survey adjusts unemployment figures for age, gender, and race by a ratio estimation technique. For example, the ratio of number of unemployed blacks to number of blacks in the workforce for a sample area can be expanded to a measure of number of unemployed blacks in a larger area by simply multiplying that sample ratio by the number of blacks in the workforce of the larger area.

The Nielsen Retail Index can provide ratios of average sales prices for the two competing brands of a product or for a single product at two points in time. The SAMI can provide total stock volume ratios for two competing brands.

Forecasting often employs a ratio estimation technique. For example, the ratio of total first-period sales for the current year to a similar total for last year can be multiplied by last year's total sales to estimate this year's total sales. Similar methods are used to forecast population growth.

In academic research, sociologists are interested in measures like the ratio of total monthly food budget to total monthly income per family or the ratio of number of children to total number of people residing in a housing unit. Medical researchers can measure the relative potency of a new drug by looking at the ratio of the average amount of new drug required to evoke a certain response to the average amount of a standard drug required for the same response.

As you can see, the possible applications of ratio estimation are endless. However, we will now shift our emphasis to the construction of estimators for μ_y, τ_y, and R, and we will provide numerical examples of each. Whenever appropriate, comparisons will be made to the estimators of these parameters presented in previous chapters.

6.3

RATIO ESTIMATION USING SIMPLE RANDOM SAMPLING

Let us assume that a simple random sample of size n is to be drawn from a finite population containing N elements. How, then, do we estimate a population mean μ_y, a total τ_y, or a ratio R, utilizing sample information on y and a subsidiary variable x?

Estimator of the population ratio R:

$$r = \frac{\sum_{i=1}^{n} y_i}{\sum_{i=1}^{n} x_i} = \frac{\bar{y}}{\bar{x}} \tag{6.1}$$

Estimated variance of r:

$$\hat{V}(r) = \hat{V}\left(\frac{\sum_{i=1}^{n} y_i}{\sum_{i=1}^{n} x_i}\right) = \left(\frac{N-n}{nN}\right)\left(\frac{1}{\mu_x^2}\right)s_r^2 \tag{6.2}$$

where

$$s_r^2 = \frac{\sum_{i=1}^{n}(y_i - rx_i)^2}{n-1} \qquad (6.3)$$

[If the population mean for x, μ_x, is unknown, we use $\bar{x}^2$ to approximate μ_x^2 in Equations (6.2) and (6.3).]

Example 6.1

In a survey to examine trends in real estate, an investigator is interested in the relative change over a two-year period in the assessed value of homes in a particular community. A simple random sample of $n = 20$ homes is selected from the $N = 1000$ homes in the community. From tax records the investigator obtains the assessed value for this year (y) and the corresponding value for two years ago (x) for each of the $n = 20$ homes included in the sample. He wishes to estimate R, the relative change in assessed value for the $N = 1000$ homes, using information contained in the sample.

The data for the real estate survey are presented in Table 6.1. We have added the $(y_i - rx_i)$ column, which is useful in the calculation of $\hat{V}(r)$.

Using the data in Table 6.1, estimate R, the relative change in real estate valuation over the given two-year period. Place a bound on the error of estimation.

Solution

A scatterplot of the data, Figure 6.1, shows evidence of a positive linear association (correlation) between y and x, which is good for ratio estimation. None of the observations deviate sharply from this pattern. The mean of the x's and the mean of the y's are used in computing r. The standard deviation s_r is simply the sample standard deviation of the values for $(y_i - rx_i)$, which are given in the third data column. (Why do these deviations average zero?) When computations are to be made by computer or graphing calculator, this spreadsheet formulation of the computations is the easiest one to use.

The estimate of R using the sample data is given by

$$r = \frac{\sum_{i=1}^{20} y_i}{\sum_{i=1}^{20} x_i} = \frac{\text{total current valuation of the 20 homes}}{\text{total valuation of the 20 homes two years ago}}$$

Using Table 6.1,

$$r = \frac{164.7}{154.5} = \frac{8.235}{7.725} = 1.066$$

TABLE 6.1 **Data and calculation for the real estate valuation survey (figures in $10,000 units)**

Home	x_i	y_i	$y_i - rx_i$
1	6.7	7.1	−0.042330
2	8.2	8.4	−0.341359
3	7.9	8.2	−0.221554
4	6.4	6.9	0.077476
5	8.3	8.4	−0.447962
6	7.2	7.9	0.224660
7	6.0	6.5	0.103884
8	7.4	7.6	−0.288544
9	8.1	8.9	0.265242
10	9.3	9.9	−0.013981
11	8.2	9.1	0.358642
12	6.8	7.3	0.051068
13	7.4	7.8	−0.088543
14	7.5	8.3	0.304854
15	8.3	8.9	0.052038
16	9.1	9.6	−0.100777
17	8.6	8.7	−0.467768
18	7.9	8.8	0.378447
19	6.3	7.0	0.284078
20	8.9	9.4	−0.087573

	N	Mean	Median	SD
x	20	7.725	7.900	0.947
y	20	8.235	8.350	0.957
$y_i - rx_i$	20	−0.0000	0.0185	0.260

FIGURE 6.1 Scatter plot for the data of Example 6.1

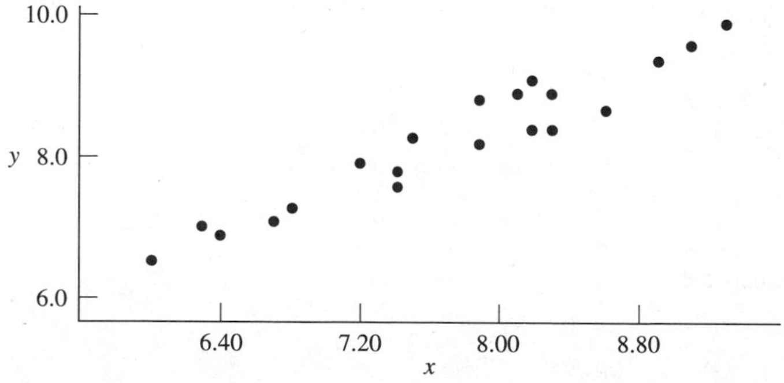

Hence, we estimate that the real estate valuation has increased approximately 7% over a two-year period in the area studied.

Using Equation (6.3) yields

$$2\sqrt{\hat{V}(r)} = 2\sqrt{\left(\frac{N-n}{nN}\right)\left(\frac{1}{\bar{x}^2}\right)s_r^2}$$

$$= 2\sqrt{\frac{1000-20}{20(1000)}\left[\frac{1}{(7.725)^2}\right](0.26)^2} = .015$$

Thus, we estimate the ratio of current real estate valuation to that of two years ago to be $r = 1.07$, and we are quite confident that the error of estimation is less than .02. That is, the true ratio R for the population should be between 1.05 and 1.09. Note that the bound on the error of estimation is quite small. Hence, r should be a fairly accurate estimate of R.

The large-sample confidence intervals based on normal distribution theory, as introduced in Chapter 2, apply in the ratio estimation case as well. Thus, for example, an approximate 90% confidence interval for the ratio R is of the form

$$r \pm 1.645\sqrt{\hat{V}(r)}$$

The estimated variance of r can be written in many forms. One that is particularly useful involves the correlation coefficient ρ between x and y. This correlation can be estimated by

$$\hat{\rho} = \frac{s_{xy}}{s_x s_y}$$

where

$$s_{xy} = \frac{1}{n-1}\sum_{i=1}^{n}(x_i - \bar{x})(y_i - \bar{y}),$$

$$s_x^2 = \frac{1}{n-1}\sum_{i=1}^{n}(x_i - \bar{x})^2,$$

$$s_y^2 = \frac{1}{n-1}\sum_{i=1}^{n}(y_i - \bar{y})^2$$

The correlation coefficient will play a key role in later discussions.

Now we can write

$$\hat{V}(r) = \frac{1-f}{n}\left(\frac{1}{\mu_x^2}\right)(s_y^2 + r^2 s_x^2 - 2r\hat{\rho}s_x s_y)$$

where $f = n/N$, the sampling fraction. This form is often used with calculations performed by standard statistical software packages. For example, MINITAB readily summarizes the data of Example 6.1 as

	N	Mean	SD
x	20	7.725	0.947
y	20	8.235	0.957

Correlation between x and y = 0.966

From this summary,

$$\hat{V}(r) = \left\{ \left(1 - \frac{20}{1000}\right)\left(\frac{1}{20}\right)\left(\frac{1}{7.725}\right)^2 \cdot [(0.957)^2 + (1.07)^2(0.947)^2 - 2(1.07)(0.966)(0.947)(0.957)] \right\}$$

$$= .0000567$$

and $2\sqrt{\hat{V}(r)} = .015$, approximately the same as the earlier calculation.

The ratio technique for estimating a population total τ_y was applied in estimating the total sugar content of a truckload of oranges. The simple estimator $N\bar{y}$ is not applicable because we do not know N, the total number of oranges in the truck. The following ratio estimation procedure can be applied in estimating τ_y whether or not N is known.

Ratio estimator of the population total $\hat{\tau}_y$:

$$\hat{\tau}_y = \frac{\sum_{i=1}^{n} y_i}{\sum_{i=1}^{n} x_i}(\tau_x) = r\tau_x \tag{6.4}$$

Estimated variance of $\hat{\tau}_y$:

$$\hat{V}(\hat{\tau}_y) = (\tau_x)^2\hat{V}(r) = \tau_x^2\left(\frac{N-n}{nN}\right)\left(\frac{1}{\mu_x^2}\right)s_r^2 \tag{6.5}$$

where μ_x and τ_x are the population mean and total, respectively, for the random variable x, and

$$s_r^2 = \frac{\sum_{i=1}^{n}(y_i - rx_i)^2}{n-1}$$

Note that although we do not need to know N or μ_x, we must know τ_x in order to estimate τ_y by use of the ratio estimation procedure.

Example 6.2

In a study to estimate the total sugar content of a truckload of oranges, a random sample of $n = 10$ oranges was juiced and weighted (see Table 6.2). The total weight of all the oranges, obtained by first weighing the truck loaded and then unloaded, was found to be 1800 pounds. Estimate τ_y, the total sugar content for the oranges, and place a bound on the error of estimation.

TABLE 6.2 **Data for Example 6.2**

Orange	*Sugar content (in pounds)*	*Weight of orange (in pounds)*	
	y	x	$y_i - rx_i$
1	0.021	0.40	−0.0016207
2	0.030	0.48	0.0028552
3	0.025	0.43	0.0006828
4	0.022	0.42	−0.0017517
5	0.033	0.50	0.0047241
6	0.027	0.46	0.0009862
7	0.019	0.39	−0.0030552
8	0.021	0.41	−0.0021862
9	0.023	0.42	−0.0007517
10	0.025	0.44	0.0001172

	n	Mean	Median	SD
y	10	0.0246	0.0240	0.0044
x	10	0.4350	0.4250	0.0354
$y_i - rx_i$	10	0.0000	−0.0003	0.0024

Solution

The sugar content of an orange is usually recorded in degrees brix, which is a measure of the number of pounds of solids (mostly sugar) per 100 pounds of juice. For our calculations, we will use the actual pounds per orange. An estimate of τ_y can be obtained by using Equation (6.4):

$$\hat{\tau}_y = r\tau_x = \frac{\sum_{i=1}^{10} y_i}{\sum_{i=1}^{10} x_i}(\tau_x) = \frac{.246}{4.35}(1800) = 101.79 \text{ pounds}$$

A bound on the error of estimation can be found if we use a modified version of Equation (6.5). Because N is unknown in this example, we assume that the finite population correction, $(N - n)/N$, is near unity. This assumption is reasonable because we expect at least $N = 4000$ oranges even in a small truckload. The sample mean $\bar{x}$ must be used in place of μ_x in Equation (6.5), because μ_x is unknown. With these adjustments, the bound becomes

$$2\sqrt{\hat{V}(\hat{\tau}_y)} = 2\sqrt{\tau_x^2\left(\frac{1}{n}\right)\left(\frac{1}{\bar{x}^2}\right)\frac{\sum_{i=1}^{n}(y_i - rx_i)^2}{n-1}}$$

$$= 2\sqrt{(1800)^2\left(\frac{1}{10}\right)\left[\frac{1}{(.435)^2}\right](.0024)^2} = 6.3$$

To summarize, the ratio estimate of the total sugar content of the truckload of oranges is $\hat{\tau}_y = 101.79$ pounds, with a bound on the error of estimation of 6.3. We are confident that the total sugar content τ_y lies in the interval

$$101.79 \pm 6.3$$

that is, the interval 95.49 to 108.09 pounds. The scatter plot, Figure 6.2, shows a strong, positive association between sugar content and weight, making the ratio estimator a reasonable choice.

FIGURE 6.2 Scatter plot for the data of Example 6.2

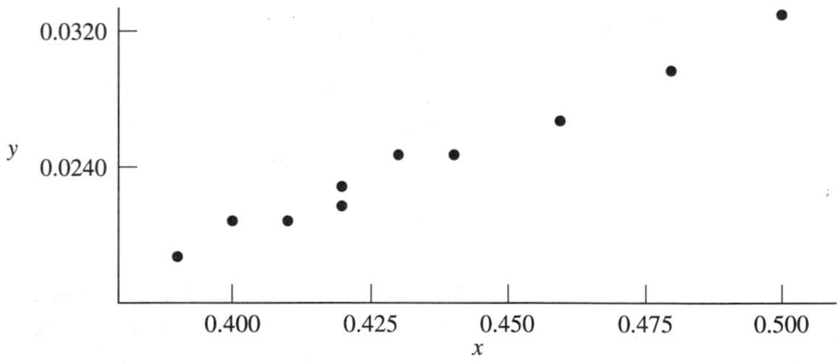

Recall that the population size N is frequently known. Consequently, the investigator must decide under what conditions use of the ratio estimator $\hat{\tau}_y = r\tau_x$ is better than use of the corresponding estimator $N\bar{y}$, where both estimators are based on simple random sampling (see Section 6.5). Generally, $r\tau_x$ possesses a smaller variance than $N\bar{y}$ when there is a strong positive correlation between x and y (where ρ, the correlation coefficient between x and y, is greater than $\frac{1}{2}$). Intuitively, this statement makes sense because in ratio estimation we are using the additional information provided by the subsidiary variable x.

If an investigator is interested in a population mean rather than a population total, the corresponding ratio estimation procedure is shown in Equations (6.6) and (6.7).

Ratio estimator of a population mean μ_y:

$$\hat{\mu}_y = \frac{\sum_{i=1}^{n} y_i}{\sum_{i=1}^{n} x_i}(\mu_x) = r\mu_x \tag{6.6}$$

Estimated variance of μ_y:

$$\hat{V}(\hat{\mu}_y) = \mu_x^2 \hat{V}(r) = \left(\frac{N-n}{nN}\right)s_r^2 \tag{6.7}$$

where

$$s_r^2 = \frac{\sum_{i=1}^{n}(y_i - rx_i)^2}{n-1}$$

Note that we do not need to know τ_x or N to estimate μ_y when using the ratio procedure; however, we must know μ_x.

Example 6.3

A company wishes to estimate the average amount of money μ_y paid to employees for medical expenses during the first three months of the current calendar year. Average quarterly reports are available in the fiscal reports of the previous year. A random sample of 100 employee records is taken from the population of 1000 employees. The sample results are summarized below. Use the data to estimate μ_y and to place a bound on the error of estimation.

$$n = 100, \qquad N = 1000$$

Total for the current quarter:

$$\sum_{i=1}^{100} y_i = 1750$$

Total for the corresponding quarter of the previous year:

$$\sum_{i=1}^{100} x_i = 1200$$

Population total τ_x for the corresponding quarter of the previous year:

$$\tau_x = 12{,}500$$

$$\sum_{i=1}^{100} y_i^2 = 31{,}650, \qquad \sum_{i=1}^{100} x_i^2 = 15{,}620, \qquad \sum_{i=1}^{100} y_i x_i = 22{,}059.35$$

Solution

The estimate of μ_y is

$$\hat{\mu}_y = r\mu_x$$

where

$$\mu_x = \frac{\tau_x}{N} = \frac{12{,}500}{1000} = 12.5$$

Then

$$\hat{\mu}_y = \frac{\sum_{i=1}^{100} y_i}{\sum_{i=1}^{100} x_i}(\mu_x) = \frac{1750}{1200}(12.5) = 18.23$$

The bound on the error of estimation can be found by expanding the quadratic term into

$$\begin{aligned}\sum_{i=1}^{100} (y_i - rx_i)^2 &= \sum_{i=1}^{100} y_i^2 + r^2 \sum_{i=1}^{100} x_i^2 - 2r \sum_{i=1}^{100} y_i x_i \\ &= 31{,}650 + (1.4583)^2(15{,}620) - (2.9166)(22{,}059.35) \\ &= 529.80\end{aligned}$$

Substituting into Equation (6.7) gives the bound on the error of estimation:

$$\begin{aligned}2\sqrt{\hat{V}(\hat{\mu}_y)} &= 2\sqrt{\left(\frac{N-n}{nN}\right)\frac{\sum_{i=1}^{n}(y_i - rx_i)^2}{n-1}} \\ &= 2\sqrt{\frac{1000-100}{100(1000)}\left(\frac{529.80}{99}\right)} = .44\end{aligned}$$

Thus, we estimate the average amount of money paid to employees for medical expenses to be \$18.23. We are very confident that the error for estimating μ_y is less than \$.44.

To remember the formulas for ratio estimation of a population mean, total, or ratio, we make the following association. The sample ratio r is given by the formula

$$r = \frac{\sum_{i=1}^{n} y_i}{\sum_{i=1}^{n} x_i} \tag{6.8}$$

The estimators of R, τ_y, and μ_y are then

$$\hat{R} = r \tag{6.9}$$

$$\hat{\tau}_y = r\tau_x \tag{6.10}$$

$$\hat{\mu}_y = r\mu_x \tag{6.11}$$

Thus, we need know only the formula for r and its relationship to $\hat{\mu}_y$ and $\hat{\tau}_y$.

Approximate variances can be obtained if you remember the basic formula,

$$\hat{V}(r) = \left(\frac{N-n}{nN}\right)\left(\frac{1}{\mu_x^2}\right)\frac{\sum_{i=1}^{n}(y_i - rx_i)^2}{n-1} \tag{6.12}$$

Then

$$\hat{V}(\hat{\tau}_y) = \tau_x^2 \hat{V}(r) \tag{6.13}$$

$$\hat{V}(\hat{\mu}_y) = \mu_x^2 \hat{V}(r) \tag{6.14}$$

6.4 SELECTING THE SAMPLE SIZE

We stated previously that the amount of information contained in the sample depends on the variation in the data (which is frequently controlled by the sample survey design) and the number of observations n included in the sample. Once the sampling procedure (design) has been chosen, the investigator must determine the number of elements to be drawn. We will consider the sample size required to estimate a population parameter R, μ_y, or τ_y to within B units for simple random sampling using ratio estimators.

Note that the procedure for choosing the sample size n is identical to that presented in Section 4.4. The number of observations required to estimate R, a population ratio, with a bound on the error of estimation of magnitude B is determined by setting 2 SD of the ratio estimator r equal to B and solving this expression for n. That is, we must solve

$$2\sqrt{V(r)} = B \tag{6.15}$$

for n. Although we have not discussed the form of $V(r)$, recall that $\hat{V}(r)$, the estimated variance of r, is given by the formula

$$\hat{V}(r) = \left(\frac{N-n}{nN}\right)\left(\frac{1}{\mu_x^2}\right)\frac{\sum_{i=1}^{n}(y_i - rx_i)^2}{n-1} \tag{6.16}$$

We can rewrite Equation (6.16) as

$$\hat{V}(r) = \left(\frac{N-n}{nN}\right)\left(\frac{1}{\mu_x^2}\right)s_r^2 \tag{6.17}$$

In this instance, we define

$$s_r^2 = \frac{\sum_{i=1}^{n}(y_i - rx_i)^2}{n-1}$$

An approximate population variance, $V(r)$, can be obtained from $\hat{V}(r)$ by replacing s^2 with the corresponding population variance σ^2. Thus, the number of observations required to estimate R with a bound B on the error of estimation is determined by solving the following equation for n:

$$2\sqrt{V(r)} = 2\sqrt{\left(\frac{N-n}{nN}\right)\left(\frac{1}{\mu_x^2}\right)\sigma^2} = B \tag{6.18}$$

Sample size required to estimate R with a bound on the error of estimation B:

$$n = \frac{N\sigma^2}{ND + \sigma^2} \tag{6.19}$$

where $D = \dfrac{B^2\mu_x^2}{4}$

In a practical situation, we are faced with a problem in determining the appropriate sample size because we do not know σ^2. If no past information is available to calculate s_r^2 as an estimate of σ^2, we take a preliminary sample of size n' and compare

$$\hat{\sigma}^2 = \frac{\sum_{i=1}^{n'}(y_i - rx_i)^2}{n'-1}$$

Then we substitute this quantity for σ^2 in Equation (6.19), and we find an *approximate* sample size. If μ_x is also unknown, it can be replaced by the sample mean $\bar{x}$, calculated from the n' preliminary observations.

Example 6.4

A manufacturing company wishes to estimate the ratio of change from last year to this year in the number of man-hours lost due to sickness. A preliminary study of $n' = 10$ employee records is made, and the results are given in Table 6.3. The company records show that the total number of man-hours lost because of sickness for the previous year was $\tau_x = 16{,}300$. Use the data to determine the sample size required to estimate R, the rate of change for the company, with a bound on the error of estimation of magnitude $B = .01$. Assume the company has 1000 employees ($N = 1000$).

TABLE 6.3 **Hours lost to sickness**

Employee	Man-hours lost in previous year, x	Man-hours lost in current year, y	$y_i - rx_i$
1	12	13	0.39326
2	24	25	−0.21349
3	15	15	−0.75843
4	30	32	0.48314
5	32	36	2.38202
6	26	24	−3.31461
7	10	12	1.49438
8	15	16	0.24157
9	0	2	2.00000
10	14	12	−2.70787

	N	Mean	Median	SD
x	10	17.80	15.00	9.99
y	10	18.70	15.50	10.36
$y_i - rx_i$	10	−0.00	0.32	1.87

Solution

From the given data, we determine

$$r = \frac{\sum_{i=1}^{10} y_i}{\sum_{i=1}^{10} x_i} = \frac{187}{178} = 1.05$$

and

$$\hat{\sigma}^2 = \frac{\sum_{i=1}^{10} (y_i - rx_i)^2}{9} = (1.86)^2 = 3.46$$

The required sample size can now be found by using Equation (6.19). Note that

$$\mu_x = \frac{\tau_x}{N} = \frac{16{,}300}{1000} = 16.3$$

and

$$D = \frac{B^2 \mu_x^2}{4} = \frac{(.01)^2 (16.3)^2}{4} = .006642$$

Thus,

$$n = \frac{N\hat{\sigma}^2}{ND + \hat{\sigma}^2} = \frac{1000(3.46)}{1000(.006642) + 3.46} = 342.5$$

Therefore, we should sample approximately 343 employee records to estimate R, the rate of change in man-hours lost due to sickness, with a bound on the error of estimation of .01 hour.

Similarly, we can determine the number of observations n needed to estimate a population mean μ_y, with a bound on the error of estimation of magnitude B. The required sample size is found by solving the following equation for n:

$$2\sqrt{V(\hat{\mu}_y)} = B \qquad (6.20)$$

Stated differently, from Equation (6.14),

$$2\mu_x \sqrt{V(r)} = B$$

The solution is shown in Equation (6.21).

Sample size required to estimate μ_y with a bound on the error of estimation B:

$$n = \frac{N\sigma^2}{ND + \sigma^2} \qquad (6.21)$$

where $D = \frac{B^2}{4}$

Note that we need not know the value of μ_x to determine n in Equation (6.21); however, we do need an estimate of σ^2, either from prior information if it is available or from information obtained in a preliminary study.

Example 6.5 An investigator wishes to estimate the average number of trees μ_y per acre on an $N = 1000$-acre plantation. She plans to sample n 1-acre plots and count the number of trees y on each plot. She also has aerial photographs of the plantation from which she can estimate the number of trees x on each plot for the entire plantation. Hence, she knows μ_x. Therefore, to use a ratio estimator of μ_y seems appropriate. Determine the sample size needed to estimate μ_y with a bound on the error of estimation of magnitude $B = 1.0$.

Solution

Assuming no prior information is available, we must conduct a preliminary study to estimate σ^2. Since an investigator can readily examine ten 1-acre plots in a day to determine the total number of trees y per plot, conducting a preliminary study of $n' = 10$ plots is convenient. The results of such a study are given in Table 6.4, with the corresponding aerial estimates x.

The scatter plot of the data, Figure 6.3, shows strong positive correlation with no unusual observations (data points).

TABLE 6.4

Plot	Aerial estimate, x	Actual number, y	$y_i - rx_i$
1	23	25	0.5625
2	14	15	0.1250
3	20	22	0.7500
4	25	24	−2.5625
5	12	13	0.2500
6	18	18	−1.1250
7	30	35	3.1250
8	27	30	1.3125
9	8	10	1.5000
10	31	29	−3.9375

	N	Mean	Median	SD
x	10	20.80	21.50	7.79
y	10	22.10	23.00	8.06
$y_i - rx_i$	10	0.000	0.41	2.06

FIGURE 6.3 Scatter plot for the data of Example 6.5

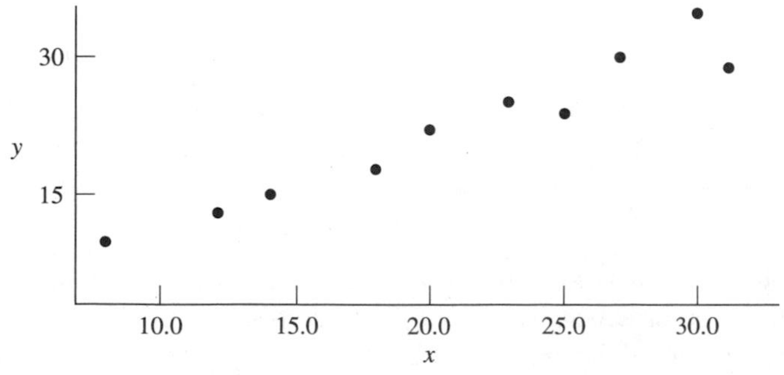

From the preliminary study,

$$r = \frac{\sum_{i=1}^{10} y_i}{\sum_{i=1}^{10} x_i} = \frac{221}{208} = \frac{22.1}{20.8} = 1.06$$

and

$$\hat{\sigma}^2 = \frac{\sum_{i=1}^{10} (y_i - rx_i)^2}{9} = 4.20$$

We now determine n from Equation (6.21), where $D = B^2/4 = \frac{1}{4}$:

$$n = \frac{N\sigma^2}{ND + \sigma^2} = \frac{10.00(4.20)}{1000(.25) + 4.20} = 16.52$$

To summarize, we need to examine approximately 17 plots to estimate μ_y, the average number of trees per 1-acre plot, with a bound on the error of estimation of $B = 1.0$. We only need 7 additional observations since we have 10 from the preliminary study.

The sample size required to estimate τ_y with a bound on the error of estimation of magnitude B can be found by solving the following expression for n:

$$2\sqrt{V(\hat{\tau}_y)} = B \tag{6.22}$$

or, from Equation (6.13),

$$2\tau_x\sqrt{V(r)} = B$$

Sample size required to estimate τ_y with a bound on the error of estimation B:

$$n = \frac{N\sigma^2}{ND + \sigma^2} \tag{6.23}$$

where $D = \dfrac{B^2}{4N^2}$

Example 6.6 An auditor wishes to compare the actual dollar value of an inventory of a hospital, τ_y, with the recorded inventory, τ_x. The recorded inventory τ_x can be summarized from computer-stored hospital records. The actual inventory τ_y could be determined by examining and counting all hospital supplies, but this process would be very time-consuming and costly. Hence, the auditor plans to estimate τ_y from a sample of n different items randomly selected from the hospital's supplies.

Records in the computer list $N = 2100$ different item types and the number of each particular item in the hospital inventory. With these data, a total value for each item, x, can be obtained by multiplying the total number of each recorded item by the unit value per item. The total dollar value of the inventory obtained from the computer is given by

$$\tau_x = \text{sum of the dollar values for the } N = 2100 \text{ items} = \sum_{i=1}^{2100} x_i$$

In this instance τ_x was found to be \$45,000. Determine the sample size (number of items) needed to estimate τ_y with a bound on the error of estimation of magnitude $B = \$500$.

Solution

Because there is no prior information available, a preliminary study must be conducted to estimate σ^2. Two people can determine the actual dollar value y for each of 15 items in a day. For this example, we will use the data from a single day's inventory ($n' = 15$) as a preliminary study to obtain a rough estimate of σ^2 and, consequently, a rough approximation of the required sample size n. Actually, the investigator would probably take a preliminary study of two or three days' inventory to provide a good approximation to σ^2 and hence n; however, to simplify computations, we will consider a preliminary study of $n' = 15$ items. These data are summarized in Table 6.5 along with the corresponding computer figures (entries in hundreds of dollars).

TABLE 6.5

Item	Dollar value from computer, x	Actual dollar value, y	$y_i - rx_i$
1	15.0	14.0	−0.72102
2	9.5	9.0	−0.32331
3	14.2	12.5	−1.43590
4	20.5	22.0	1.88128
5	6.7	6.3	−0.27539
6	9.8	8.4	−1.21773
7	25.7	28.5	3.27799
8	12.6	10.0	−2.36565
9	15.1	14.4	−0.41916
10	30.9	28.2	−2.12529
11	7.3	15.5	8.33577
12	28.6	26.3	−1.76807
13	14.7	13.1	−1.32660
14	20.5	19.5	−0.61872
15	10.9	9.8	−0.89727

	N	Mean	Median	SD
x	15	16.13	14.70	7.57
y	15	15.83	14.00	7.38
$y_i - rx_i$	15	0.00	−0.72	2.73

From these data,

$$r = \frac{\bar{y}}{\bar{x}} = \frac{15.83}{16.13} = .9814 \approx .98$$

and

$$\hat{\sigma}^2 = \frac{\sum_{i=1}^{15} (y_i - rx_i)^2}{14} = (2.73)^2 = 7.45$$

The required sample size now can be found by using Equation (6.23). We have

$$D = \frac{B^2}{4N^2} = \frac{(500)^2}{4(2100)^2} = .01417$$

and hence

$$n = \frac{N\sigma^2}{ND + \sigma^2} = \frac{2100(7.45)}{2100(.01417) + 7.45} = 420.2$$

Thus, the auditor must sample approximately 421 items to estimate τ_y, the actual dollar value of the inventory, to within $B = \$500$.

It appears that solutions to ratio estimation problems, as in Example 6.6, are neatly solved by the techniques just presented, but let's take a closer look at the sensitivity of these calculations to a few particular data points. The scatter plot for the data of Example 6.6 is shown in Figure 6.4. Note that most of the points lie close to a straight line with positive slope, so there is strong positive correlation between y and x. One of the points, however, lies quite far from the straight-line pattern generated by the rest of the points. On another scale, Figure 6.5 plots the deviations $(y_i - rx_i)$ against the x values. Again, the one observation shows up as having an extremely large deviation and hence makes an unusually large contribution to the variance.

FIGURE 6.4 Scatter plot for the data of Example 6.6

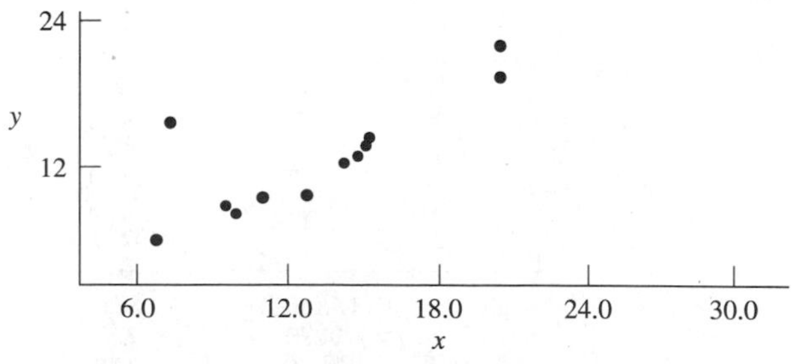

FIGURE 6.5 Scatter plot of the deviations ($y_i - rx_i$) for Example 6.6

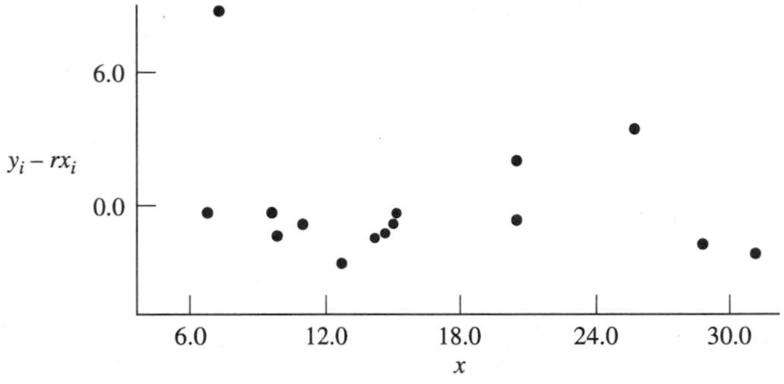

This seemingly unusual data value should be checked for accuracy. Perhaps someone misread the y value or the x value. If it is correct, perhaps it represents a very unusual case (such as old items that have increased greatly in value since being purchased) and can be removed from the database with justification for separate handling.

What happens to the analysis if this one data point is removed? Figure 6.6 shows that the scatter plot is now well behaved, and Figure 6.7 shows that the deviations are now more randomly distributed around zero. These deviations seem to increase in variability as x increases, but that is actually a good case for ratio estimation, as we will see later. The summary calculations show that

$$r = \frac{15.86}{16.76} = 0.95$$

FIGURE 6.6 Scatter plot for the data of Example 6.6, with the large deviation removed

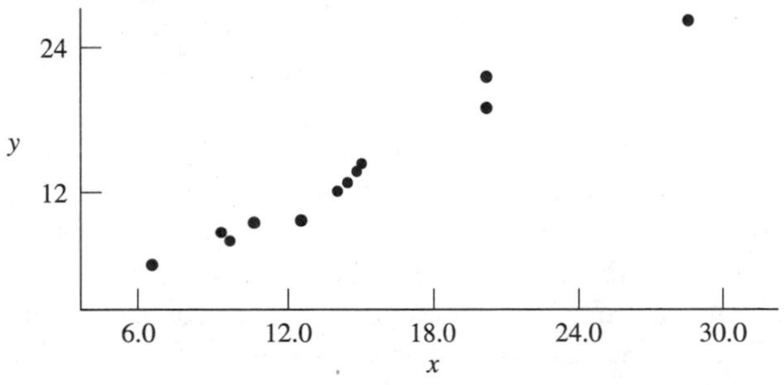

	N	Mean	Median	SD
x	14	16.76	14.85	7.44
y	14	15.86	13.55	7.65
y - rx	14	-0.000	-0.349	1.58

FIGURE 6.7 Scatter plot of the deviations $(y_i - rx_i)$ for Example 6.6, with the large deviation removed

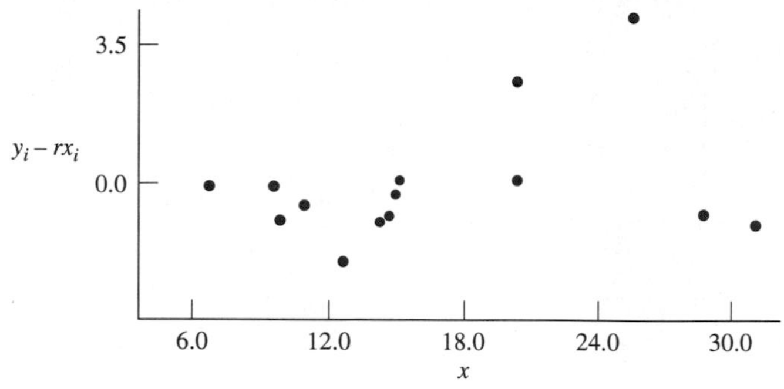

and

$$\hat{\sigma}^2 = (1.58)^2 = 2.50$$

These values produce a sample-size estimate of 163 as compared with the original estimate of 421! So, the estimates in ratio estimation, particularly estimates of variance and sample size, are quite sensitive to data points that do not fit the ideal pattern for that technique, so-called influential observations. It is very important to plot the data and look for these unusual data points before proceeding with an analysis.

6.5

RATIO ESTIMATION IN STRATIFIED RANDOM SAMPLING

For the same reasons indicated in Chapter 5, stratifying the population before using a ratio estimator is sometimes advantageous. We will assume that we can take a large enough sample of both x's and y's in each stratum for the variance approximations to work fairly well.

There are two different methods for constructing estimators of a ratio in stratified sampling. One is to estimate the ratio of μ_y to μ_x within each stratum and then form a weighted average of these *separate* estimates as a single estimate of the population ratio. The result of this procedure is called a *separate ratio estimator.*

The other method involves first estimating μ_y by the usual $\bar{y}_{st}$ and similarly estimating μ_x by $\bar{x}_{st}$. Then $\bar{y}_{st}/\bar{x}_{st}$ can be used as an estimator of μ_y/μ_x. This estimator is called a *combined ratio estimator.*

We will not introduce a general (and cumbersome) notation for these estimators but will illustrate their use by a numerical example. The derivation in the appendix shows that the variance of a sum of random variables is the sum of the variances, if the variables are independent. This fact will allow us to use a sum of terms similar to those in Equation (6.7) for the variance of either the combined or the separate ratio estimator. The next two examples illustrate the techniques used.

Example 6.7 Refer to Example 6.4. Treat the 10 observations given there on man-hours lost due to sickness as a simple random sample from company A. Thus, $n_A = 10$, $\bar{y}_A = 18.7$, $\bar{x}_A = 17.8$, $r_A = 1.05$, $N_A = 1000$, and $\tau_{xA} = 16{,}300$.

A simple random sample of $n_B = 10$ measurements was taken from company B within the same industry. (Assume companies A and B together form the population of workers of interest in this problem.) The data are given in Table 6.6. It is known that $N_B = 1500$ employees and $\tau_{xB} = 12{,}800$. Find the separate ratio estimate of μ_y and its estimated variance.

TABLE 6.6

Employee	Man-hours lost in previous year, x_B	Man-hours lost in current year, y_B
1	10	8
2	8	0
3	0	4
4	14	6
5	12	10
6	6	0
7	4	2
8	0	4
9	8	4
10	16	8
	78	46

Solution

The ratio estimator of μ_{yA} is $(\bar{y}_A/\bar{x}_A)(\mu_{xA})$ [see Equation (6.6)], and its estimated variance is given by Equation (6.7). The corresponding estimator of μ_{yB} is $(\bar{y}_B/\bar{x}_B)(\mu_{xB})$, with a similar estimated variance.

To obtain an estimator of μ_y, the population mean of the y's, we need to average the estimators together choosing weights proportional to the stratum sizes, as in Chapter 5. Thus, $\hat{\mu}_{yRS}$, given by

$$\hat{\mu}_{yRS} = \left(\frac{N_A}{N}\right)\left(\frac{\bar{y}_A}{\bar{x}_A}\right)(\mu_{xA}) + \left(\frac{N_B}{N}\right)\left(\frac{\bar{y}_B}{\bar{x}_B}\right)(\mu_{xB})$$

will be the estimator of μ_y, with estimated variance

$$\hat{V}(\hat{\mu}_{yRS}) = \left(\frac{N_A}{N}\right)^2 \left(\frac{N_A - n_A}{N_A n_A}\right) \frac{\sum_{i=1}^{n_A} (y_i - r_A x_i)^2}{n_A - 1} + \left(\frac{N_B}{N}\right)^2 \left(\frac{N_B - n_B}{N_B n_B}\right) \frac{\sum_{i=1}^{n_B} (y_i - r_B x_i)^2}{n_B - 1}$$

The observed value of $\hat{\mu}_{yRS}$ from the data is

$$\left(\frac{1000}{2500}\right)\left(\frac{18.7}{17.8}\right)(16.3) + \left(\frac{1500}{2500}\right)\left(\frac{4.6}{7.8}\right)(8.53) = 9.87$$

Summarizing all data, including the key components of the two variances of ratios, in a spreadsheet formation yields the array shown in Table 6.7.

TABLE 6.7

Row	x_A	y_A	$y_A - r_A x_A$	x_B	y_B	$y_B - r_B x_B$
1	12	13	0.39326	10	8	2.10256
2	24	25	−0.21349	8	0	−4.71795
3	15	15	−0.75843	0	4	4.00000
4	30	32	0.48314	14	6	−2.25641
5	32	36	2.38202	12	10	2.92308
6	26	24	−3.31461	6	0	−3.53846
7	10	12	1.49438	4	2	−0.35897
8	15	16	0.24157	0	4	4.00000
9	0	2	2.00000	8	4	−0.71795
10	14	12	−2.70787	16	8	−1.43590

	n	Mean	Median	SD
x_A	10	17.80	15.00	9.99
y_A	10	18.70	15.50	10.36
$y_A - r_A x_A$	10	−0.000	0.317	1.86
x_B	10	7.80	8.00	5.45
y_B	10	4.60	4.00	3.41
$y_B - r_B x_A$	10	0.000	−0.538	3.12

From this, it is relatively easy to construct the estimated variance to be

$$\hat{V}(\hat{\mu}_{yRS}) = \left(\frac{1000}{2500}\right)^2 \left(\frac{990}{1000}\right)\left(\frac{1}{10}\right)(1.86)^2 + \left(\frac{1500}{2500}\right)^2 \left(\frac{1490}{1500}\right)\left(\frac{1}{10}\right)(3.12)^2 = 0.40$$

Example 6.8 Refer to the data of Example 6.7 and find a combined ratio estimate of μ_y.

Solution

Here we use $\bar{y}_{st}$ to estimate μ_y, $\bar{x}_{st}$ to estimate μ_x, and

$$\hat{\mu}_{yRC} = \frac{\bar{y}_{st}}{\bar{x}_{st}} (\mu_x)$$

as the combined ratio estimator of μ_y. If we denote $(\bar{y}_{st}/\bar{x}_{st})$ by r_C, the estimated variance of $\hat{\mu}_{yRC}$ is

$$\hat{V}(\hat{\mu}_{yRC}) = \left(\frac{N_A}{N}\right)^2 \left(\frac{N_A - n_A}{N_A n_A}\right) s_{rA}^2 + \left(\frac{N_B}{N}\right)^2 \left(\frac{N_B - n_B}{N_B n_B}\right) s_{rB}^2$$

where s_{rA}^2 is the variance of the terms $(y_i - r_C x_i)$ for data set A and s_{rB}^2 is the variance of the corresponding deviations for data set B.

For the data given,

$$\bar{y}_{st} = (.4)(18.7) + (.6)(4.6) = 10.24$$
$$\bar{x}_{st} = (.4)(17.8) + (.6)(7.8) = 11.80$$
$$r_C = .868$$
$$\mu_x = \frac{16{,}300 + 12{,}800}{2500} = 11.64$$

Hence, the observed value of $\hat{\mu}_{yRC}$ is

$$\frac{10.24}{11.80}(11.64) = 10.10$$

The spreadsheet formulation of the data and the key components of the variance calculations, along with the statistical summary, is shown in Table 6.8. From these components, the variance is estimated to be

$$\hat{V}(\hat{\mu}_{yRC}) = \left(\frac{1000}{2500}\right)^2 \left(\frac{990}{1000}\right)\left(\frac{1}{10}\right)(2.39)^2 + \left(\frac{1500}{2500}\right)^2 \left(\frac{1490}{1500}\right)\left(\frac{1}{10}\right)(4.00)^2 = 0.66$$

TABLE 6.8

Row	x_A	y_A	$y_A - r_C x_A$	x_B	y_B	$y_B - r_C x_B$
1	12	13	2.58400	10	8	−0.68000
2	24	25	4.16800	8	0	−6.94400
3	15	15	1.98000	0	4	4.00000
4	30	32	5.96000	14	6	−6.15200
5	32	36	8.22400	12	10	−0.41600
6	26	24	1.43200	6	0	−5.20800
7	10	12	3.32000	4	2	−1.47200
8	15	16	2.98000	0	4	4.00000
9	0	2	2.00000	8	4	−2.94400
10	14	12	−0.15200	16	8	−5.88800

	n	Mean	Median	SD
x_A	10	17.80	15.00	9.99
y_A	10	18.70	15.50	10.36
$y_A - r_C x_A$	10	3.25	2.78	2.39
x_B	10	7.80	8.00	5.45
y_B	10	4.60	4.00	3.41
$y_B - r_C x_B$	10	−2.17	−2.21	4.00

On comparing Examples 6.7 and 6.8, we see that the combined ratio estimator gives the larger estimated variance. This result is generally the case, and so we should employ the separate ratio estimator most of the time. However, the separate ratio estimator may have a larger bias since each stratum ratio estimate contributes to that bias. In summary, if the stratum sample sizes are large enough (say, 20 or so) so that the separate ratios do not have large biases and so that the variance approximations work adequately, then use the separate ratio estimator. If stratum sample sizes are very small, or if the within-stratum ratios are all approximately equal, then the combined ratio estimator may perform better.

Of course, an estimator of the population total can be found by multiplying either of the estimators above by the population size N, and the variances can be adjusted accordingly. Thus, we might use the notation

$$\hat{\tau}_{yRS} = N\hat{\mu}_{yRS}$$

6.6 REGRESSION ESTIMATION

We saw in Section 6.5 that the ratio estimator is most appropriate when the relationship between y and x is linear through the origin. If there is evidence of a linear relationship between the observed y's and x's, but not necessarily one that would pass through the origin, then this extra information provided by the auxiliary variable x may be taken into account through a regression estimator of the mean μ_y. One must still have knowledge of μ_x before the estimator can be employed, as it was in the case of ratio estimation of μ_y.

The underlying line that shows the basic relationship between the y's and x's is sometimes referred to as the *regression line of y upon x*. Thus, the subscript L in the ensuing formulas is used to denote *linear regression*.

The estimator given next assumes the x's to be fixed in advance and the y's to be random variables. We can think of the x value as something that has already been observed, like last year's first-quarter earnings, and the y response as a random variable yet to be observed, such as the current quarterly earnings of a company for which x is already known. The probabilistic properties of the estimator then depend only on y for a given set of x's.

Regression estimator of a population mean μ_y:

$$\hat{\mu}_{yL} = \bar{y} + b(\mu_x - \bar{x}) \tag{6.24}$$

where

$$b = \frac{\sum_{i=1}^{n}(y_i - \bar{y})(x_i - \bar{x})}{\sum_{i=1}^{n}(x_i - \bar{x})^2}$$

Estimated variance of $\hat{\mu}_{yL}$:

$$\hat{V}(\hat{\mu}_{yL}) = \left(\frac{N-n}{Nn}\right)\left(\frac{1}{n-2}\right)\left[\sum_{i=1}^{n}(y_i - \bar{y})^2 - b^2\sum_{i=1}^{n}(x_i - \bar{x})^2\right] \tag{6.25}$$

$$= \left(\frac{N-n}{Nn}\right)\text{MSE} \tag{6.26}$$

where MSE is the mean square error from the standard simple linear regression of y on x.

The calculations necessary for regression estimation of a mean, illustrated in Example 6.9, are similar to those performed when fitting a straight line to paired data by the least-squares method.

Example 6.9

A mathematics achievement test was given to 486 students prior to their entering a certain college. From these students a simple random sample of $n = 10$ students was selected and their progress in calculus observed. Final calculus grades were then reported, as given in the accompanying table. It is known that $\mu_x = 52$ for all 486 students taking the achievement test. Estimate μ_y for this population and place a bound on the error of estimation.

Student	Achievement test score, x	Final calculus grade, y
1	39	65
2	43	78
3	21	52
4	64	82
5	57	92
6	47	89
7	28	73
8	75	98
9	34	56
10	52	75

Solution

The first step of the analysis is to look at a scatter plot of the data. Figure 6.8 shows that there is a strong positive association between y and x, and a straight line looks like a reasonable model for this relationship. Fitting a straight line through these data by the least-squares method produces a line of the form

$$\hat{y}_i = a + bx_i$$

where a is the y-intercept at $x = 0$ and b is the slope. Using the least-squares solution for the intercept changes the form of this equation to

$$\hat{y}_i = \bar{y} + b(x_i - \bar{x})$$

and the estimator μ_{yL} is obtained by substituting μ_x for x_i. The observed value of μ_{yL} is then

$$\bar{y} + b(\mu_x - \bar{x}) = 76 + (.766)(52 - 46) = 80.6$$

Notice that the regression estimator of μ_y inflates the value of $\bar{y}$ since $\bar{x}$ turns out to be less than μ_x and b is positive.

FIGURE 6.8 Scatter plot for Example 6.9

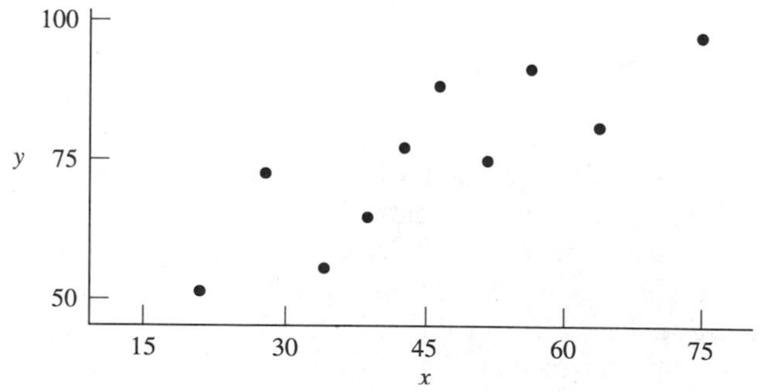

Deviations between the data point y_i and the value $\hat{y}_i$ of the regression line at x_i are called *residuals*. The MSE is based on the sum of the squared residuals and can be written

$$\text{MSE} = \frac{\sum_{i=1}^{n} (y_i - \hat{y}_i)^2}{n-2}$$

Standard computer printout for fitting a simple linear model to these data is shown below. The value for b is shown as the coefficient of x in the regression equation; in this case, $b = 0.766$. MSE is the value opposite "Error" and under "MS" in the table, 75.8 in this case. The value labeled "s" is the square root of MSE, and the value labeled "R-sq" is the square of the correlation coefficient between y and x.

```
The regression equation is
y=40.8+0.766x

s=8.704 R-sq=70.5%

Analysis of Variance

SOURCE          DF        SS         MS
Regression       1    1450.0     1450.0
Error            8     606.0       75.8
Total            9    2056.0
```

Thus,

$$\hat{V}(\hat{\mu}_{yL}) = \left(\frac{N-n}{Nn}\right)\text{MSE}$$
$$= \frac{476}{486(10)}(75.8) = 7.42$$

and

$$2\sqrt{\hat{V}(\hat{\mu}_{yL})} = 5.45$$

which is very close to the value obtained in the previous calculations.

It is good practice to plot the residuals to see if any are unusually large deviations or if an emerging pattern emerges that suggests the simple linear model is not a good one. For a good model of y versus x, the residuals should simply be a random scattering of points. Figure 6.9 shows the residuals in this case to be fairly random; no particularly obvious pattern seems to be apparent.

A close examination of the data on sugar content and weight of oranges given in Example 6.2 might suggest that a regression estimator is more appropriate than a ratio estimator. (A plot of the points will show that the regression line does not appear to go

FIGURE 6.9 Residual plot for Example 6.9

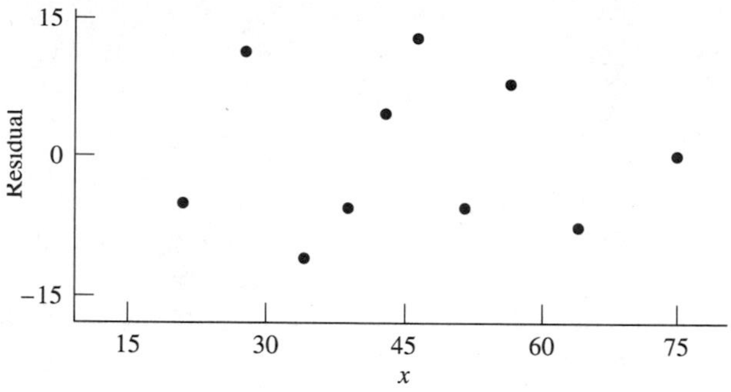

through the origin.) However, the regression estimator of a total is of the form $N\hat{\mu}_{yL}$, specifically requiring knowledge of N. Since the ratio estimator also works well in this case, determining the number of oranges in the truckload may not be worth the extra cost and time. In other cases, N may be known or easily found. Thus, one should carefully consider the choice between ratio and regression estimators when estimating population means or totals. There will be more on this in Section 6.8.

6.7

DIFFERENCE ESTIMATION

The difference method of estimating a population mean or total is similar to the regression method in that it adjusts the $\bar{y}$ value up or down by an amount depending on the difference $(\mu_x - \bar{x})$. However, the regression coefficient b is not computed. In effect, b is set equal to unity.

The difference method is, then, easier to employ than the regression method and frequently works just as well. It is commonly employed in auditing procedures, and we will consider such an example in this section.

The following formulas hold provided that simple random sampling was employed.

Difference estimator of a population μ_y:

$$\hat{\mu}_{yD} = \bar{y} + (\mu_x - \bar{x}) = \mu_x + \bar{d} \tag{6.27}$$

where $\bar{d} = \bar{y} - \bar{x}$

Estimated variance of $\hat{\mu}_{yD}$:

$$\hat{V}(\hat{\mu}_{yD}) = \left(\frac{N-n}{Nn}\right)\frac{\sum_{i=1}^{n}(d_i-\bar{d})^2}{n-1} \tag{6.28}$$

where $d_i = y_i - x_i$

Example 6.10

Auditors are often interested in comparing the audited value of items with the book value. Generally, book values are known for every item in the population, and audit values are obtained for a sample of these items. The book values can then be used to obtain a good estimate of the total or average audit value for the population.

Suppose a population contains 180 inventory items with a stated book value of \$13,320. Let x_i denote the book value and y_i the audit value of the ith item. A simple random sample of $n = 10$ items yields the results shown in the accompanying table. A plot of these data, Figure 6.10 shows them to lie along a straight line through the origin. Estimate the mean audit value of μ_y by the difference method and estimate the variance of $\hat{\mu}_{yD}$.

Sample	Audit value, y_i	Book value, x_i	d_i
1	9	10	−1
2	14	12	+2
3	7	8	−1
4	29	26	+3
5	45	47	−2
6	109	112	−3
7	40	36	+4
8	238	240	−2
9	60	59	+1
10	170	167	+3

Solution

Since $\bar{y} = 72.1$, $\bar{x} = 71.7$, and $\mu_x = 74.0$,

$$\hat{\mu}_{yD} = \mu_x + \bar{d} = 74.0 + (72.1 - 71.7) = 74.4$$

Also,

$$\left(\frac{1}{n-1}\right)\sum_{i=1}^{n}(d_i-\bar{d})^2 = \left(\frac{1}{n-1}\right)\left[\sum_{i=1}^{n} d_i^2 - n\bar{d}^2\right]$$

$$= \frac{58 - 10(.4)^2}{9} = 6.27$$

FIGURE 6.10 Plot of y versus x for Example 6.10

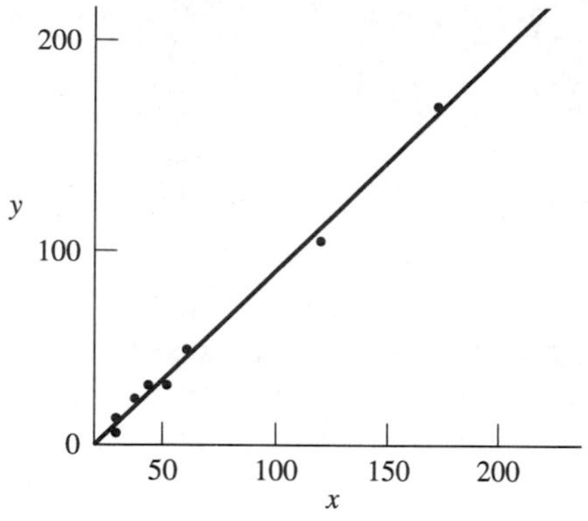

Thus,

$$\hat{V}(\hat{\mu}_{yD}) = \left(\frac{N-n}{Nn}\right) \frac{\sum_{i=1}^{n}(d_i - \bar{d})^2}{n-1} = \left[\frac{180-10}{(180)10}\right](6.27) = .59$$

6.8

RELATIVE EFFICIENCY OF ESTIMATORS

We have seen that the sample mean, ratio estimator, regression estimator, and difference estimator can all be used as estimators of a population mean μ_y. How do we tell which one is best for a particular sampling situation? Actually, we cannot always answer definitively, but there are some guidelines that compare the properties of estimators. One such guideline can be expressed in terms of *relative efficiency* of estimators.

Suppose we have two estimators E_1 and E_2 for a mean μ. If both E_1 and E_2 are unbiased, or nearly unbiased, estimators of μ, then we should generally choose the one with the smaller variance as the better estimator. This will produce the narrower confidence interval estimate for μ. Variances usually decrease as the sample size increases, so we must compare the variances of E_1 and E_2 assuming equal sample sizes for both estimators. It is convenient to describe the relative size of the two variances by looking at their ratio. This ratio will be called the *relative* efficiency, denoted by RE, for the two estimators. We set up the relative efficiency ratio so that a *large* value is favorable to the estimator named first. Thus, the relative efficiency of E_1 to E_2 (or E_1 with respect to E_2) is given by

$$\text{RE}\left(\frac{E_1}{E_2}\right) = \frac{V(E_2)}{V(E_1)}$$

If RE (E_1/E_2) is large (greater than 1), then $V(E_2)$ will be larger than $V(E_1)$, which is favorable to E_1 as an estimator of μ. Remember, the sample sizes for E_1 and E_2 must be equal in this calculation.

Suppose $\text{RE}(E_1/E_2)=2$. This implies that $V(E_2)=2V(E_1)$, which is a favorable case for E_1. Another way to make this comparison is to state that the sample size for E_2 would have to be twice that for E_1 in order to make E_1 and E_2 equivalent in terms of variance. Thus, RE can be thought of in terms of equivalent sample sizes (or sampling effort or sampling cost). An RE of 1 implies that the two estimators are equivalent; it does not matter which one we use.

RE is usually defined in terms of theoretical variances. However, we have given mostly estimated variances in this text. So, we will continue in that spirit and define

$$\widehat{\text{RE}}\left(\frac{E_1}{E_2}\right)=\frac{\hat{V}(E_2)}{\hat{V}(E_1)}$$

We now must be careful in the interpretation; $\widehat{\text{RE}}(E_1/E_2) > 1$ does not necessarily mean that $V(E_2) > V(E_1)$, since we are only dealing with estimators of variances, which could change from sample to sample. However, if we have large samples (and hence good estimators of the variances), a value of $\widehat{\text{RE}}(E_1/E_2)$ considerably greater than 1 would certainly imply strongly that E_1 is probably the better estimator.

We now want to compute estimated REs for various combinations of the four estimators of μ_y mentioned above. First, however, we must consider the question of bias, since it is inappropriate to compare variances for biased estimators. The mean $\bar{y}$ of a simple random sample is always an unbiased estimator of μ_y, so there is no bias problem (theoretically, at least) in dealing with this estimator.

On the other hand, the ratio estimator of μ_y is generally biased, since $r=\bar{y}/\bar{x}$ is generally a biased estimator of $R=\mu_y/\mu_x$. The bias becomes negligible if the relationship between y and x falls along a straight line that runs through the origin. An approximation to the relative bias of r is given by

$$\frac{E(r)-R}{R}\approx\left(\frac{N-n}{Nn}\right)\left(\frac{s_x^2}{\bar{x}^2}-\hat{\rho}\frac{s_y}{\bar{y}}\cdot\frac{s_x}{\bar{x}}\right)$$

where $\hat{\rho}$ is the sample correlation coefficient between x and y.

As to the bias of the other two estimators, the regression estimator is unbiased if the relationship between y and x (the regression of y on x) falls along a straight line, not necessarily through the origin. The difference estimator is always unbiased in simple random sampling.

On comparing the ratio estimator to the simple mean per element $\bar{y}$, we have

$$\widehat{\text{RE}}\left(\frac{\hat{\mu}_y}{\bar{y}}\right)=\frac{\hat{V}(\bar{y})}{\hat{V}(\hat{\mu}_y)}$$

$$=\frac{s_y^2}{s_y^2+r^2s_x^2-2r\hat{\rho}s_xs_y}$$

Now, $\widehat{\text{RE}}(\hat{\mu}_y/\bar{y}) > 1$ if

$$s_y^2 + r^2 s_x^2 - 2r\hat{\rho}s_x s_y < s_y^2$$

or

$$r^2 s_x^2 < 2r\hat{\rho}s_x s_y$$

or

$$rs_x^2 < 2\hat{\rho}s_x s_y \qquad \text{(assuming } r > 0\text{)}$$

or

$$\hat{\rho} > \frac{1}{2}\frac{rs_x}{s_y} = \frac{1}{2}\frac{s_x/\bar{x}}{s_y/\bar{y}}$$

The quantity $s_x/\bar{x}$ is sometimes called the coefficient of variation. For situations in which the ratio estimator is commonly used, the y's are updated values of the x's (first-quarter earnings in one year compared to first-quarter earnings in the previous year, audit value versus book value, and the like). In such cases, the coefficient of variation for the y's should be quite close to that for the x's. Thus, in such situations, the ratio estimator is more efficient than the simple mean per element estimator if $\hat{\rho} > \frac{1}{2}$. Overall, the ratio estimator will be more efficient than $\bar{y}$ if the variation among the x's is small relative to the variation among the y's and the correlation between x and y is a high positive value. If the experimenter has options as to how to choose the x-values, she should choose them to be nearly constant.

A simple comparison of the regression estimator with the mean per element $\bar{y}$ and the ratio estimator $\hat{\mu}_y$ requires some modification of the estimated variance. Recall that

$$\hat{\mu}_{yL} = \bar{y} + b(\mu_x - \bar{x})$$

where b is the usual estimator of the slope of a regression line. The estimated variance of $\hat{\mu}_{yL}$ was given [see Equation (6.25)] as

$$\hat{V}(\hat{\mu}_{yL}) = \left(\frac{N-n}{Nn}\right)\left(\frac{1}{n-2}\right)\left[\sum_{i=1}^{n}(y_i - \bar{y})^2 - b^2\sum_{i=1}^{n}(x_i - \bar{x})^2\right]$$

If we make the slight change of replacing $(n-2)$ by $(n-1)$ in the denominator, we have

$$\hat{V}(\hat{\mu}_{yL}) \approx \left(\frac{N-n}{Nn}\right)[s_y^2 - b^2 s_x^2]$$

and, since

$$b = \hat{\rho}\,\frac{s_y}{s_x}$$

this becomes

$$\hat{V}(\hat{\mu}_{yL}) \approx \left(\frac{N-n}{Nn}\right)s_y^2(1 - \hat{\rho}^2)$$

This approximation to $V(\hat{\mu}_{yL})$ is good as long as n is reasonably large; the $(n - 2)$ was used in the denominator to prevent serious underestimation of the variance in small-sample situations.

Using the variance approximation simplified above,

$$\widehat{\mathrm{RE}}\left(\frac{\hat{\mu}_{yL}}{\bar{y}}\right) = \frac{s_y^2}{s_y^2(1-\hat{\rho})^2} = \frac{1}{1-\hat{\rho}^2}$$

which will always be greater than unity if $\hat{\rho}$ differs from zero. In fact, the RE can become infinitely large as $\hat{\rho}$ approaches unity. Thus $\hat{\mu}_{yL}$ is always more efficient than $\bar{y}$ as an estimator of μ_y. (However, recall that $\hat{\mu}_{yL}$ could have serious bias problems unless the regression of y on x is truly linear.)

When comparing regression to ratio estimation,

$$\widehat{\mathrm{RE}}\left(\frac{\hat{\mu}_{yL}}{\hat{\mu}_y}\right) = \frac{s_y^2 + r^2 s_x^2 - 2r\hat{\rho}s_x s_y}{s_y^2(1-\hat{\rho}^2)}$$

In this case, $\widehat{\mathrm{RE}} > 1$ implies

$$r^2 s_x^2 - 2r\hat{\rho}s_x s_y > -\hat{\rho}^2 s_y^2$$

or

$$(\hat{\rho}s_y - rs_x)^2 > 0$$

Since $\hat{\rho}s_y = bs_x$, this can be written

$$(bs_x - rs_x)^2 > 0$$

which implies

$$(b - r)^2 > 0$$

Thus, the regression estimator is more efficient than the ratio estimator unless $b = r$, in which case they are equivalent. The case $b = r$ will occur when the regression of y on x is linear through the origin *and* the variance of y is proportional to x. Such a situation is demonstrated in Figure 6.11, where case receipts for a specified sales period are related to the number of customers. Note that the scatter of receipt values widens as x increases.

The difference estimator

$$\hat{\mu}_{yD} = \bar{y} + (\mu_x - \bar{x})$$

is always an unbiased estimator of μ in simple random sampling, and its estimated variance

$$\hat{V}(\hat{\mu}_{yD}) = \left(\frac{N-n}{Nn}\right)\frac{\sum_{i=1}^{n}(d_i - \bar{d})^2}{n-1}$$

FIGURE 6.11 Plot of cash receipts versus number of customers

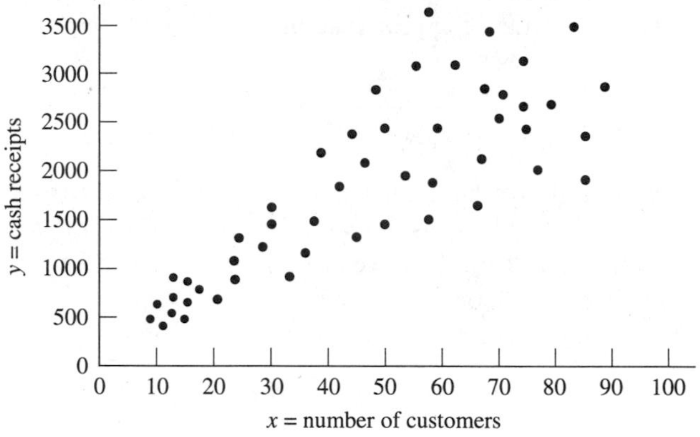

can be written

$$\hat{V}(\hat{\mu}_{yD}) = \left(\frac{N-n}{Nn}\right)\left(\frac{1}{n-1}\right)\sum_{i=1}^{n}[(y_i - \bar{y}) - (x_i - \bar{x})]^2$$
$$= \left(\frac{N-n}{Nn}\right)[s_y^2 + s_x^2 - 2\hat{\rho}s_x s_y]$$

On comparing the difference estimator with the sample mean per element, we have

$$\widehat{\mathrm{RE}}\left(\frac{\hat{\mu}_{yD}}{\bar{y}}\right) = \frac{s_y^2}{s_y^2 + s_x^2 - 2\hat{\rho}s_x s_y}$$

which is greater than unity when

$$2\hat{\rho}s_x s_y > s_x^2$$

or

$$\hat{\rho} > \frac{s_x}{2s_y}$$

If the variation in x's and y's is about the same, the difference estimator will be more efficient than $\bar{y}$ when the correlation between x and y is greater than $\frac{1}{2}$.

On comparing the regression estimator with the difference estimator, we have

$$\widehat{\mathrm{RE}}\left(\frac{\hat{\mu}_{yL}}{\hat{\mu}_{yD}}\right) = \frac{s_y^2 + s_x^2 - 2\hat{\rho}s_x s_y}{s_y^2(1-\hat{\rho}^2)}$$

which is greater than unity when

$$s_x^2 - 2\hat{\rho}s_x s_y > -\hat{\rho}^2 s_y^2$$

or

$$(s_x - \hat{\rho} s_y)^2 > 0$$

Since $bs_x = \hat{\rho} s_y$, the regression estimator will be equivalent to the difference estimator when $b = 1$. Otherwise, the regression estimator will be more efficient than the difference estimator.

We will now look at some numerical values of REs for data analyzed earlier in this chapter. The data from Table 6.1 on real estate valuation is plotted in Figure 6.12. Notice that the data points tend to fall along a straight line with slope near unity (in fact, $b = 0.977$ for the regression estimator) and y-intercept near zero. For this case,

$$\frac{E(r) - R}{R} \approx 0.0053$$

so the relative bias of the ratio estimator is not serious.

For these data,

$$\widehat{\text{RE}}\left(\frac{\hat{\mu}_{y\text{L}}}{\hat{\mu}_y}\right) = 1.13 \qquad \text{and} \qquad \widehat{\text{RE}}\left(\frac{\hat{\mu}_{y\text{D}}}{\hat{\mu}_y}\right) = 1.01$$

so that all three estimators, regression, ratio, and difference, are about equivalent in terms of the size of the estimated variance. Any of the three estimators would work well for the problem of estimating μ_y or τ_y with these data. But

$$\widehat{\text{RE}}\left(\frac{\hat{\mu}_{y\text{L}}}{\bar{y}}\right) = 14.96$$

so that $\bar{y}$ would be a poor estimator of μ_y in comparison to any one of the three estimators making use of the x's. In other words, it would take about 15 times as many observations to achieve the same variance with $\bar{y}$ as is achieved with $\hat{\mu}_{y\text{L}}$.

FIGURE 6.12 Data from Table 6.1

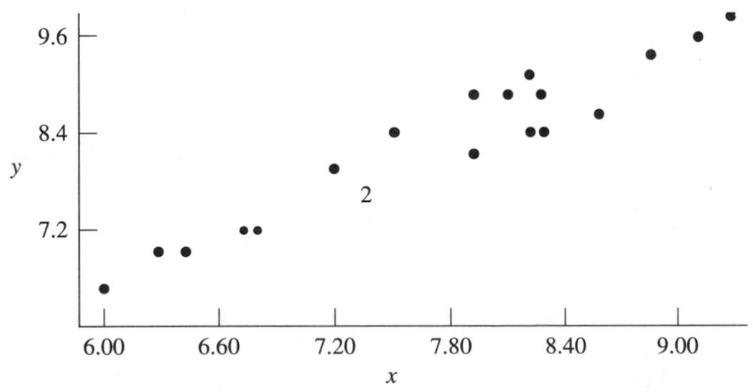

The data from Table 6.2 on sugar content versus weight of oranges, is plotted in Figure 6.13. Here, the data points fall along a straight line, but the slope is not close to unity (in fact, $b = 0.123$) and the y-intercept is significantly different from zero. The relative bias of r is -0.00077, still not serious, but

$$\widehat{\text{RE}}\left(\frac{\hat{\mu}_{y\text{L}}}{\hat{\mu}_y}\right) = 16.79$$

This implies that better accuracy could be achieved in the estimation of μ_y or τ_y by employing the regression estimator rather than the ratio estimator. Differences do not make sense in this problem.

From Example 6.9, we have data on final calculus grades versus achievement test scores, plotted in Figure 6.14. The slope of this line is not close to unity ($b = 0.766$) and the y-intercept is far from zero. Calculations show that

$$\widehat{\text{RE}}\left(\frac{\hat{\mu}_{y\text{L}}}{\hat{\mu}_y}\right) = 4.84 \qquad \text{and} \qquad \widehat{\text{RE}}\left(\frac{\hat{\mu}_{y\text{L}}}{\hat{\mu}_{y\text{D}}}\right) = 1.22$$

Regression estimation is considerably better than ratio estimation here, but a difference estimator could have been used with little loss of efficiency.

For the data of Example 6.10, all three methods, ratio, regression, and difference, are approximately equivalent; the difference estimator is the easiest to calculate, and so it is a reasonable choice.

In summary, the analysis of bivariate data should always begin by plotting the points. If the points fall along a straight line through the origin and if it is appropriate to take differences, any of the three methods involving x's and y's is a possibility. For the difference estimator to work well, the slope of the line must be close to 1. If the points fall along a straight line not through the origin, then the regression estimator is likely to

FIGURE 6.13 Data from Table 6.2

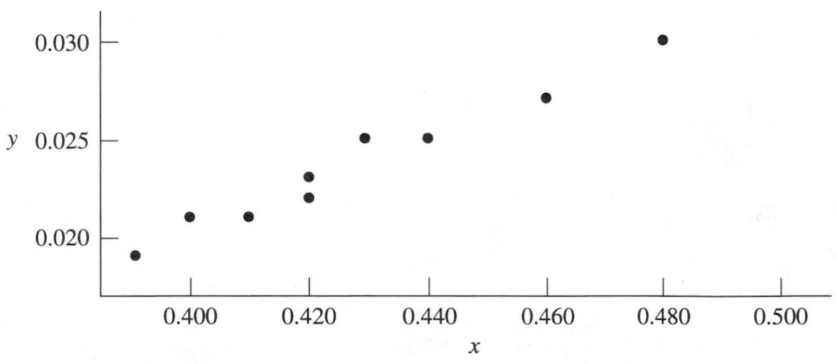

FIGURE 6.14 Data from Example 6.9

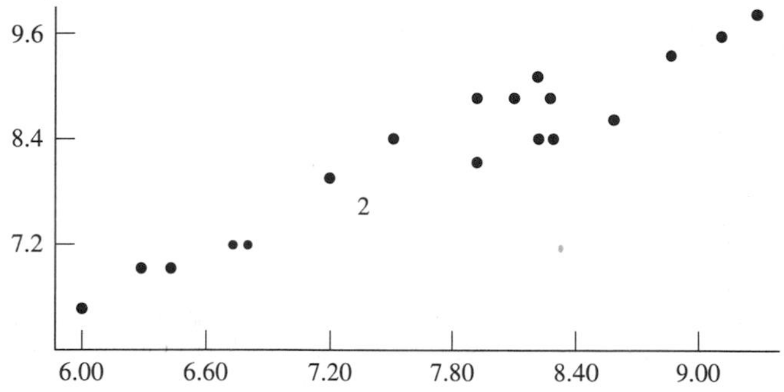

be best, in terms of variance. If the points do not fall along a straight line, perhaps $\bar{y}$ should be used (or a more detailed regression analysis undertaken).

6.9

SUMMARY

This chapter has briefly presented ratio estimation of a population mean, total, and ratio for simple random sampling. By measuring a variable y and a subsidiary variable x on each element in the sample, we obtain additional information for estimating the population parameter of interest. When a strong positive correlation exists between the variables x and y, the ratio estimation procedure usually provides more precise estimators of μ_y and τ_y than do the standard techniques presented in Chapter 4.

Sample-size requirements were presented for estimating μ_y, τ_y, and R with a bound on the error of estimation equal to B. In each case, one must obtain an estimate of σ^2 from prior information or from a preliminary study to approximate the required sample size.

Regression estimation is another technique for incorporating information on a subsidiary variable. This method is usually better than ratio estimation if the relationship between the y's and x's is a straight line not through the origin.

Although these methods can be employed with any sampling design, we have concentrated on simple random sampling, while mentioning stratified random sampling for the ratio case.

The method of difference estimation is similar in principle to regression estimation. It works well when the plot of y versus x reveals points lying uniformly close to a straight line with unit slope.

CASE STUDY REVISITED

THE ESTIMATION OF INVENTORY ERROR

An auditor can use the method of ratio estimation to obtain an estimate of total audited amount, which can then be used to estimate the total inventory error, as outlined at the beginning of this chapter.

An electronics firm reports that their inventory of $N = 100$ computer terminals amounts to \$150,000. An auditor decides to estimate the total error on this item by sampling five terminals and determining their actual value. The sample data and pertinent calculations are as follows (in units of \$1000):

Book amount, x_i	Audit amount, y_i	rx_i	$y_i - rx_i$	$(y_i - rx_i)^2$
1.3	1.1	1.17	−.07	.0049
1.2	1.3	1.08	.22	.0484
1.5	1.4	1.35	.05	.0025
1.7	1.5	1.53	−.03	.0009
1.3	1.0	1.17	−.17	.0289
7.0	6.3			.0856

$$r = \frac{\sum_{i=1}^{n} y_i}{\sum_{i=1}^{n} x_i} = \frac{6.3}{7.0} = .9$$

$$\bar{x} = \frac{7.0}{5} = 1.4$$

The estimate of τ_y, the total audited amount, is

$$\hat{\tau}_y = r\tau_x = (.9)(150) = 135$$

(in \$1000 units), and the bound on the error is

$$2\sqrt{(\tau_x)^2\left(\frac{N-n}{nN}\right)\left(\frac{1}{\bar{x}}\right)^2\left(\frac{1}{n-1}\right)\sum_{i=1}^{n}(y_i - rx_i)^2}$$

$$= 2\sqrt{(150)^2\left(\frac{95}{500}\right)\left(\frac{1}{1.4}\right)^2\left(\frac{1}{4}\right)(.0856)} = 14$$

Thus, we estimate the total audited amount to be between $135 - 14 = 121$ and $135 + 14 = 149$. The total error, $\tau_e = \tau_x - \tau_y$, is then estimated to be between

$$150 - 149 = 1 \qquad \text{and} \qquad 150 - 121 = 29$$

Note that it is quite likely that the total error is positive.

EXERCISES

6.1 A forester is interested in estimating the total volume of trees in a timber sale. He records the volume for each tree in a simple random sample. In addition, he measures the basal area for each tree marked for sale. He then uses a ratio estimator of total volume.

The forester decides to take a simple random sample of $n = 12$ from the $N = 250$ trees marked for sale. Let x denote basal area and y the cubic-foot volume for a tree. The total basal area for all 250 trees, τ_x, is 75 square feet. Use the data in the accompanying table to estimate τ_y, the total cubic-foot volume for those trees marked for sale, and place a bound on the error of estimation.

Tree sampled	Cubic-foot basal area, x	Volume, y	Tree sampled	Cubic-foot basal area, x	Volume, y
1	.3	6	7	.6	12
2	.5	9	8	.5	9
3	.4	7	9	.8	20
4	.9	19	10	.4	9
5	.7	15	11	.8	18
6	.2	5	12	.6	13

6.2 Use the y data in Exercise 6.1 to compute an estimate of τ_y, using $N\bar{y}$. Place a bound on the error of estimation. Compare your results with those obtained in Exercise 6.1. Why is the estimate $N\bar{y}$, which does not use any basal-area data, much larger than the ratio estimate? (Look at μ_x and $\bar{x}$. Can you speculate about the reason for this discrepancy?)

6.3 A consumer survey was conducted to determine the ratio of the money spent on food to the total income per year for households in a small community. A simple random sample of 14 households was selected from 150 in the community. Sample data are given in the accompanying table. Estimate R, the population ratio, and place a bound on the error of estimation. Which points have greatest influence on the estimate?

Household	Total income, x	Amount spent on food, y	Household	Total income, x	Amount spent on food, y
1	25,100	3800	8	28,200	3600
2	32,200	5100	9	34,600	3800
3	29,600	4200	10	32,700	4100
4	35,000	6200	11	31,500	4500
5	34,400	5800	12	30,600	5100
6	26,500	4100	13	27,700	4200
7	28,700	3900	14	28,500	4000

6.4 A corporation is interested in estimating the total earnings from sales of color television sets at the end of a three-month period. The total earnings figures are available for all districts within the corporation for the corresponding three-month period of the previous year. A simple random sample of 13 district offices is selected from the 123 offices within the corporation. Using a ratio estimator, estimate τ_y and place a bound on the error of estimation. Use the data in the accompanying table and take $\tau_x = 128{,}200$.

Office	Three-month data from previous year, x_i	Three-month data from current year, y_i	Office	Three-month data from previous year, x_i	Three-month data from current year, y_i
1	550	610	8	1200	1440
2	720	780	9	1350	1570
3	1500	1600	10	1750	2210
4	1020	1030	11	670	980
5	620	600	12	729	865
6	980	1050	13	1530	1710
7	928	977			

6.5 Use the data in Exercise 6.4 to estimate the mean earnings for offices within the corporation. Place a bound on the error of estimation.

6.6 An investigator has a colony of $N = 763$ rats that have been subjected to a standard drug. The average length of time to thread a maze correctly under influence of the standard drug was found to be $\mu_x = 17.2$ seconds. The investigator now would like to subject a random sample of 11 rats to a new drug. Estimate the average time required to thread the maze while under the influence of the new drug. (See the data in the accompanying table.) Place a bound on the error of estimation. [Hint: To employ a ratio estimator for μ_y is reasonable if we assume that the rats will react to the new drug in much the same way as they reacted to the standard drug.] What is the most unusual feature of the scatter plot for these data?

Rat	Standard drug, x_i	New drug, y_i	Rat	Standard drug, x_i	New drug, y_i
1	14.3	15.2	7	17.6	17.5
2	15.7	16.1	8	14.3	14.1
3	17.8	18.1	9	14.9	15.2
4	17.5	17.6	10	17.9	18.1
5	13.2	14.5	11	19.2	19.5
6	18.8	19.4			

6.7 A group of 100 rabbits is being used in a nutrition study. A prestudy weight is recorded for each rabbit. The average of these weights is 3.1 pounds. After two months, the experimenter wants to obtain a rough approximation of the average weight of the rabbits. She selects $n = 10$ rabbits at random and weighs them. The original weights and current weights are presented in the accompanying table. Estimate the average current weight, and place a bound on the error of estimation. What does the scatter plot suggest concerning the relationship between current weight and original weight for these rabbits?

Rabbit	Original weight	Current weight	Rabbit	Original weight	Current weight
1	3.2	4.1	6	3.1	4.1
2	3.0	4.0	7	3.0	4.2
3	2.9	4.1	8	3.2	4.1
4	2.8	3.9	9	2.9	3.9
5	2.8	3.7	10	2.8	3.8

6.8 A social worker wants to estimate the ratio of the average number of rooms per apartment to the average number of people per apartment in an urban ghetto area. He selects a simple random sample of 25 apartments from the 275 in the ghetto area. Let x_i denote the number of people in apartment i and let y_i denote the number of rooms in apartment i. From a count of the number of rooms and number of people in each apartment, the following data are obtained:

$$\bar{x} = 9.2, \qquad \bar{y} = 169.0$$

$$\sum_{i=1}^{25} x_i^2 = 2240, \qquad \sum_{i=1}^{25} x_i y_i = 522, \qquad \sum_{i=1}^{25} y_i^2 = 169.0$$

Estimate the ratio of average number of rooms to average number of people for this area and place a bound on the error of estimation.

6.9 A forest resource manager is interested in estimating the number of dead fir trees in a 300-acre area of heavy infestation. Using an aerial photo, she divides the area into 200 $1\frac{1}{2}$-acre plots. Let x denote the photo count of dead firs and y the actual ground count for a simple random sample of $n = 10$ plots. The total number of dead fir trees obtained from the photo count is $\tau_x = 4200$. Use the sample data in the accompanying table to estimate τ_y, the total number of dead firs in the 300-acre area. Place a bound on the error of estimation.

Plot sampled	Photo count, x_i	Ground count, y_i	Plot sampled	Photo count, x_i	Ground count, y_i
1	12	18	6	30	36
2	30	42	7	12	14
3	24	24	8	6	10
4	24	36	9	36	48
5	18	24	10	42	54

6.10 Members of a teachers' association are concerned about the salary increases given to high school teachers in a particular school system. A simple random sample of $n = 15$ teachers is selected from an alphabetical listing of all high school teachers in the system. All 15 teachers are interviewed to determine their salaries for this year and the previous year (see the accompanying table). Use these data to estimate R, the rate of change, for $N=750$ high school teachers in the community school system. Place a bound on the error of estimation. What pattern of grouping is suggested by the scatter plot?

Teacher	Past year's salary	Present year's salary	Teacher	Past year's salary	Present year's salary
1	20,400	21,500	9	20,416	21,420
2	21,700	22,600	10	20,397	21,600
3	22,792	23,920	11	23,152	24,560
4	24,956	26,400	12	21,436	22,750
5	21,355	22,020	13	24,192	25,800
6	20,108	21,308	14	22,006	23,300
7	22,891	24,100	15	22,311	23,920
8	20,216	21,320			

6.11 An experimenter was investigating a new food additive for cattle. Midway through the two-month study, he was interested in estimating the average weight for the entire herd of $N=500$ steers. A simple random sample of $n=12$ steers was selected from the herd and weighed. These data and prestudy weights are presented in the accompanying table for all cattle sampled. Assume μ_x, the prestudy average, was 880 pounds. Estimate μ_y, the average weight for the herd, and place a bound on the error of estimation. Which points have greatest influence on the estimate? Do these points look like they might be in error? Why?

Steer	Prestudy weight (in pounds)	Present weight (in pounds)	Steer	Prestudy weight (in pounds)	Present weight (in pounds)
1	815	897	7	1323	1428
2	919	992	8	1067	1152
3	690	752	9	789	875
4	984	1093	10	573	642
5	500	768	11	834	909
6	560	828	12	1049	1122

6.12 An advertising firm is concerned about the effect of a new regional promotional campaign on the total dollar sales for a particular product. A simple random sample of $n=20$ stores is drawn from the $N=452$ regional stores in which the product is sold. Quarterly sales data are obtained for the current three-month period and the three-month period prior to the new campaign. Use these data (see the accompanying table) to estimate τ_y, the total sales for the current period, and place a bound on the error of estimation. Assume $\tau_x=216{,}256$. Does it look like these stores may be in two different size groups? (Check the scatter plot.)

Store	Precampaign sales	Present sales	Store	Precampaign sales	Present sales
1	208	239	11	599	626
2	400	428	12	510	538
3	440	472	13	828	888
4	259	276	14	473	510
5	351	363	15	924	998
6	880	942	16	110	171
7	273	294	17	829	889
8	487	514	18	257	265
9	183	195	19	388	419
10	863	897	20	244	257

6.13 Use the data of Exercise 6.12 to determine the sample size required to estimate τ_y, with a bound on the error of estimation equal to $3800.

6.14 Refer to Exercises 6.4 and 6.5. By using a regression estimator, estimate the mean earnings μ_y and place a bound on the error of estimation. Compare your answer with that of Exercise 6.5. Are there any advantages to using the regression estimator here?

6.15 Show how to adjust Equations (6.24) and (6.25) for estimating a total τ_y rather than a mean μ_y.

6.16 Refer to Exercise 6.9. Estimate τ_y by using a regression estimator and place a bound on the error of estimation. Do you think the regression estimator is better than the ratio estimator for this problem?

6.17 Traders on the futures market are interested in relative prices of certain commodities rather than specific price levels; these relative prices can be presented in terms of a ratio. One such important ratio in agriculture is the cattle/hog ratio. From 64 trading days in the first quarter of 1977, the cattle and hog prices were sampled on 18 days, with the results as shown in the accompanying table. Estimate the true value of (μ_y/μ_x) for this period and place a bound on the error of estimation.

Cattle, y_i	Hogs, x_i	Cattle, y_i	Hogs, x_i
42.40	47.80	39.65	49.40
41.40	48.60	38.45	44.30
39.60	48.20	37.80	43.90
39.45	46.75	37.20	42.70
37.00	46.50	37.60	43.25
37.80	45.40	37.50	44.55
38.55	47.30	36.90	45.10
38.60	48.20	37.30	45.00
38.80	49.40	38.60	45.25

Which data points have greatest influence on the answer? What happens to the estimated ratio when they are removed?

6.18 Under what conditions should one employ a ratio estimator of τ_y rather than an estimator of the form $N\bar{y}$?

6.19 Discuss the relative merits of ratio, regression, and different estimation.

6.20 Exercise 5.4, page 176, gives data on the typical sales price and typical size of houses for certain MSAs and CMSAs in the United States. Treating these data as coming from a stratified random sample with MSAs and CMSAs as the two strata, estimate the average price per square foot for new one-family homes in the United States. Place a bound on the error of estimation.

6.21 A traditional audit expresses retail sales as opening inventory plus store purchases minus closing inventory. Thus, such an audit will look at these three items for a retail store over a period of time (say, six weeks) in order to report total sales. Such data combined from several stores and collected for a variety of competing brands allow one to estimate market shares (percentage of the total market held by a certain brand).

Faster methods of estimating market shares are the weekend selldown and store purchase audit methods. The first eliminates the store purchases, since purchases are minimal on a weekend, but uses a shorter time frame and is subject to distortion by weekend specials. The second uses only purchase information to compute market share and involves no audit of inventories.

Data on market shares calculated by the three methods, traditional (T), weekend (W), and purchases (P), are given in the accompanying table for one brand of beer. Observations were taken in six different time periods within a year.

T	W	P
15	16	12
18	17	14
16	17	20
14	16	11
13	12	8
16	18	15

(a) Estimate the ratio of the average market share calculated by the weekend method to that calculated by the traditional method. Place a bound on the error of estimation.
(b) Estimate the ratio of the average market share calculated by the purchase method to that calculated by the traditional method. Place a bound on the error of estimation.
(c) Which of the less costly methods (W or P) compares more favorably with the traditional method?
(d) What problems for this analysis show up in the scatter plots?

6.22 From the data given in the accompanying table on expenditures from six different areas of the U.S. health care field, estimate the ratio of health care expenditures in 1991 to those for 1980, and place a bound on the error of estimation. What are the shortcomings of this estimate of the true ratio of health care expenditures in the United States? (Figures are in billions of dollars.)

Area	1991	1980
Hospital care	288.6	102.4
Physicians' services	142.0	41.9
Dentists' services	37.1	14.4
Home health care	9.8	1.3
Drugs	60.7	21.6
Nursing home care	59.9	20.0

Source: U.S. Bureau of the Census, *Statistical Abstract of the United States, 1993–94,* Washington, D.C., 199.

6.23 National income from manufacturing industries is to be estimated for 1989 from a sample of 6 of the 19 industry categories that reported figures early for that year. Incomes from all 19 industries are known for 1980 and the total is $674 billion. From the data provided, estimate the total national income from manufacturing in 1989, with a bound on the error. All figures are in billions of constant (1982) dollars.

Industry	1980	1989
Lumber and wood products	21	26
Electric and electronic equipment	63	91
Motor vehicles and equipment	35	47
Food and kindred products	60	70
Textile mill products	16	17
Chemicals and allied products	50	76

Source: U.S. Bureau of the Census, *Statistical Abstract of the United States, 1993–94,* Washington, D.C., 199.

(a) Find a ratio estimator of the 1989 total income, and place a bound on the error of estimation.
(b) Find a regression estimator of the 1989 total income, and place a bound on the error of estimation.
(c) Find a difference estimator of the 1989 total income, and place a bound on the error of estimation.
(d) Which of the three methods, parts (a), (b), or (c), is most appropriate in this case? Why?

6.24 The sales manager of a firm wants to measure the relationship between monthly sales and monthly advertising costs. What parameter would you suggest he estimate? Why? What data should he collect?

6.25 A certain manufacturing firm produces a product that is packaged under two brand names, for marketing purposes. These two brands serve as strata for estimating potential sales volume for the next quarter. A simple random sample of customers for each brand is contacted and asked to provide a potential sales figure y (in number of units) for the coming quarter. Last year's true sales figure, for the same quarter, is available for each of the sampled customers and is denoted by x. The data are given in the accompanying table. The sample for brand I was taken from a list of 120 customers for whom the total sales in the same quarter of last year was 24,500 units. The brand II sample came from 180 customers with a total quarterly sales last year of 21,200 units. Find a ratio estimate of the total potential sales for next quarter. Estimate the variance of your estimator.

Brand I		***Brand II***	
x_i	y_i	x_i	y_i
204	210	137	150
143	160	189	200
82	75	119	125
256	280	63	60
275	300	103	110
198	190	107	100
		159	180
		63	75
		87	90

6.26 For Exercises 6.1 and 6.2, a regression estimator could be employed. Compute the relative efficiency of
(a) ratio estimation to simple random sampling.
(b) regression estimation to simple random sampling.
(c) regression estimation to ratio estimation.
Can you give practical reasons for the results in parts (a), (b), and (c)?

6.27 For Exercise 6.6 compute the estimated relative efficiencies of
(a) regression estimation to ratio estimation.
(b) regression estimation to difference estimation.
(c) ratio estimation to difference estimation.
Which method of estimation would you recommend?

6.28 Follow the instructions given in Exercise 6.27 for the data of Exercises 6.9 and 6.11.

6.29 It is desired to estimate the average amount of overdue accounts for a certain firm. A simple random sample of accounts will yield an observation y_i on the current amount overdue. A check of records will give the overdue amount x_i for the same time period last year. Also, the population mean μ_x for last year can be determined. Discuss how you would determine whether to use
(a) the sample mean of the y_i's,
(b) a ratio estimator,
(c) a regression estimator,
(d) a different estimator
as an estimator of the mean amount overdue.

6.30 We have seen that it is often of interest to compare audit values (y_i) with book values (x_i), and that the difference $d_i = y_i - x_i$ is sometimes a good basis from which to make this comparison. The ratio of this difference to the book value is called the *taint* of a particular item, denoted by T_i. Thus, for item i,

$$T_i = \frac{d_i}{x_i}$$

Using the data on audit and book value in Example 6.10, estimate the true ratio of average difference to average book value (which is one way to describe the average taint).

6.31 It is desired to estimate the average amount of overdue accounts for a firm with two branches. The branches are treated as strata. In each stratum, a ratio estimator is used, because exact information on the overdue amounts for this time last year can be determined. A simple random sample of accounts within each stratum will yield an observation y_i on the current amount overdue. A check of records will give the overdue amount x_i for this time last year. Also, the population means μ_x for last year can be determined. The information available is then as follows:

Branch 1	Branch 2
N_1	N_2
n_1	n_2
$\bar{y}_1$	$\bar{y}_2$
$\bar{x}_1$	x_2
μ_{x_1}	μ_{x_2}

Suggest an estimator of the average amount of overdue accounts for the firm and also suggest an approximate variance.

6.32 A problem of both economic and social concern in the United States is the importation and sale of illicit drugs. The data shown below are a part of a larger body of data collected by the Florida Attorney General's Office in an attempt to relate the incidence of drug seizures and drug arrests to the characteristics of the Florida counties. They show the number y of drug arrests per county, the density x_1 of the county (population per square mile), and the number x_2 of law enforcement employees.

The data shown below are from a sample of 10 of Florida's 67 counties.

(a) Estimate the ratio of number of arrests to population density for the state. Place a bound on the error of estimation.

(b) Estimate the ratio of number of arrests to number of law enforcement employees for the state. Place a bound on the error of estimation.

(c) Find the most influential data point in each case. What happens to the estimates when this point is removed?

County	1	2	3	4	5	6	7	8	9	10
Population density, x_1	169	68	278	842	18	42	112	529	276	613
Number of law enforcement employees, x_2	498	35	772	5788	18	57	300	1762	416	520
Number of arrests, y	370	44	716	7416	25	50	189	1097	256	432

6.33 The Florida Game and Freshwater Fish Commission is interested in estimating weights of alligators from much more easily observed lengths. Data on the lengths (L) and weights (W) of 25 alligators are given in the accompanying table. Estimate the average weight of a population of alligators for which the average length is 100 inches.

Alligator	L	W
1	94	130
2	74	51
3	147	640
4	58	28
5	86	80
6	94	110
7	63	33
8	86	90
9	69	36
10	72	38
11	128	366
12	85	84
13	82	80
14	86	83
15	88	70
16	72	61
17	74	54
18	61	44
19	90	106
20	89	84
21	68	39
22	76	42
23	114	197
24	90	102

Weight is in pounds; length in inches.

6.34 In building a model to study automobile fuel consumption, Biggs and Akcelik (*Journal of Transportation Engineering*, vol. 113, no. 1, January 1987, pp. 101–106) begin by looking at the relationship between idle fuel consumption and engine capacity. Suppose the data are as follows:

Idle fuel consumption (mL/s)	Engine size (L)
0.18	1.2
0.21	1.2
0.17	1.2
0.31	1.8
0.34	1.8
0.29	1.8
0.42	2.5
0.39	2.5
0.45	2.5
0.52	3.4

(continued)

(continued)

Idle fuel consumption (mL/s)	Engine size (L)
0.61	3.4
0.44	3.4
0.62	4.2
0.65	4.2
0.59	4.2

For a population of automobiles with engine size 2.5 liters, estimate the average idle fuel consumption, with a bound on the error of estimation.

EXPERIENCES WITH REAL DATA

6.1 The data set TEMPS in the appendix (and on the data disk) shows normal temperature (T) and amount of precipitation (P) for weather stations around the United States. Use the January and March precipitation data for a sample of n stations to determine the following estimates.
(a) Estimate the ratio of the average March precipitation to the average January precipitation.
(b) Estimate the average March precipitation for all stations, making use of the January and March data. Choose one of the three possible estimators from this chapter and give reasons for your choice.
(c) Select an appropriate sample size and place a bound on the error of estimation in both parts (a) and (b).

6.2 Data from the 1990 United States census is given on the data disk and in the appendix. Using the four regions of the country as strata, form a stratified ratio estimate of the 1990 population total making use of the 1980 data for states and for the United States as a whole. Choose an appropriate sample size and allocation. Place a bound on the error of estimation.

6.3 An interesting project is to estimate what proportion of the money spent for entertainment by students in your community goes to a specific type of entertainment, such as movie theaters. You can obtain this estimate by listing a simple random sample of n students, calling them on the telephone (or interviewing them personally), and recording the total amount spent on entertainment (x_i) as well as the amount spent on movies (y_i). Then estimate the ratio (μ_y/μ_x) and place a bound on the error.

Think about sample size before you begin the study. Also, concentrating on students in one locality, such as an apartment building or group of fraternity houses, rather than students at large may be most convenient. Nonresponse is always a problem when dealing with human populations, so try to think of ways to minimize it.

6.4 Ratio estimation is often a convenient method of estimating properties of physical objects that are difficult to measure directly. Gather a box of rocks or other irregularly shaped objects. You wish to estimate the total volume of the rocks. Volume of irregularly shaped objects is somewhat difficult to measure, but volume is related to weight, which is quite easy to measure. Thus, volume can be estimated by using the ratio of volume to weight.

Select a sample of n rocks. Measure the weight and the volume for each rock in the sample. (You may want to use water displacement as a method of measuring volume.) Then obtain the total weight of all rocks in the box. Use these data to construct an estimate, with a bound on the error, of the total volume of the rocks.

6.5 The data set SCHOOLS in the appendix and on the data disk contains information on various aspects of education for all 50 states and the District of Columbia. The variables are the following:

ExpPP90 = Expenditures per pupil for 1990 (dollars)
ExpPC90 = Expenditures per capita for 1990 (dollars)
TeaSal90 = Average teacher salary for 1990 (in thousands of dollars)
%Compl = % of residents over the age of 25 who completed high school
%Dropout = % of 16 to 19-year-olds who dropped out of school as measured in 1990
ExpPP92 = Expenditures per pupil for 1992 (dollars)
ExpPC92 = Expenditures per capita for 1992 (dollars)
TeaSal92 = Average teacher salary for 1992 (in thousands of dollars)
Region = 1 for northeast, 2 for midwest, 3 for south, 4 for west
Population = Approximate population in 1992 (millions)
Enroll = Number of students enrolled in 1990 (thousands)
Teachers = Number of teachers as of 1990 (thousands)

For parts (a)–(d), ExpPP90 and ExpPC90 are assumed known for all states; the others are assumed to be known only for a sample of states that you select. All estimates should be based on a sample of 10 states.

(a) Estimate the mean per-capita education expenditure for the United States in 1992 by using data on the mean per-capita expenditure per state in 1990 and employing

(i) ratio estimation,
(ii) regression estimation,
(iii) difference estimation.

Discuss which of the three techniques you think is best. Use at least one plot to justify your selection.

(b) Estimate the mean teacher salary for the United States in 1992 by making use of either per-pupil or per-capita expenditures for 1990. Choose any method of estimation you like. Justify your choice of variable and method, making use of at least one plot in your justification.
(c) Estimate the student to teacher ratio for the United States.
(d) Discuss how you would estimate the average dropout rate based on the variables provided.

C A S E S T U D Y

IS QUALITY BEING MAINTAINED BY THE MANUFACTURER?

Consumers and manufacturers are both concerned about the quality of items purchased or produced. Consumers want assurances that they are buying a product that will perform according to specifications, and manufacturers want evidence that their products are meeting certain standards. Such evidence is most often provided by quality control sampling plans within the manufacturing operation. These sampling plans often involve selecting items for inspection from a continuously moving production line. The selection process may require the sampling of every 100th manufactured item, one item every hour, or some similar systematic plan. The observed quality characteristic for each item may be a measurement, such as weight or time to failure, or simply a classification into the categories of "conforming" and "nonconforming." The average value of the observed characteristic is then compared with a standard value to see whether quality is being maintained.

In the manufacture of certain hydraulic equipment, one important component is a bronze casting. The main quality characteristic of the casting is the percentage of copper it contains. In the manufacturing process, one casting is selected after each half-hour of production, and the data on percentage copper are accumulated over an eight-hour day. Thus, 16 measurements are obtained. For one day's production the percentages of copper averaged 87, with a variance of 18. The standard for the bronze was 90% copper. Is the standard being met? Techniques presented in this chapter will help us answer the question.

CHAPTER 7

SYSTEMATIC SAMPLING

7.1 INTRODUCTION

As we have seen in Chapters 4 and 5, both simple random sampling and stratified random sampling require very detailed work in the sample selection process. Sampling units on an adequate frame must be numbered (or otherwise identified) so that a randomization device, such as a random number table, can be used to select specific units for the sample. A sample survey design that is widely used primarily because it simplifies the sample selection process is called *systematic sampling*.

The basic idea of systematic sampling is as follows: Suppose a sample of n names is to be selected from a long list. A simple way to make this selection is to choose an appropriate interval and to select names at equal intervals along the list. Thus, every tenth name might be selected, for example. If the starting point for this regular selection process is random, the result is a systematic sample.

Definition 7.1

A sample obtained by randomly selecting one element from the first k elements in the frame and every kth element thereafter is called a *1-in-k systematic sample*, with a random start.

As in previous chapters, we present methods for estimating a population mean, total, and proportion. We will also discuss appropriate bounds on the error of estimation and sample-size requirements.

Systematic sampling provides a useful alternative to simple random sampling for the following reasons:

1. Systematic sampling is easier to perform in the field and hence is less subject to selection errors by field-workers than are either simple random samples or stratified random samples, especially if a good frame is not available.
2. Systematic sampling can provide greater information per unit cost than simple random sampling can provide.

In general, systematic sampling involves random selection of one element from the first k elements and then selection of every kth element thereafter. This procedure is easier to perform and usually less subject to interviewer error than is simple random sampling. For example, using simple random sampling to select a sample of $n = 50$ shoppers on a city street corner would be difficult. The interviewer could not determine which shoppers to include in the sample, because the population size N would not be known until all shoppers had passed the corner. In contrast, the interviewer could take a systematic sample (say, 1 in 20 shoppers) until the required sample size was obtained. This procedure would be an easy one for even an inexperienced interviewer.

Besides being easier to perform and less subject to interviewer error, systematic sampling *frequently* provides more information per unit cost than does simple random sampling. A systematic sample is generally spread more uniformly over the entire population and thus may provide more information about the population than an equivalent amount of data contained in a simple random sample. Consider the following illustration: We wish to select a 1-in-5 systematic sample of travel vouchers from a stack of $N = 1000$ (that is, sample $n = 200$ vouchers) to determine the proportion of vouchers filed incorrectly. A voucher is drawn at random from the first 5 vouchers (for example, no. 3), and every fifth voucher thereafter is included in the sample. (See the accompanying table.)

Suppose that most of the first 500 vouchers have been correctly filed, but because of a change in clerks, the second 500 have all been incorrectly filed. Simple random sampling could accidentally select a large number (perhaps all) of the 200 vouchers from either the first or the second 500 vouchers and hence yield a very poor estimate of p. In contrast, systematic sampling would select an equal number of vouchers from each of the two groups and would give a very accurate estimate of the fraction of vouchers incorrectly filed.

Voucher	Voucher sampled
1	
2	
3	3
4	
5	
6	
7	
8	8
9	
10	
⋮	⋮
996	
997	
998	998
999	
1000	

Additional examples are discussed in Section 7.3 to illustrate how to choose between systematic and simple random sampling in a given situation. Note, however, that the accuracy of estimates from systematic sampling depend on the *order* of the sampling units in the frame. If the incorrect vouchers had been randomly dispersed among all vouchers, then the advantage of systematic sampling would have been lost.

Systematic sampling is very commonly used in a wide variety of contexts. The U.S. census directs only a minimal number of questions to every resident, but it gathers much more information from a systematic sample of residents. In the 1980 census, 14 items were on the short form distributed to all residents. Another 42 items were on the long form that was distributed to, approximately, a 1-in-5 systematic sample of residents.

The Gallup poll begins its sampling process by listing 200,000 election districts in the United States and then systematically selecting 300 for a follow-up study of households. The households, or dwellings, within a sampled district may again be selected systematically—by choosing the second dwelling in every other block when moving east to west, for example.

Industrial quality control sampling plans are most often systematic in structure. An inspection plan for manufactured items moving along an assembly line may call for inspecting every 50th item. An inspection of cartons of products stored in a warehouse may suggest sampling the 2nd carton from the left in the 3rd row down from the top in every 5th stack. In the inspection of work done at fixed stations, the inspection plan may call for walking up and down the rows of workstations and inspecting the machinery at every 10th station. The time of day is often important in assessing quality of worker performance, and so an inspection plan may call for sampling the output of a workstation at systematically selected times throughout the day.

Auditors are frequently confronted with the problem of sampling a list of accounts to check compliance with accounting procedures or to verify dollar amounts. The most natural way to sample these lists is to choose accounts systematically.

Market researchers and opinion pollsters who sample people on the move very often employ a systematic design. Every 20th customer at a checkout counter may be asked his or her opinion on the taste, color, or texture of a food product. Every 10th person boarding a bus may be asked to fill out a questionnaire on bus service. Every 100th car entering an amusement park may be stopped and the driver questioned on various advertising policies of the park or on ticket prices. All of these samples are systematic samples.

Crop-yield estimates often result from systematic samples of fields and small plots within fields. Similarly, foresters may systematically sample field plots to estimate the proportion of diseased trees or may systematically sample trees themselves to study growth patterns.

Thus, systematic sampling is a popular design. We will now investigate the construction of these designs and the properties of resulting estimators of means, totals, and proportions.

7.2

HOW TO DRAW A SYSTEMATIC SAMPLE

Although simple random sampling and systematic sampling both provide useful alternatives to one another, the methods of selecting the sample data are different. A simple random sample from a population is selected by using a table of random numbers, as noted in Section 4.3. In contrast, various methods are possible in systematic sampling. The investigator can select a 1-in-3, a 1-in-5, or, in general, a 1-in-k systematic sample. For example, a medical investigator is interested in obtaining information about the average number of times 15,000 specialists prescribed a certain drug in the previous year ($N = 15{,}000$). To obtain a simple random sample of $n = 1600$ specialists, we would use the methods of Section 4.3 and refer to a table of random numbers; however, this procedure would require a great deal of work. Alternatively, we could select one name (specialist) at random from the first $k = 9$ names appearing on the list and then select every ninth name thereafter until a sample of size 1600 is selected. This sample is called a *1-in-9 systematic sample*.

Perhaps you wonder how k is chosen in a given situation. If the population size N is known, we can determine an approximate sample size n for the survey (see Section 7.5) and then choose k to achieve that sample size. There are $N = 15{,}000$ specialists in the population for the medical survey. Suppose the required sample size is $n = 100$. We must then choose k to be 150 or less. For $k = 150$, we will obtain exactly $n = 100$ observations, while for $k < 150$, the sample size will be greater than 100.

In general, for a systematic sample of n elements from a population of size N, k must be less than or equal to N/n (that is, $k \leq N/n$). Note in the preceding illustration that $k \leq 15{,}000/100$; that is, $k \leq 150$.

We cannot accurately choose k when the population size is unknown. We can determine an approximate sample size n, but we must guess the value of k needed to achieve a sample of size n. If too large a value of k is chosen, the required sample size n will not be obtained by using a 1-in-k systematic sample from the population. This result presents no problem if the experimenter can return to the population and conduct another 1-in-k systematic sample until the required sample size is obtained. However, in some situations, obtaining a second systematic sample is impossible. For example, conducting another 1-in-20 systematic sample of shoppers is impossible if the required sample of $n = 50$ shoppers is not obtained at the time they pass the corner.

7.3 ESTIMATION OF A POPULATION MEAN AND TOTAL

As we have repeatedly stressed, the objective of most sample surveys is to estimate one or more population parameters. We can estimate a population mean μ from a systematic sample by using the sample mean $\bar{y}$. This outcome is shown in Equation (7.1).

Estimator of the population mean μ:

$$\hat{\mu} = \bar{y}_{sy} = \frac{\sum_{i=1}^{n} y_i}{n} \tag{7.1}$$

where the subscript sy signifies that systematic sampling was used.

Estimated variance of $\bar{y}_{sy}$:

$$\hat{V}(\bar{y}_{sy}) = \frac{s^2}{n}\left(\frac{N-n}{N}\right) \tag{7.2}$$

assuming a randomly ordered population.

If N is unknown, we eliminate the fpc, $(N - n)/N$, in Equation (7.2).

You will recognize that the estimated variance of $\bar{y}_{sy}$ given in Equation (7.2) is identical to the estimated variance of $\bar{y}$ obtained by using simple random sampling (Section 4.3). This result does not imply that the true variance of $\bar{y}_{sy}$ is the same as that of $\bar{y}$. The variance of $\bar{y}$ is given by

$$V(\bar{y}) = \frac{\sigma^2}{n}\left(\frac{N-n}{N-1}\right) \tag{7.3}$$

Similarly, the variance of $\bar{y}_{sy}$ is given by

$$V(\bar{y}_{sy}) = \frac{\sigma^2}{n}[1 + (n-1)\rho] \tag{7.4}$$

where ρ is a measure of the correlation between pairs of elements within the same systematic sample, which will be discussed in Section 7.7. If ρ is close to 1, then the elements within the sample are all quite similar with respect to the characteristic being measured, and systematic sampling will yield a higher variance of the sample mean than will simple random sampling. If ρ is negative, then systematic sampling may be better than simple random sampling. The correlation may be negative if elements within the systematic sample tend to be extremely different. (Note that ρ cannot be so large negatively that the variance expression becomes negative.) For ρ close to 0 and N fairly large, systematic sampling is roughly equivalent to simple random sampling.

An unbiased estimate of $V(\bar{y}_{sy})$ cannot be obtained by using the data from only one systematic sample. This statement does not imply that we can never obtain an estimate of $V(\bar{y}_{sy})$. When systematic sampling is equivalent to simple random sampling, we can take $V(\bar{y}_{sy})$ to be approximately equal to the estimated variance of $\bar{y}$ based on simple random sampling.

For which populations does this relationship occur? To answer this question, we must consider the following three types of populations:

Random population
Ordered population
Periodic population

Definition 7.2

A population is *random* if the elements of the population are in random order.

Suppose we are interested in determining the average daily sales volume for a chain of grocery stores. The population of daily sales is clearly periodic, with peak sales occurring toward the end of each week. The effectiveness of a 1-in-k sample depends on the value we choose for k. If we sample daily sales every Wednesday, we would probably underestimate the true average daily sales volume. Similarly, if we sample sales every Friday, we would probably overestimate the true average sales. We might sample every ninth workday to avoid consistently sampling either the low- or high-sales days.

Elements of a systematic sample drawn from a periodic population can be homogeneous (that is, $\rho > 0$). For example, the elements within a systematic sample of daily sales taken every Wednesday would be fairly homogeneous. It can be shown, using Equations (7.3) and (7.4), that when N is large and $\rho > 0$,

$$V(\bar{y}_{sy}) > V(\bar{y})$$

Thus, in this case, systematic sampling provides less information per unit cost than does simple random sampling. As in the preceding situations, $V(\bar{y}_{sy})$ cannot be estimated directly by using a single systematic sample. We can approximate its value by using

$\hat{V}(\bar{y})$, as for simple random sampling. In general, this estimator should underestimate the true variance of $\bar{y}_{sy}$.

To avoid this problem that occurs with systematic sampling from a periodic population, the investigator could change the random starting point several times. This procedure would reduce the possibility of choosing observations from the same relative position in a period population. For example, when a 1-in-10 systematic sample is being drawn from a long list of file cards, a card is randomly selected from the first 10 cards (for example, no. 2) and every tenth card thereafter. This procedure can be altered by randomly selecting a card from the first 10 (for example, no. 2) and every tenth card thereafter for perhaps 15 selections to obtain the numbers

$$2, 12, 22, \ldots, 152$$

At this point, another random starting point can be selected from the next 10 numbers:

$$153, 154, 155, \ldots, 162$$

If 156 is selected, we then proceed to select every tenth number thereafter for the next 15 selections. This entire process is repeated until the desired sample size is obtained.

The process of selecting a random starting point several times throughout the systematic sample has the effect of shuffling the elements of the population and then drawing a systematic sample. Hence, we can assume that the sample obtained is equivalent to a systematic sample drawn from a random population. The variance of $\bar{y}_{sy}$ can then be approximated by using

$$\hat{V}(\bar{y}_{sy}) = \frac{s^2}{n}\left(\frac{N-n}{N}\right)$$

We will assume a random population for our examples in the next few sections. Alternatives are given in Sections 7.6 and 7.7.

Example 7.1

The federal government keeps track of various indicators on the performance of manufacturing industries in the country by collecting annual data on variables such as number of employees and payroll. The Standard Industrial Classification (SIC) system divides the manufacturing industry into 79 groups. Table 7.1 shows data for 1987 and 1991 on number of employees (in thousands) and annual payroll (in millions of dollars) for a sample of 16 industrial groups selected systematically from a somewhat arbitrary ordering of the 79 groups. (This type of sample could be used for a preliminary report to be released before data from all groups were available.) The other variable reported is value added by manufacturer (in millions of dollars), which will be used later.

A. Estimate the mean 1991 payroll per industrial group, with a bound on the error of estimation.
B. Estimate the mean difference in number of employees between 1987 and 1991, with a bound on the error of estimation.

TABLE 7.1

		1987		1991		
Row	Industry	Employees	Payroll	Employees	Payroll	Value added
1	Fluid milk	72.4	1,681	65.5	1,840	5,656
2	Cookies	45.4	1,010	49.3	1,247	4,991
3	Food preparation	57.9	1,110	54.7	1,193	5,666
4	Paper mills	129.1	4,597	130.3	5,224	15,510
5	Periodicals	110.0	2,983	110.6	3,661	13,794
6	Business forms	53.2	1,276	46.3	1,269	3,840
7	Soap	31.7	956	36.6	1,221	8,165
8	Ag-chemicals	16.1	18	16.4	620	4,951
9	Plastics	384.7	7,286	388.9	8,540	20,121
10	Fabricated metal	80.9	1,881	77.0	2,064	3,900
11	Int. comb. engines	64.0	2,043	56.3	1,909	4,677
12	Computers	151.9	4,953	126.0	5,344	16,878
13	Telephone, telegraph	112.3	3,178	94.1	3,468	9,503
14	Motor vehicle parts	389.1	11,947	369.9	12,587	25,213
15	Space vehicles	166.7	6,415	135.8	6,026	13,551
16	Photographic equipment	88.0	2,878	78.0	3,044	14,603

	N	Mean	Median	SD
1991 payroll	16	3,704	2,554	3,230
$E_{91} - E_{87}$	16	−7.36	−5.40	10.95

Source: U.S. Bureau of the Census, *Statistical Abstract of the United States, 1993–94* (113 ed.), Washington, D.C., 1994.

Solution

A. Since the data are taken in order from a list, plotting the data in the order they were taken is a good idea. This will show any indication of trends that could cause trouble in estimating the variance of the sample mean. Figure 7.1 shows the plot of the 1991 payroll values versus the ordered sample numbers. There is a slight tendency for the values taken later in the list to be a little larger and more variable than those taken earlier in the list, but the trend does not appear to be serious. Estimating the variance by methods of simple random sampling should produce fairly good results here.

From the statistical summaries given below the plot, it is easy to see that

$$\bar{y}_{sy} = 3704,$$

$$\hat{V}(\bar{y}_{sy}) = \left(\frac{79-16}{79}\right)\left(\frac{1}{16}\right)(3230)^2$$

$$2\sqrt{\hat{V}(\bar{y}_{sy})} = 2\sqrt{\left(\frac{79-16}{79}\right)\left(\frac{1}{4}\right)}(3230) = 1442$$

FIGURE 7.1 Plot of 1991 payroll data for Example 7.1

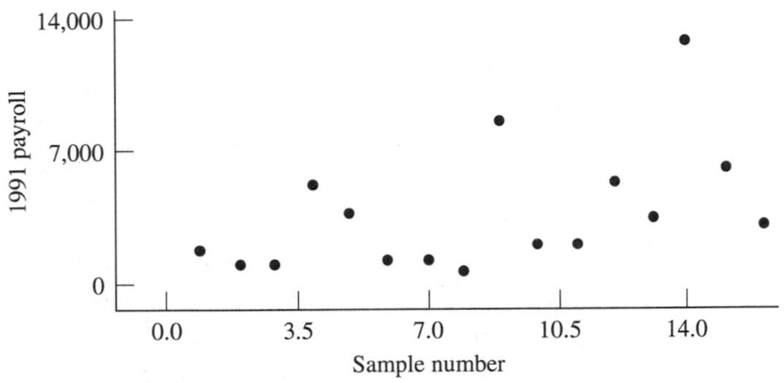

This is quite a large bound on the error! It is due to the fact that a small sample is taken from a population that is highly variable. A better design might be to stratify the industry groups into those with large payrolls and those with small payrolls before doing the sampling. Nevertheless, the use of the formula for simple random sampling is probably not inflating the variance estimate very much in this case.

B. Figure 7.2 shows the plot of the differences between the 1991 and 1987 numbers of employees for the sampled industrial groups. The differences show considerably less variability than the numbers of employees would show. There is a slight decreasing trend toward the later sample numbers but, again, this does not appear to be too serious. The variance approximation from simple random sampling may be a slight overestimate of the true variance, but probably not by much.

FIGURE 7.2 Plot of the differences in number of employees for Example 7.1

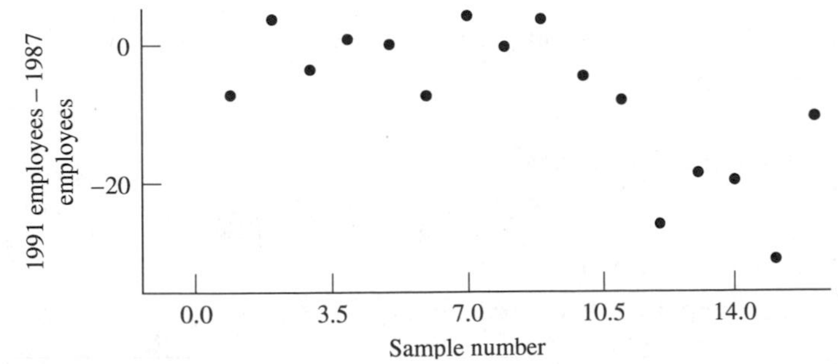

From the statistical summary below the data set, it follows that, for estimating the mean difference, with the estimator denoted by $\bar{x}_{sy}$, we have

$$\bar{x}_{sy} = -7.36$$

$$\hat{V}(\bar{x}_{sy}) = \left(\frac{79-16}{79}\right)\left(\frac{1}{16}\right)(10.95)^2$$

$$2\sqrt{\hat{V}(\bar{x}_{sy})} = 2\sqrt{\left(\frac{79-16}{79}\right)}\left(\frac{1}{4}\right)(10.95) = 4.89$$

Thus, $\bar{x}_{sy} \pm 2\sqrt{\hat{V}(\bar{x}_{sy})}$ yields an interval of (− 12.25, − 2.47). Note that the resulting interval estimate of the true difference in mean number of employees per group is negative, suggesting that there was a significant drop in average number of employees per industry group over the period from 1987 to 1991.

Recall that estimation of a population total requires knowledge of the total number of elements N in the population when we are using the procedures of Chapters 4 and 5. For example, we use

$$\hat{\tau} = N\bar{y}$$

as an estimator of τ from simple random sampling. Also, we use

$$\hat{\tau}_{st} = \sum_{i=1}^{L} N_i\bar{y}_i$$

where

$$\sum_{i=1}^{L} N_i = N$$

as an estimator of τ from stratified random sampling from L strata (Section 5.3). Similarly, we need to know N to estimate τ when we are using systematic sampling.

The population size is unknown in many practical situations, which suggests using systematic sampling; however, when N is known, we can estimate τ by using Equations (7.5) and (7.6).

Estimator of the population total τ:

$$\hat{\tau} = N\bar{y}_{sy} \tag{7.5}$$

Estimated variance of τ:

$$\hat{V}(N\bar{y}_{sy}) = N^2\hat{V}(\bar{y}_{sy}) = N^2\left(\frac{s^2}{n}\right)\left(\frac{N-n}{N}\right) \tag{7.6}$$

assuming a randomly ordered population.

Note that the results presented in Equations (7.5) and (7.6) are identical to those presented for estimating a population total under simple random sampling. This result does not imply that the variance of $N\bar{y}_{sy}$ is the same as the variance of $N\bar{y}$. Again, we cannot obtain an unbiased estimator of $V(N\bar{y}_{sy})$ from the data in a single systematic sample. However, in certain circumstances, as noted earlier, systematic sampling is equivalent to simple random sampling, and we can use the result presented in Section 4.3.

Example 7.2

Returning to the data from the systematic sample of 16 industry groups from the population of 79, as shown in Example 7.1, it is now of interest to estimate the total value added by manufacture for the 79 groups. (This is one of the important measures used in assessing taxes on manufactured items.) From the data provided, estimate this total for 1991 and find a bound for the error of estimation.

Solution

Beginning with a plot of the data against sample number, Figure 7.3, we see that the data may have a slight increasing trend as we move toward the later values on the list, but this trend is not pronounced. The variance approximation from simple random sampling should work well.

FIGURE 7.3 Plot of the value-added data for Example 7.2

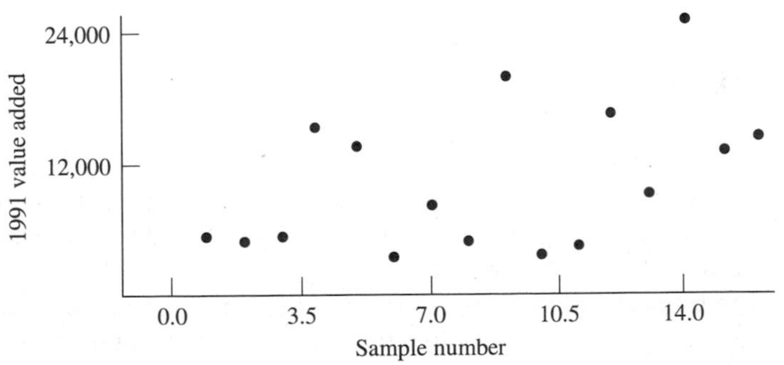

Using the statistical summary provided below, it is easy to write down the estimate of the total and the bound on the error.

	N	Mean	Median	SD
1991 value added	16	10,689	8834	6570

$$\hat{\tau} = N\bar{y}_{sy} = 79(10{,}689) = 844{,}431$$

$$2\sqrt{\hat{V}(\hat{\tau})} = 2N\sqrt{\left(\frac{N-n}{N}\right)}\left(\frac{1}{\sqrt{n}}\right)s$$

$$= 2(79)\sqrt{\left(\frac{79-16}{79}\right)}\left(\frac{1}{4}\right)(6570)$$

$$= 231{,}750$$

Notice that this bound is quite large. Again, we are attempting to estimate a total from highly variable data with a small sample; the precision of the result is not great in this case. To achieve greater precision, the sample size should be increased or the design changed, or both.

If stratifying the populations is advantageous, systematic sampling can be used within each stratum in place of simple random sampling. Using the estimator of Equation (7.1) with its estimated variance (7.2) within each stratum, the resulting estimator of the population mean will look similar to Equation (5.1), with an estimated variance given by Equation (5.2). Such a situation might arise if we were to stratify an industry by plants and then take a systematic sample of the records within each plant to estimate average accounts receivable, time lost to accidents, and so on.

7.4

ESTIMATION OF A POPULATION PROPORTION

An investigator frequently wishes to use data from a systematic sample to estimate a population proportion. For example, to determine the proportion of registered voters in favor of an upcoming bond issue, the investigator might use a 1-in-k systematic sample from the voter registration list.

The estimator of the population proportion p obtained from systematic sampling is denoted by $\hat{p}_{sy}$. As in the simple random sampling (Section 4.5), the properties of $\hat{p}_{sy}$ parallel those of the sample mean $\bar{y}_{sy}$ if the response measurements are defined as follows: Let $y_i = 0$ if the ith element sampled does not possess the specified characteristic and $y_i = 1$ if it does. The estimator $\hat{p}_{sy}$ is then the average of the 0 and 1 values from the sample.

Estimator of the population proportion p:

$$\hat{p}_{sy} = \bar{y}_{sy} = \frac{\sum_{i=1}^{n} y_i}{n} \tag{7.7}$$

Estimated variance of $\hat{p}_{sy}$:

$$\hat{V}(\hat{p}_{sy}) = \frac{\hat{p}_{sy}\hat{q}_{sy}}{n-1}\left(\frac{N-n}{N}\right) \tag{7.8}$$

where $\hat{q}_{sy} = 1 - \hat{p}_{sy}$ assuming a randomly ordered population.

We can ignore the fpc, $(N - n)/N$, in Equation (7.8) if the population size N is unknown but can be assumed large relative to n.

We again note that the estimated variance of $\hat{p}_{sy}$ (or $\bar{y}_{sy}$) is identical to the estimated variance of $\hat{p}$ (or $\bar{y}$), using simple random sampling (Section 4.5). This result does not imply that the corresponding population variances are equal; however, if N is large and if the observations within a systematic sample are unrelated (that is, $\rho = 0$), the two population variances will be equal.

Example 7.3

A 1-in-6 systematic sample is obtained from a voter registration list to estimate the proportion of voters in favor of the proposed bond issue. Several different random starting points are used to ensure that the results of the sample are not affected by periodic variation in the population. The coded results of this preelection survey are as shown in the accompanying table. Estimate p, the proportion of the 5775 registered voters in favor of the proposed bond issue ($N = 5775$). Place a bound on the error of estimation.

Voter	Response
4	1
10	0
16	1
⋮	⋮
5760	0
5766	0
5772	1
	$\sum_{i=1}^{962} y_i = 652$

Solution

The sample proportion is given by

$$\hat{p}_{sy} = \frac{\sum_{i=1}^{962} y_i}{962} = \frac{652}{962} = .678$$

Since N is large and several random starting points were chosen in drawing the systematic sample, we can assume that

$$\hat{V}(\hat{p}_{sy}) = \frac{\hat{p}_{sy}\hat{q}_{sy}}{n-1}\left(\frac{N-n}{N}\right)$$

provides a good estimate of $V(\hat{p}_{sy})$.

The bound on the error of estimation is

$$2\sqrt{\hat{V}(\hat{p}_{sy})} = 2\sqrt{\frac{\hat{p}_{sy}\hat{q}_{sy}}{n-1}\left(\frac{N-n}{N}\right)}$$

$$= 2\sqrt{\frac{(.678)(.322)}{961}\left(\frac{5775-962}{5775}\right)} \approx .028$$

Thus, we estimate .678 (67.8%) of the registered voters favor the proposed bond issue. We are relatively confident that the error of estimation is less than .028 (2.8%).

7.5 SELECTING THE SAMPLE SIZE

Now let us determine the number of observations necessary to estimate μ to within B units. The required sample size is found by solving the following equation for n:

$$2\sqrt{V(\bar{y}_{sy})} = B \tag{7.9}$$

The solution to Equation (7.9) involves both σ^2 and ρ, which must be known (at least approximately) in order to solve for n. Although these parameters sometimes can be estimated if data from a prior survey are available, we do not discuss this method in this text. Instead, we use the formula for n for simple random sampling. This formula could give an extra large sample for ordered populations and too small a sample for periodic populations. As noted earlier, the variances of $\bar{y}_{sy}$ and $\bar{y}$ are equivalent if the population is random.

Sample size required to estimate μ with a bound B on the error of estimation:

$$n = \frac{N\sigma^2}{(N-1)D + \sigma^2} \tag{7.10}$$

where $D = \dfrac{B^2}{4}$

Example 7.4

The management of a large utility company is interested in the average amount of time delinquent bills are overdue. A systematic sample will be drawn from an alphabetical list of $N = 2500$ overdue customer accounts. In a similar survey conducted the previous year, the sample variance was found to be $s^2 = 100$ days. Determine the sample size required to estimate μ, the average amount of time utility bills are overdue, with a bound on the error of estimation of $B = 2$ days.

Solution

A reasonable assumption is that the population is random; hence, $\rho \approx 0$. Then we can use Equation (7.10) to find the approximate sample size. Replacing σ^2 by s^2 and setting

$$D = \frac{B^2}{4} = \frac{4}{4} = 1$$

we have

$$n = \frac{N\sigma^2}{(N-1)D + \sigma^2} = \frac{2500(100)}{2499(1) + 100} = 96.19$$

Thus, management must sample approximately 97 accounts to estimate the average amount of time delinquent bills are overdue, to within two days.

To determine the sample size required to estimate τ with a bound on the error of estimation of magnitude B, we use the corresponding method presented in Section 4.4.

The sample size required to estimate p to within B units is found by using the sample-size formula for estimating p under simple random sampling.

Sample size required to estimate p with a bound B on the error of estimation:

$$n = \frac{Npq}{(N-1)D + pq} \tag{7.11}$$

where $q = 1 - p$ and $D = \dfrac{B^2}{4}$

In a practical situation, we do not know p. We can find an approximate sample size by replacing p with an estimated value. If no prior information is available to estimate p, we can obtain a conservative sample size by setting $p = .5$.

Example 7.5

An advertising firm is starting a promotion campaign for a new product. The firm wants to sample potential customers in a small community to determine customer acceptance.

To eliminate some of the costs associated with personal interviews, the investigators decide to run a systematic sample from $N = 5000$ names listed in a community registry and collect the data via telephone interviews. Determine the sample size required to estimate p, the proportion of people who consider the product "acceptable," with a bound on the error of estimation of magnitude $B = .03$ (that is, 3%).

Solution

The required sample size can be found by using Equation (7.11). Although no previous data are available on this new product, we can still find an approximate sample size. Set $p = .5$ in Equation (7.11) and

$$D = \frac{B^2}{4} = \frac{(.03)^2}{4} = .000225$$

Then the required sample size is

$$n = \frac{Npq}{(N-1)D + pq} = \frac{5000(.5)(.5)}{4999(.000225) + (.5)(.5)} = 909.240$$

Hence, the firm must interview 910 people to determine consumer acceptance to within 3%.

7.6

REPEATED SYSTEMATIC SAMPLING

We stated in Section 7.3 that we cannot estimate the variance of $\bar{y}_{sy}$ from information contained in a single systematic sample unless the systematic sampling generates, for all practical purposes, a random sample. When this result occurs, we can use the random sampling estimation procedures outlined in Section 4.3. However, in most cases, systematic random sampling is *not* equivalent to simple random sampling. An alternate method must be used to estimate $V(\bar{y}_{sy})$. Repeated systematic sampling is one such method.

As the name implies, repeated systematic sampling requires the selection of more than one systematic sample. For example, 10 1-in-50 systematic samples, each containing 6 measurements, could be acquired in approximately the same time as 1 1-in-5 systematic sample containing 60 measurements. Both procedures yield 60 measurements for estimating the population mean μ, but the repeated sampling procedure allows us to estimate $V(\bar{y}_{sy})$ by using the square of the deviations of the $n_s = 10$ individual sample means about their mean. The average $\hat{\mu}$ of the 10 sample means will estimate the population mean μ.

To select n_s repeated systematic samples, we must space the elements of each sample further apart. Thus, 10 1-in-50 samples ($n_s = 10$, $k' = 50$) of 6 measurements each contain the same number of measurements as does a single 1-in-5 sample ($k = 5$) containing $n = 60$ measurements. The starting point for each of the n_s systematic samples is randomly selected from the first k' elements. The remaining elements in each sample are acquired by adding k', $2k'$, and so forth, to the starting point until the total number per sample, n/n_s, is obtained.

A population consists of $N = 960$ elements, which we can number consecutively. To select a systematic sample of size $n = 60$, we choose $k = N/n = 16$ and a random number between 1 and 16 as a starting point. What procedure do we follow to select 10 repeated systematic samples in place of the 1 systematic sample? First, we choose $k' = 10k = 10(16) = 160$. Next, we select 10 random numbers between 1 and 160. Finally, the constant 160 is added to each of these random starting points to obtain 10 numbers between 161 and 320; the process of adding the constant is continued until 10 samples of size 6 are obtained.

A random selection of 10 integers between 1 and 160 gives the following:

$$73, 42, 81, 145, 6, 21, 86, 17, 112, 102$$

These numbers form the random starting points for 10 systematic samples, as shown in Table 7.2. The second element in each sample is found by adding 160 to the first, the third by adding 160 to the second, and so forth.

TABLE 7.2 **Selection of repeated systematic samples**

Random starting point	Second element in sample	Third element in sample		Sixth element in sample
6	166	326	...	806
17	177	337	...	817
21	181	341	...	821
42	202	362	...	842
73	233	393	...	873
81	241	401	...	881
86	246	406	...	886
102	262	422	...	902
112	272	432	...	912
145	305	465	...	945

We frequently select $n_s = 10$ to allow us to obtain enough sample means to acquire a satisfactory estimate of $V(\hat{\mu})$. We choose k' to give the same number of measurements as would be obtained in a single 1-in-k systematic sample; thus,

$$k' = kn_s$$

The formulas for estimating μ from n_s systematic samples are shown in Equations (7.12) and (7.13).

Estimator of the population mean μ using n_s 1-in-k' systematic samples:

$$\hat{\mu} = \sum_{i=1}^{n_s} \frac{\bar{y}_i}{n_s} \tag{7.12}$$

where $\bar{y}_i$ represents the average of the ith systematic sample.

Estimated variance of $\hat{\mu}$:

$$\hat{V}(\hat{\mu}) = \left(\frac{N-n}{N}\right) \frac{\sum_{i=1}^{n_s} (\bar{y}_i - \hat{\mu})^2}{n_s(n_s - 1)} \tag{7.13}$$

We can also use repeated systematic sampling to estimate a population total τ, if N is known. The necessary formulas are given in Equations (7.14) and (7.15).

Estimator of the population total τ using n_s 1-in-k' systematic samples:

$$\hat{\tau} = N\hat{\mu} = N \sum_{i=1}^{n_s} \frac{\bar{y}_i}{n_s} \tag{7.14}$$

Estimated variance of $\hat{\tau}$:

$$\hat{V}(\hat{\tau}) = N^2 \hat{V}(\hat{\mu}) = N^2 \left(\frac{N-n}{N}\right) \frac{\sum_{i=1}^{n_s} (\bar{y}_i - \hat{\mu})^2}{n_s(n_s - 1)} \tag{7.15}$$

Example 7.6

A state park charges admission by carload rather than by person, and a park official wants to estimate the average number of persons per car for a particular summer holiday. She knows from past experience that there should be about 400 cars entering the park, and she wants to sample 80 cars. To obtain an estimate of the variance, she uses repeated systematic sampling with 10 samples of 8 cars each. Using the data given in Table 7.3, estimate the average number of persons per car and place a bound on the error of estimation.

TABLE 7.3 Data on number of persons per car (the responses y_i are in parentheses)

Random starting point	Second element	Third element	Fourth element	Fifth element	Sixth element	Seventh element	Eighth element	$\bar{y}_i$
2 (3)	52 (4)	102 (5)	152 (3)	202 (6)	252 (1)	302 (4)	352 (4)	3.75
5 (5)	55 (3)	105 (4)	155 (2)	205 (4)	255 (2)	305 (3)	355 (4)	3.38
7 (2)	57 (4)	107 (6)	157 (2)	207 (3)	257 (2)	307 (1)	357 (3)	2.88
13 (6)	63 (4)	113 (6)	163 (7)	213 (2)	263 (3)	313 (2)	363 (7)	4.62
26 (4)	76 (5)	126 (7)	176 (4)	226 (2)	276 (6)	326 (2)	376 (6)	4.50
31 (7)	81 (6)	131 (4)	181 (4)	231 (3)	281 (6)	331 (7)	381 (5)	5.25
35 (3)	85 (3)	135 (2)	185 (3)	235 (6)	285 (5)	335 (6)	385 (8)	4.50
40 (2)	90 (6)	140 (2)	190 (5)	240 (5)	290 (4)	340 (4)	390 (5)	4.12
45 (2)	95 (6)	145 (3)	195 (6)	245 (4)	295 (4)	345 (5)	395 (4)	4.25
46 (6)	96 (5)	146 (4)	196 (6)	246 (3)	296 (3)	346 (5)	396 (3)	4.38

Solution

For one systematic sample,

$$k = \frac{N}{n} = \frac{400}{80} = 5$$

Hence, for $n_s = 10$ samples,

$$k' = 10k = 10(5) = 50$$

The following ten random numbers between 1 and 50 are drawn:

$$13, 35, 2, 40, 26, 7, 31, 45, 5, 46$$

Cars with these numbers form the random starting points for the systematic samples.

For Table 7.3, the quantity $\bar{y}_1$ is the average for the first row, $\bar{y}_2$ is the average for the second row, and so forth. The estimate of μ is

$$\hat{\mu} = \frac{1}{n_s} \sum_{i=1}^{n_s} \bar{y}_i = \frac{1}{10}(3.75 + 3.38 + \cdots + 4.38) = 4.16$$

The following identity can be established:

$$\sum_{i=1}^{n_s} (\bar{y}_i - \hat{\mu})^2 = \sum_{i=1}^{n_s} \bar{y}_i^2 - \frac{1}{n_s}\left(\sum_{i=1}^{n_s} \bar{y}_i\right)^2$$

Substituting, we obtain

$$\sum_{i=1}^{10} (\bar{y}_i - \hat{\mu})^2 = 177.410 - \frac{1}{10}(1733.06) = 4.104$$

Thus, the estimated variance of $\hat{\mu}$ is

$$\hat{V}(\hat{\mu}) = \left(\frac{N-n}{N}\right) \frac{\sum_{i=1}^{n_s} (\bar{y}_i - \hat{\mu})^2}{n_s(n_s - 1)} = \left(\frac{400 - 80}{400}\right)\left[\frac{4.104}{10(9)}\right] = .0365$$

The estimate of μ with a bound on the error of estimation is

$$\hat{\mu} \pm 2\sqrt{\hat{V}(\hat{\mu})} \qquad \text{or} \qquad 4.163 \pm 2\sqrt{.0365} \qquad \text{or} \qquad 4.16 \pm .38$$

Therefore our best estimate of the average of persons per car is 4.16. The error of estimation should be less than .38 with probability approximately .95.

7.7

FURTHER DISCUSSION OF VARIANCE ESTIMATORS

To learn more about the behavior of $V(\bar{y}_{sy})$, we may view a systematic sample of n elements as a single cluster sample selected from k possible cluster samples in the population under study. Schematically, think of the population as being arranged in a rectangular array as shown in Table 7.4. Here, $N = nk$. Systematic sampling, as discussed earlier, involves randomly selecting *one* of the k clusters (rows) and hence one of the k possible sample means. In other words, $\bar{y}_{sy}$ can take on the values $\bar{y}_1$, $\bar{y}_2, \ldots, \bar{y}_k$ with equal probabilities.

We know from Equation (7.4) that

$$V(\bar{y}_{sy}) = \frac{\sigma^2}{n}[1 + (n-1)\rho]$$

where ρ measures the correlation among elements in the same cluster (the intracluster correlation) and must lie between $-[1/(n-1)]$ and 1.

If we had a population of measurements, as indicated in Table 7.4, we could make analysis-of-variance-type calculations of between cluster mean square (MSB), within-cluster mean square (MSW), and total sum of squares (SST). These are defined as

$$\text{MSB} = \frac{n}{k-1}\sum_{i=1}^{k}(\bar{y}_i - \bar{\bar{y}})^2 \tag{7.16}$$

$$\text{MSW} = \frac{1}{k(n-1)}\sum_{i=1}^{k}\sum_{j=1}^{n}(y_{ij} - \bar{y}_i)^2 \tag{7.17}$$

$$\text{SST} = \sum_{i=1}^{k}\sum_{j=1}^{n}(y_{ij} - \bar{\bar{y}})^2 \tag{7.18}$$

TABLE 7.4 **A population of k clusters, each of size n**

Cluster number	Sample number 1	2	3	...	n	Mean
1	y_{11}	y_{12}	y_{13}	...	y_{1n}	$\bar{y}_1$
2	y_{21}	y_{22}	y_{23}	...	y_{2n}	$\bar{y}_2$
3	y_{31}	y_{32}	y_{33}	...	y_{3n}	$\bar{y}_3$
⋮	⋮	⋮	⋮	⋮	⋮	
k	y_{k1}	y_{k2}	y_{k3}	...	y_{kn}	$\bar{y}_k$

where $\bar{\bar{y}}$ is the overall mean per element. Using these terms,

$$\rho = \frac{(k-1)n\ \text{MSB} - \text{SST}}{(n-1)\text{SST}} \tag{7.19}$$

which, for large $N = nk$, is approximately

$$\rho \approx \frac{\text{MSB} - \text{MST}}{(n-1)\text{MST}} \tag{7.20}$$

where $\text{MST} = \text{SST}/(nk-1)$.

Looking at (7.20), we can see that MSB plays a key role in the behavior of $V(\bar{y}_{sy})$. If MSB is small compared to MST, ρ will be negative, and $V(\bar{y}_{sy})$ will be smaller than the corresponding $V(\bar{y})$ from simple random sampling. If, on the other hand, MSB is larger than MST, ρ will be positive, and $V(\bar{y}_{sy})$ will be larger than $V(\bar{y})$. A small MSB implies that the cluster means are very nearly equal; thus, we should choose clusters to have similar means but as much internal variation as possible.

Some calculations on actual populations of measurements will help us understand these concepts. Table 7.5 shows a population of 20 random digits divided into 4 clusters, each of size 5. The analysis of variance calculations, with $k = 4$ clusters of $n = 5$ measurements yield

TABLE 7.5 **Clusters of random digits**

Cluster number	Population (N = 20) 8152	5722	9504	8399	6423	$\bar{y}_i$	$\hat{V}(\bar{y})$
1	8	5	9	8	6	7.2	.405
2	1	7	5	3	4	4.0	.750
3	5	2	0	9	2	3.6	1.845
4	2	2	4	9	3	4.0	1.275

Analysis of Variance

Source	DF	SS	MS
Factor	3	42.20	14.07
Error	16	114.00	7.13
Total	19	156.20	

from which we identify MSB = 14.07, MSW = 7.13, and SST = 156.20. Then, using (7.19),

$$\rho = \frac{3(5)(14.07) - 156.20}{4(156.20)} = .088$$

We should not be surprised that ρ is close to zero, since the data are simply random digits. Each of the k clusters should behave like a simple random sample of n observations.

TABLE 7.6 **Clusters of ordered integers**

Cluster number	Population (N = 20): 1 2 3 4	5 6 7 8	9 10 11 12	13 14 15 16	17 18 19 20	$\bar{y}$	$\hat{V}(\bar{y})$
1	1	5	9	13	17	9.0	6.0
2	2	6	10	14	18	10.0	6.0
3	3	7	11	15	19	11.0	6.0
4	4	8	12	16	20	12.0	6.0

For purposes of studying a nonrandom situation, look at Table 7.6, which displays a population of the first 20 ordered integers broken into 4 clusters of 5 observations each. The analysis of variance calculations yield

Analysis of Variance

Source	DF	SS	MS
Factor	3	25.0	8.3
Error	16	640.0	40.0
Total	19	665.0	

from which

$$\rho = \frac{3(5)(8.3) - 665.0}{4(665)} = -.203$$

Since this is a *very* ordered population, producing ordered systematic samples, we should expect the large negative value of ρ. For ordered populations, systematic sampling produces more precise estimates than does simple random sampling.

Either by direct calculation or by using Equation (7.4), we can find the true variance of $\bar{y}_{sy}$ for the situations in which the population is known. Referring to Table 7.5,

$$V(\bar{y}_{sy}) = \frac{1}{4}\left[\sum_{i=1}^{4} \bar{y}_i^2 - \frac{1}{4}\left(\sum_{i=1}^{4} \bar{y}_i\right)^2\right] = 2.11$$

Using Equation (7.4),

$$\begin{aligned} V(\bar{y}_{sy}) &= \frac{\sigma^2}{n}[1 + (n-1)\rho] \\ &= \frac{\text{SST}}{n(kn)}[1 + (n-1)\rho] \\ &= \frac{156.2}{5(20)}[1 + (4)(.088)] = 2.11 \end{aligned}$$

The right-hand column of Table 7.5 shows the estimated variance of the individual systematic samples (clusters) that could have been obtained using the formula from simple random sampling given as Equation (7.2). Notice that all four of these estimated variances are smaller than the true variance but are reasonably close.

Similar calculations for the data of Table 7.6 show $V(\bar{y}_{sy}) = 1.25$. Now, the four sample estimates of $V(\bar{y}_{sy})$ are quite far away from the true value. (All four estimates equal 6.0.) This is consistent with earlier discussions; in ordered populations, the simple random sample formula will overestimate the true variance of $\bar{y}_{sy}$.

What can we do about this state of affairs? Can we find another estimator of $V(\bar{y}_{sy})$ that might perform better than the estimator used in simple random sampling? Fortunately, we can produce a rather simple estimator that does perform well in most cases, especially in the case of linear trends in the population. That new estimator is based on the following idea. Suppose $y_1, y_2, \ldots, y_n$ is a random sample with $E(y_i) = \mu$ and $V(y_i = \sigma^2)$. The usual estimator of σ^2 is based on $\sum (y_i - \bar{y})^2$, but if we knew that $\mu = 0$, the estimator of σ^2 would be based on $\sum y_i^2$, and $\sum y_i^2/n$ would be an unbiased estimator of σ^2.

Now, suppose μ is not zero. Choose two sample values, y_i and y_j, and construct $d_i = y_i - y_j$. It follows that $E(d_i) = 0$ and $V(d_i) = 2\sigma^2$. If we make up n_d such differences, then $\sum_{i=1}^{n_d} d_i^2/n_d$ is an estimator of $2\sigma^2$, from which an estimator of σ^2 is easily obtained. Since we want to estimate a mean $\bar{y}_{sy}$ with a sample of n measurements from a population of N measurements, the estimator based on d_i becomes

$$\hat{V}_d(\bar{y}_{sy}) = \frac{N-n}{Nn}\frac{1}{2n_d}\sum_{i=1}^{n_d} d_i^2 \tag{7.21}$$

Instead of making arbitrary differences, we usually take successive differences of the form

$$d_i = y_{i+1} - y_i, \qquad i = 1, \ldots, (n-1)$$

A sample of size n yields $n-1$ such successive differences, and so Equation (7.21) becomes

$$\hat{V}_d(\bar{y}_{sy}) = \frac{N-n}{Nn} \frac{1}{2(n-1)} \sum_{i=1}^{n-1} d_i^2 \tag{7.22}$$

For the random digit data of Table 7.5, the successive differences are given in Table 7.7. The right-hand column gives $\hat{V}_d(\bar{y}_{sy})$ calculated from Equation (7.22). These estimated variances are quite close to the ones produced by the simple random sampling formula (see Table 7.5) and are in the neighborhood of the true $V(\bar{y}_{sy}) = 2.11$.

Moving to the ordered integer data of Table 7.6, we see that the successive differences in all four possible systematic samples are (4, 4, 4, 4). Based on these data, $\hat{V}_d(\bar{y}_{sy}) = 1.20$ for each systematic sample. Note here that $\hat{V}_d(\bar{y}_{sy})$ is very close to the true $V(\bar{y}_{sy}) = 1.25$. Also, the simple random sampling estimate (6, in this case) does *not* work well at all.

A good rule for practical work in sampling seems evident—we should use the variance estimate based on successive differences whenever we suspect the population elements to be other than purely randomly ordered. We should use the variance estimate from simple random sampling *only* when we have good reason to believe that the population elements are in purely random order.

The same methods and principles apply to estimation of a proportion. To demonstrate this, we took the random digit data of Table 7.5 and coded each digit as even or odd. (The objective here would be to estimate the proportion of even numbers in the population.) The results are shown in Table 7.8. The calculations for $\hat{V}(\hat{p})$ come from simple random sampling formula (7.8) and those for $\hat{V}_d(\hat{p})$ from Equation (7.22). The true $V(\hat{p}_{st}) = .030$ in this case, so both methods perform reasonably well. That is to be expected for purely random numbers such as these.

Table 7.9 shows similar calculations for the data of Table 7.6 with integers ≤ 10 coded as 0 and integers > 10 coded as 1. (The objective, here, would be to estimate the proportion of integers greater than 10 in the population.) For this situation, $V(\hat{p}_{sy}) = .01$, and we see that the variance estimates based on successive differences are superior to

TABLE 7.7 **Successive differences from Table 7.5**

Cluster number	Differences (absolute value)				$\hat{V}_d(\bar{y}_{sy})$
1	3	4	1	2	.562
2	6	2	2	1	.844
3	3	2	9	7	2.681
4	0	2	5	6	1.219

TABLE 7.8 **Even–odd random numbers**

Cluster number	1001	0011	0011	1000	1110	$\hat{p}$	$\hat{V}(\hat{p})$	$\hat{V}_d(\hat{p})$
1	1	0	0	1	1	.6	.045	.047
2	0	0	0	0	1	.2	.030	.023
3	0	1	1	0	1	.6	.045	.070
4	1	1	1	0	0	.6	.045	.023

(1 = even)

TABLE 7.9 **Ordered integers 1–20**

Cluster number	0000	0000	0011	1111	1111	$\hat{p}$	$\hat{V}(\hat{p})$	$\hat{V}_d(\hat{p})$
1	0	0	0	1	1	.4	.045	.023
2	0	0	0	1	1	.4	.045	.023
3	0	0	1	1	1	.6	.045	.023
4	0	0	1	1	1	.6	.045	.023

(0 indicates ≤ 10, 1 indicates > 10)

those based on simple random sampling results. Again, if the data in the population do not come about in random order, then the successive differences should be used in estimating the variance of the sample proportion.

We now discuss two examples of a more realistic type, in which the population as a whole is not under study.

Example 7.7

The data set TEMPS in the appendix and on the data disk shows an alphabetical listing of 88 weather stations across the United States. It is desired to estimate the average January precipitation across the U.S. by systematically sampling $n = 8$ of these stations. (A reporter may only have time to call 8 stations.) In this case, $k = 11$ so a random start was selected among the first 11 stations on the list. (It turned out to be number 9.) The January precipitation (in inches) for the 8 sampled stations were

$$0.5, 1.8, 1.9, 4.7, 1.7, 0.7, 1.7, 2.8$$

Estimate the average January precipitation for all 88 stations and place a bound on the error of estimation.

Solution

The plot of these data versus sample number, Figure 7.4, shows a small cyclical trend as the systematic selection moves across the alphabetical listing. If there is such a trend in

FIGURE 7.4 Plot of the precipitation data for Example 7.7

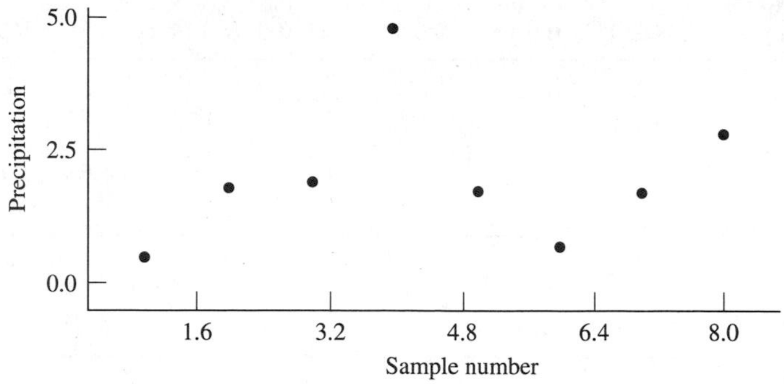

the population, this sample seems to have captured both high and low values, so the variance approximation from simple random sampling should still be acceptable.

Now,

$$\bar{y}_{sy} = 1.98$$

and

$$s^2 = 1.728$$

Hence,

$$\hat{V}(\bar{y}_{sy}) = \frac{N-n}{Nn} s^2 = \frac{88-8}{(88)(8)}(1.728)$$
$$= .196$$

and

$$2\sqrt{\hat{V}(y_{sy})} = .89$$

If we had some doubts about the randomness of the order of the measurements on the list, we could use

$$\hat{V}_d(\bar{y}_{sy}) = \frac{N-n}{Nn}\frac{1}{2(n-1)}\sum_{i=1}^{n-1} d_i^2$$
$$= \frac{80}{88(8)}\frac{1}{2(7)}(21.75) = .176$$

and

$$2\sqrt{\hat{V}_d(\bar{y}_{sy})} = .84$$

These estimated variances are quite close to each other, and either will work well with this somewhat cyclical data. The one based on differences, though, is slightly smaller because of the pattern in the sample data.

Example 7.8 The data set RIVER in the appendix and on the data disk shows the mean daily flow rates for a certain Florida river. Suppose we choose to estimate the average flow rate for the fall (October, November, December) of 1977 by looking at every tenth day. Choosing a random start (4, in this case) between 1 and 10 yields $n = 9$ measurements during this three-month period. They are (in cubic feet per second):

$$38, 24, 17, 11, 4.7, 7.5, 4.0, 2.6, 4.9$$

Estimate the average daily flow for the period under study.

Solution

A plot of these data, Figure 7.5, shows a definite and pronounced decreasing trend as one moves from the earlier to the later observations. The variance approximation from simple random sampling will overestimate the true variance by a considerable amount, if this trend is also apparent in the population.

FIGURE 7.5 Plot of the flow-rate data for Example 7.8

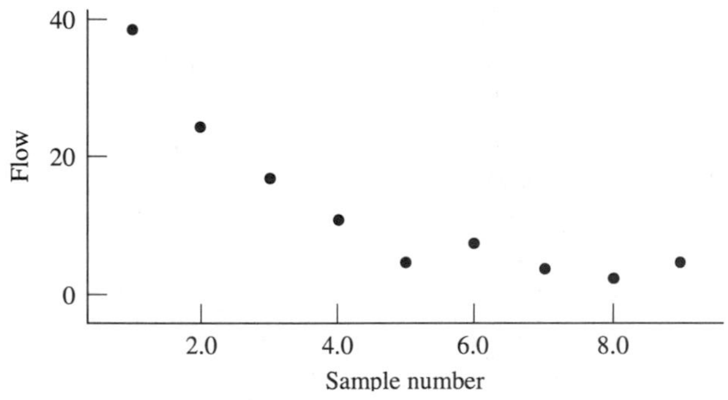

Equations (7.1) and (7.2) yield

$$\bar{y}_{sy} = 12.63$$

$$\hat{V}(\bar{y}_{sy}) = \frac{83}{92(9)}(139.838) = 14.018$$

and

$$2\sqrt{\hat{V}(\bar{y}_{sy})} = 7.49$$

The successive difference estimator of the variance gives

$$\hat{V}_d(\bar{y}_{sy}) = \frac{83}{92(9)}\frac{348.03}{2(8)} = 2.180$$

and

$$2\sqrt{\hat{V}_d(\bar{y}_{sy})} = 2.95$$

Obviously, the difference method produced a much smaller variance estimate. But is it realistic? The sample data do seem to have a linearly decreasing trend across time, which could reflect a similar trend in the population. Also, Florida is fairly dry in the fall months (but not in the summer), so it would not be surprising to see the flow rates decreasing during the fall. All things considered, the successive difference method seems to be the better one. Our sample mean of 12.63 should lie within 2.95 units of the true population mean.

7.8 SUMMARY

Systematic sampling is presented as an alternative to simple random sampling. Systematic sampling is easier to perform and therefore less subject to interviewer errors than simple random sampling. In addition, systematic sampling often provides more information per unit cost than does simple random sampling.

We considered estimation of a population mean, total, and proportion using the estimators $\bar{y}_{sy}$, $N\bar{y}_{sy}$, and $\hat{p}_{sy}$, respectively. The corresponding bounds on the errors of estimation were given for these estimators.

We must first consider the type of population under investigation in order to choose between systematic and simple random sampling. For example, when N is large and $\rho < 0$, the variance of $\bar{y}_{sy}$ is smaller than the corresponding variance of $\bar{y}$ based on simple random sampling. A systematic sample is preferable when the population is ordered and N is large. When the population is random, the two sampling procedures are equivalent, and either design can be used. Care must be used in applying systematic sampling to periodic populations.

Sample-size requirements for estimating μ, τ, and p are determined by using formulas presented for simple random sampling.

Repeated systematic sampling and use of successive differences allow the experimenter to estimate the population mean or total and the variance of the estimator without making any assumptions about the nature of the population.

CASE STUDY REVISITED

THE ASSESSMENT OF PRODUCT QUALITY

The quality control problem involving percentage of copper in bronze castings, given at the beginning of this chapter, showed a systematic sample of 16 measurements with $\bar{y} = 87$ and $s^2 = 18$. Even though the sample was selected systematically, we can estimate the population mean by

$$\bar{y} \pm 2\sqrt{\frac{s^2}{n}}$$

assuming N is large compared with n. Thus, we have

$$87 \pm 2\sqrt{\frac{18}{16}}$$

$$87 \pm 2$$

or 85 to 89 as the best estimate of the true mean of the production process. Since the standard is 90, apparently the process is not performing up to the advertised standard on this day. The supervisor in charge will want to look into possible causes for this failure.

In this case, systematic sampling is reasonable since it forces the sample to cover the entire day of production. If quality tends to decrease (or increase) during the day, this sampling plan may detect it. A simple random sample could concentrate all sampled items in the morning (or afternoon) hours.

EXERCISES

7.1 Suppose that a home mortgage company has N mortgages numbered serially in the order that they were granted over a period of 20 years. There is a generally increasing trend in the unpaid balances because of the rising cost of housing over the years. The company wishes to estimate the total amount of unpaid balances. Would you employ a systematic or a simple random sample? Why?

7.2 A corporation lists employees by income brackets (alphabetically within brackets) from highest to lowest. If the objective is to estimate average income per employee, should systematic, stratified, or simple random sampling be used? Assume that costs are equivalent for the three methods and that you can stratify on income brackets. Discuss the advantages and disadvantages of the three methods.

7.3 A retail store with four departments has charge accounts arranged by department, with past-due accounts at the front of each departmental list. Suppose the departments average around 10 accounts each, with approximately 40% past due. On a given day the accounts might appear as shown in the accompanying table (with account numbers 1 through 40). The store wishes to estimate the proportion of past-due accounts by systematic sampling.

	Department			
Account numbers	1–11	12–20	21–28	29–40
Delinquent accounts	1, 2, 3, 4	12, 13, 14	21, 22, 23, 24, 25	29, 30, 31, 32

(a) List all possible 1-in-10 systematic samples and compute the exact variance of the sample proportion. (Note that there are 10 possible values, not all distinct, for the sample proportion, each with probability $\frac{1}{10}$ of occurring.)
(b) List all possible 1-in-5 systematic samples and compute the exact variance of the sample proportion.
(c) Compare the result in part (a) with an approximate variance that would have been obtained in a simple random sample of size $n=4$ from this population. Similarly, compare the result in part (b) with what would have been obtained from a simple random sample with $n=8$. What general conclusions can you make?

7.4 The management of a particular company is interested in estimating the proportion of employees favoring a new investment policy. A 1-in-10 systematic sample is obtained from employees leaving the building at the end of a particular workday. Use the data in the accompanying table to estimate p, the proportion in favor of the new policy, and place a bound on the error of estimation. Assume $N=2000$.

Employee sampled	Response
3	1
13	0
23	1
⋮	⋮
1993	1
	$\sum_{i=1}^{200} y_i = 132$

7.5 For the situation outlined in Exercise 7.4, determine the sample size required to estimate p to within .01 unit. What type of systematic sample should be run?

7.6 The quality control section of an industrial firm uses systematic sampling to estimate the average amount of fill in 12-ounce cans coming off an assembly line. The data in the accompanying table represent a 1-in-50 systematic sample of the production in one day. Estimate μ and place a bound on the error of estimation. Assume $N = 1800$.

Amount of fill (in ounces)					
12.00	11.97	12.01	12.03	12.01	11.80
11.91	11.98	12.03	11.98	12.00	11.83
11.87	12.01	11.98	11.87	11.90	11.88
12.05	11.87	11.91	11.93	11.94	11.89
11.75	11.93	11.95	11.97	11.93	12.05
11.85	11.98	11.87	12.05	12.02	12.04

7.7 Use the data of Exercise 7.6 to determine the sample size required to estimate μ to within .03 unit.

7.8 Soil experts want to determine the amount of exchangeable calcium (in parts per million) in a plot of ground. So that the sampling scheme is simplified, a rectangular grid is superimposed on the field. Soil samples are taken at each point of intersection on the grid (see the diagram). Use the following data to determine the average amount of exchangeable calcium on the plot of ground. Place a bound on the error of estimation.

$$n = 45$$

$$\sum y_i = 90{,}320 \qquad \text{exchangeable calcium}$$

$$\sum y_i^2 = 184{,}030{,}000$$

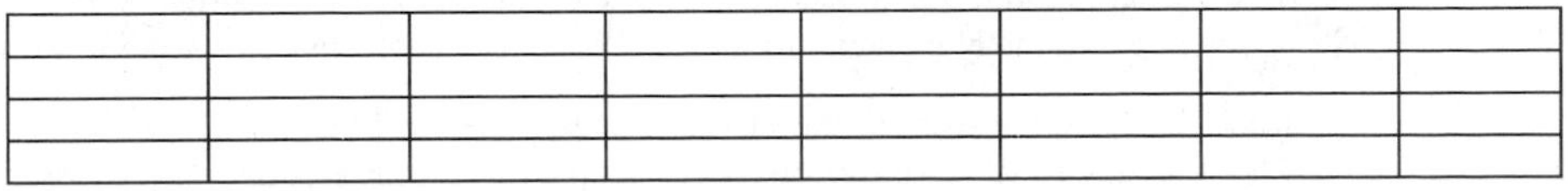

7.9 The highway patrol of a particular state is concerned about the proportion of motorists who carry their licenses. A checkpoint is set up on a major highway, and the driver of every seventh car is questioned. Use the data in the accompanying table to estimate the proportion of drivers carrying their licenses. Place a bound on the error of estimation. Assume that $N = 2800$ cars pass the checkpoint during the sampling period.

Car	Response, y_i
1	1
2	1
3	0
⋮	⋮
400	1
	$\sum y_i = 324$

7.10 The highway patrol expects at least $N = 3000$ cars to pass the checkpoint. Determine the sample size required to estimate p to within $B = .015$ unit.

7.11 A college is concerned about improving its relations with a neighboring community. A 1-in-150 systematic sample of the $N = 4500$ students listed in the directory is taken to estimate the total amount of money spent on clothing during one quarter of the school year. The results of the sample are listed in the accompanying table. Use these data to estimate τ and place a bound on the error of estimation.

Student	Amount spent (in dollars)	Student	Amount spent (in dollars)	Student	Amount spent (in dollars)
1	30	11	29	21	9
2	22	12	21	22	15
3	10	13	13	23	6
4	62	14	15	24	93
5	28	15	23	25	21
6	31	16	32	26	20
7	40	17	14	27	13
8	29	18	29	28	12
9	17	19	48	29	29
10	51	20	50	30	38

7.12 What sample size is needed to estimate τ in Exercise 7.11 with a bound on the error of estimation approximately equal to \$10,000? What systematic sampling scheme would you recommend?

7.13 Refer to the data on industry groups provided in Example 7.1. Estimate the true difference between the mean payrolls in 1987 and 1991, with a bound on the error. Does this produce the same type of change as the difference in the mean number of employees? Does systematic sampling appear to work well for estimating this difference?

7.14 A group of guidance counselors is concerned about the average yearly tuition for out-of-state students in 371 junior colleges. From an alphabetical list of these colleges, a 1-in-7 systematic sample is drawn. Data concerning out-of-state tuition expenses for an academic year (September to June) are obtained for each college sampled. Let y_i be the amount of tuition required for the ith college sampled. Use the following data to estimate μ and place a bound on the error of estimation.

$$\sum_{i=1}^{53} y_i = \$11{,}950 \qquad \text{and} \qquad \sum_{i=1}^{53} y_i^2 = \$2{,}731{,}037$$

7.15 Museum officials are interested in the total number of persons who visit their museum during a 180-day period when an expensive antique collection is on display. Since monitoring the museum traffic each day is too costly, officials decide to obtain these data every tenth day. The information from this 1-in-10 systematic sample is summarized in the accompanying table. Use these data to estimate τ, the total number of persons visiting the museum during the specified period. Place a bound on the error of estimation.

Day	Number of people visiting the museum
3	160
13	350
23	225
⋮	⋮
173	290
	$\sum_{i=1}^{18} y_i = 4868$
	$\sum_{i=1}^{18} y_i^2 = 1{,}321{,}450$

7.16 Foresters are interested in determining the mean timber volume per acre for 520 one-acre plots ($N = 520$). A 1-in-25 systematic sample is conducted. Using the data presented in the accompanying table, estimate μ, the average timber volume per plot, and place a bound on the error of estimation.

Plot sampled	Volume (in board feet)	Plot sampled	Volume (in board feet)
4	7030	279	7540
29	6720	304	6720
54	6850	329	6900
79	7210	354	7200
104	7150	379	7100
129	7370	404	6860
154	7000	429	6800
179	6930	454	7050
204	6570	479	7420
229	6910	504	7090
254	7380		

7.17 The officers of a certain professional society wish to determine the proportion of the membership that favors several proposed revisions in refereeing practices. They conduct a 1-in-10 systematic sample from an alphabetical list of the $N = 650$ registered members. Let $y_i = 1$ if the ith person sampled favors the proposed changes and $y_i = 0$ if he opposes the changes. Use the following sample data to estimate p, the proportion of members in favor of the proposed changes. Place a bound on the error of estimation.

$$\sum_{i=1}^{65} y_i = 48$$

7.18 In a sociological survey a 1-in-50 systematic sample is drawn from city tax records to determine the total number of families in the city who rent their homes. Let $y_i = 1$ if the family in the ith household sampled rents and let $y_i = 0$ if the family does not. There are $N = 15{,}200$ households in the community. Use the following to estimate τ, the total number of families who rent. Place a bound on the error of estimation.

$$\sum_{i=1}^{304} y_i = 88$$

[*Hint*: If $\hat{p}$ = estimated fraction who rent, then $N\hat{p}$ is an estimate of the total number who rent; $\hat{V}(N\hat{p}) = N^2 \hat{V}(\hat{p})$.]

7.19 A farmer wishes to estimate the total weight of fruit to be produced in a field of zucchini (squash) by sampling just prior to harvest. The plot consists of 20 rows with 400 plants per row. The manufacturer of the seeds says that each plant can yield up to 8 pounds of fruit. Outline an appropriate systematic sampling plan for this problem so as to estimate the total weight of fruit to within 2000 pounds.

7.20 The following table shows the number of births (in thousands) and the birth rate (in number of births per thousand of population) in the United States for a systematic sample of years between 1950 and 1990.

Year	1950	1955	1960	1965	1970	1975	1980	1985	1990
Births	3632	4097	4258	3760	3731	3144	3612	3761	4158
Rate	24.1	25.0	23.7	19.4	18.4	14.6	15.9	15.8	16.7

Source: U.S. Bureau of the Census, *Statistical Abstract of the United States, 1993–94* (113 ed.), Washington, D.C., 1994.

(a) Estimate the total number of births during this 41-year period. Find an appropriate estimate of the variance.

(b) Estimate the mean birth rate during this period and find an appropriate estimator of the variance. Do you think this mean would be a good predictor of the birth rate for 1995? Explain.

7.21 The following table shows the number of divorces (in thousands) in the United States for a systematic sample of years between 1950 and 1990. Estimate the total number of divorces for this period and find an appropriate variance approximation for your estimate.

Year	1950	1955	1960	1965	1970	1975	1980	1985	1990
Divorces	385	377	393	479	708	1036	1189	1190	1175

Source: U.S. Bureau of the Census, *Statistical Abstract of the United States, 1993–94* (113 ed.), Washington, D.C., 1994.

Would the mean number of divorces over this period be a good predictor of the number of divorces for 1995? Explain.

7.22 A quality control inspector must sample silicon wafers, from which computer chips will be made, after they are baked in an oven. Slotted trays containing many wafers are put through the oven, one after another, all day long. Position on the tray and time of day may have important bearings on the quality of the wafer. Suggest a sampling plan, with the goal being to estimate the proportion of defective wafers.

7.23 A warehouse contains stacks of automobile batteries that must be sampled for quality inspection. Each stack has a different production date code, and the stacks are arranged chronologically. The stacks are of approximately equal size. Suggest a sampling plan for estimating the proportion of defective batteries.

7.24 An auditor is confronted with a long list of accounts receivable for a firm. She must verify the amounts on 10% of these accounts and estimate the average difference between the audited and book values.
(a) Suppose the accounts are arranged chronologically, with the older accounts tending to have smaller values. Would you choose a systematic or a simple random sampling design to select the sample?
(b) Suppose the accounts are arranged randomly. Would you choose a systematic or a simple random sampling design to select the sample?
(c) Suppose the accounts are grouped by department and then listed chronologically within departments. The older accounts again tend to have smaller values. Would you choose a systematic or a simple random sampling design to select the sample?

7.25 The market share for a certain food product is to be estimated by recording store purchases of the product for certain weeks selected throughout the year. Discuss the advantages and disadvantages of a systematic selection of the weeks for this study.

7.26 Crop yield for a large field of wheat is to be estimated by sampling small plots within the field while the grain is ripening. The field is on sloping land, with higher fertility toward the lower side.
(a) Suggest a systematic sampling design for the small plots.
(b) Could other sampling designs be used effectively in this case?

7.27 The variance of estimators from simple random sampling can be calculated using simple random sampling results or using successive differences. Does the use of successive differences produce a better estimator of the variance in
(a) Exercise 7.6?
(b) Exercise 7.11?
(c) Exercise 7.16?
(d) Exercise 7.20?
(e) Exercise 7.21?

7.28 A certain bank has 20,000 home mortgages numbered serially in the order that they were granted, over a period of years. There is a tendency for unpaid balances to increase from the lowest to highest number on the list, because housing costs have been rising. It is desired to estimate the total unpaid balance. A sample of size 200 is to be used, and four clerks are available to do the work. How would you design the sampling scheme and estimate the desired total? Construct an approximate variance from your estimator.

7.29 We have available an ordered list of addresses of individual family dwellings along a long street that intersects a city. For estimating the proportion of households containing an unemployed adult, would you use systematic or simple random sampling? Why?

EXPERIENCES WITH REAL DATA

7.1 Locate the stock price summaries for the week in your local weekend newspaper. These summaries usually list the high and low prices of each stock for the week, along with the difference between the closing price for the current week and that for the previous week.
(a) Select a systematic sample of stocks, and estimate the proportion that have a lower closing price this week than they had the previous week. Place a bound on the error of estimation.
(b) How do you think systematic sampling would compare with simple or stratified random sampling in this case?

7.2 The data set TEMPS in the appendix and on the data disk shows normal monthly temperatures and amounts of precipitation reported by weather stations around the United States. For the month of March, estimate the mean temperature and the mean amount of precipitation across these weather stations by using
(a) A single systematic sample of size 30, with an appropriate variance approximation.
(b) Three repeated systematic samples of size 10 each, with an appropriate variance approximation.
Which method do you prefer, and why?

7.3 The data set RIVER in the appendix and on the data disk shows the daily discharge rates of a certain Florida river over a two-year period. Estimate the mean discharge rate for the entire period by using
(a) A single systematic sample of 100 observations, with an appropriate variance approximation.
(b) Five repeated systematic samples of 20 observations each, with an appropriate variance approximation.
(c) A simple random sample of 100 observations, with an appropriate variance approximation.
Discuss the relative merits of these three methods.

7.4 From a list of names, like those in a student directory, select a systematic sample and interview the selected persons to find out whether they favor a certain issue of current importance (such as a proposed government action or a pending campus decision). Estimate the population proportion favoring the issue and place a bound on the error of estimation.

Repeat the procedure just outlined three more times so that four independent systematic samples are available. Compare the results from the individual samples with the combined result of the four samples analyzed according to the methods of Section 7.6. If you prefer to work with something other than lists of people, use other listed records in a similar way. For example, you could systematically sample names of employees from a file and estimate average age, income, and so on.

C A S E   S T U D Y

WHAT ARE THE CHARACTERISTICS OF THE PEOPLE LIVING IN YOUR NEIGHBORHOOD?

Suppose a firm wants to locate a business in your neighborhood. How can it find information on the characteristics of the people living there without conducting its own survey? One way is to consult the block statistics data from the U.S. Census Bureau. *Block statistics* give demographic information—such as total number of residents, number in certain minority groups, number over the age of 65, and number of owners and renters—on very small regions that often conform to city blocks. These data are used by market researchers, housing and transportation planners, and community associations, among others.

The business that is considering a location in your neighborhood caters to those aged 65 and over. Thus, it wants to estimate the proportion of residents in this age category who live in a 40-block area. The firm decides to sample 5 of the 40 blocks and obtain the data from block statistics. The sampled blocks form clusters of people, and hence the techniques of cluster sampling must be used. (This problem is a scaled-down version of a real problem. Usually, the number of blocks and the sample size would be much larger.)

CHAPTER 8

CLUSTER SAMPLING

8.1

INTRODUCTION

Recall that the objective of sample survey design is to obtain a specified amount of information about a population parameter at minimum cost. Stratified random sampling is often better suited for this than is simple random sampling for the three reasons indicated in Section 5.1. Systematic sampling often gives results at least as accurate as those from simple random sampling, and it is easier to perform, as discussed in Section 7.1. This chapter introduces a fourth design, cluster sampling, which sometimes gives more information per unit cost than do any of the other three designs discussed previously.

Definition 8.1

A *cluster sample* is a probability sample in which each sampling unit is a collection, or cluster, of elements.

Cluster sampling is less costly than simple or stratified random sampling if the cost of obtaining a frame that lists all population elements is very high or if the cost of obtaining observations increases as the distance separating the elements increases.

To illustrate, suppose we wish to estimate the average income per household in a large city. How should we choose the sample? If we use simple random sampling, we will need a frame listing all households (elements) in the city, and this frame may be very costly or impossible to obtain. We cannot avoid this problem by using stratified random sampling because a frame is still required for each stratum in the population. Rather than draw a simple random sample of *elements*, we could divide the city into regions such as blocks (or clusters of elements) and select a simple random sample of blocks from the population. This task is easily accomplished by using a frame that lists all city blocks. Then the income of every household within each sampled block could be measured.

To illustrate the second reason for using cluster sampling, suppose that a list of households in the city is available. We could select a simple random sample of households, which probably would be scattered throughout the city. The cost of conducting interviews in the scattered households would be large owing to the interviewer travel time and other related expenses. Stratified random sampling could lower these expenses, but using cluster sampling is a more effective method of reducing travel costs. Elements within a cluster should be close to each other geographically, and hence travel expenses should be reduced. Obviously, travel within a city block would be minimal when compared with the travel associated with simple random sampling of households within the city.

To summarize, cluster sampling is an effective design for obtaining a specified amount of information at minimum cost under the following conditions:

1. A good frame listing population elements either is not available or is very costly to obtain, while a frame listing clusters is easily obtained.
2. The cost of obtaining observations increases as the distance separating the elements increases.

City blocks are frequently used as clusters of households or people because the U.S. Census Bureau reports very detailed block statistics. In census data, a block may be a standard city block or an irregularly shaped area with identifiable political or geographic boundaries. Block statistics are reported for all urban areas and for all other places with concentrations of 10,000 or more people. In total, block statistics cover 77% of the nation's population. Data reported for each block include total population, racial mix, and number of housing units, and they may include the dollar value of the property, whether the unit is owned or rented by the inhabitants, and whether the unit has complete plumbing facilities.

Block statistics from the Census Bureau are widely used in cluster sampling by market research firms, which may want to estimate the potential market for a product, the potential sales if a new store were to open in the area, or the potential number of clients for a new service, such as an emergency medical facility.

State and local governments sample blocks (clusters of housing units or people) to plan new transportation methods and facilities and to plan housing developments. Similarly, community organizations, such as churches, use block statistics to determine optimal sites for expansion.

There are many other common examples of the uses of cluster sampling. Housing units themselves are clusters of people and may form convenient sampling units when sampling college students, for example. Hospitals form convenient clusters of patients with certain illnesses for studies of the average length of time hospitalized or average number of recurrences of these illnesses.

Elements other than people are often sampled in clusters. An automobile forms a nice cluster of four tires for studies of tire wear and safety. A circuit board manufactured for a computer forms a cluster of semiconductors for testing. An orange tree forms a cluster of oranges for investigating an insect infestation. A plot in a forest contains a cluster of trees for estimating timber volume or proportions of diseased trees. As you can see, the list of possible clusters that are convenient units for sampling is endless.

We will now discuss the details of selecting a cluster sample.

8.2

HOW TO DRAW A CLUSTER SAMPLE

The first task in cluster sampling is to specify appropriate clusters. Elements within a cluster are often physically close together and hence tend to have similar characteristics. Stated another way, the measurement on one element in a cluster may be highly correlated with the measurement on another. Thus, the amount of information about a population parameter may not be increased substantially as new measurements are taken within a cluster. Since measurements cost money, an experimenter would waste money by choosing too large a cluster size. However, situations may arise in which elements within a cluster are very different from one another. In such cases, a sample containing a few large clusters could produce a very good estimate of a population parameter, such as the mean.

For example, suppose clusters are formed by boxes of components coming off production lines, one cluster of components per line. If all lines have approximately the same rate of defects, then the components in each cluster (box) are about as variable with respect to quality as the population as a whole. In this situation, a good estimate of the proportion of defectives produced could be obtained from one or two clusters.

In contrast, suppose school districts are specified as clusters of households for estimating the proportion of households that favor a rezoning plan. Since the clusters contain many households, resources allow only a small number of clusters—say, two or three—to be sampled. In this case, most households in one district may be happy with their schools and not favor rezoning, whereas most households in another district may be unhappy with their schools and strongly favor rezoning. A small sample of school districts may miss one or the other of these groups entirely, thereby yielding a very poor estimate. More information may be obtained by sampling a larger number of clusters of smaller size.

The problem of choosing an appropriate cluster size can be even more difficult when an infinite number of possible cluster sizes are available, as in the selection of forest plots to estimate the proportion of diseased trees. If there is variability in the density of diseased trees across the forest, then many small plots (clusters), randomly or systematically located, would be desirable. However, to randomly locate a plot in a forest is quite time-consuming, and once it is located, sampling many trees in that one plot is economically desirable. Thus, many small plots are advantageous for controlling variability, but a few large plots are advantageous economically. A balance between size and number of plots must be achieved. There are no good rules that always hold for making this decision. Each problem must be studied on its own, but pilot surveys with various plot sizes might help point the experimenter in the correct direction.

Notice the main difference between the optimal construction of strata (Chapter 5) and the construction of clusters. Strata are to be as homogeneous (alike) as possible within, but one stratum should differ as much as possible from another with respect to the characteristic being measured. Clusters, on the other hand, should be as heterogeneous (different) as possible within, and one cluster should look very much like another in order for the economic advantages of cluster sampling to pay off.

Once appropriate clusters have been specified, a frame that lists all clusters in the population must be composed. A simple random sample of clusters is then selected from this frame by using the methods of Section 4.2. We illustrate with the following example.

Example 8.1

A sociologist wants to estimate the per-capita income in a certain small city. No list of resident adults is available. How should he design the sample survey?

Solution

Cluster sampling seems to be the logical choice for the survey design because no lists of elements are available. The city is marked off into rectangular blocks, except for two industrial areas and three parks that contain only a few houses. The sociologist decides that each of the city blocks will be considered one cluster, the two industrial areas will be considered one cluster, and, finally, the three parks will be considered one cluster. The clusters are numbered on a city map, with the numbers from 1 to 415. The experimenter has enough time and money to sample $n = 25$ clusters and to interview every household within each cluster. Hence, 25 random numbers between 1 and 415 are selected from Appendix Table 2, and the clusters having these numbers are marked on the map. Interviewers are then assigned to each of the sampled clusters.

8.3

ESTIMATION OF A POPULATION MEAN AND TOTAL

Cluster sampling is simple random sampling with each sampling unit containing a number of elements. Hence, the estimators of the population mean μ and total τ are similar to those for simple random sampling. In particular, the sample mean $\bar{y}$ is a good estimator of the population mean μ. An estimator of μ and two estimators of τ are discussed in this section.

The following notation is used in this chapter:

N = The number of *clusters* in the population

n = The number of clusters selected in a simple random sample

m_i = The number of elements in cluster i, $i = 1, \ldots, N$

$\bar{m} = \frac{1}{n}\sum_{i=1}^{n} m_i$ = the average cluster size for the sample

$M = \sum_{i=1}^{N} m_i$ = the number of elements in the population

$\bar{M} = M/N$ = the average cluster size for the population

y_i = The total of all observations in the ith cluster

The estimator of the population mean μ is the sample mean $\bar{y}$, which is given by

$$\bar{y} = \frac{\sum_{i=1}^{n} y_i}{\sum_{i=1}^{n} m_i}$$

Thus, $\bar{y}$ takes the form of a ratio estimator, as developed in Chapter 6, with m_i taking the place of x_i. Then the estimated variance of $\bar{y}$ has the form of the variance of a ratio estimator, given by Equation (6.2).

Estimator of the population mean μ:

$$\bar{y} = \frac{\sum_{i=1}^{n} y_i}{\sum_{i=1}^{n} m_i} \tag{8.1}$$

Estimated variance of $\bar{y}$:

$$\hat{V}(\bar{y}) = \left(\frac{N-n}{Nn\bar{M}^2}\right)s_r^2 \tag{8.2}$$

where

$$s_r^2 = \frac{\sum_{i=1}^{n}(y_i - \bar{y}m_i)^2}{n-1} \tag{8.3}$$

Here $\bar{M}$ can be estimated by $\bar{m}$ if M is unknown.

The estimated variance in Equation (8.2) is biased and a good estimator of $V(\bar{y})$ only if n is large—say, $n \geq 20$. The bias disappears if the cluster sizes $m_1, m_2, \ldots, m_N$ are equal.

Let us illustrate the use of the formulas with an example.

Example 8.2 Interviews are conducted in each of the 25 blocks sampled in Example 8.1. The data on incomes are presented in Table 8.1. Use the data to estimate the per-capita income in the city and place a bound on the error of estimation.

TABLE 8.1 **Per-capita income**

Cluster i	Number of residents, m_i	Total income per cluster, y_i	Cluster i	Number of residents, m_i	Total income per cluster, y_i
1	8	\$96,000	14	10	\$49,000
2	12	121,000	15	9	53,000
3	4	42,000	16	3	50,000
4	5	65,000	17	6	32,000
5	6	52,000	18	5	22,000
6	6	40,000	19	5	45,000
7	7	75,000	20	4	37,000
8	5	65,000	21	6	51,000
9	8	45,000	22	8	30,000
10	3	50,000	23	7	39,000
11	2	85,000	24	3	47,000
12	6	43,000	25	8	41,000
13	5	54,000			
				$\sum_{i=1}^{25} m_i =$ 151	$\sum_{i=1}^{25} y_i =$ \$1,329,000

Solution

Since the estimator of the mean per element is a ratio estimator, computations proceed exactly as they did for ratio estimators in Chapter 6. In a spreadsheet on a calculator or computer, a column of values for $(y_i - \bar{y}m_i)$ is established, and the basic component of the estimated variance is the sample variance of these differences. A summary of the basic statistics for these data is as follows.

	N	Mean	Median	SD
Resident	25	6.040	6.000	2.371
Income	25	53,160	49,000	21,784
$y_i - \bar{y}m_i$	25	0	993	25,189

The best estimate of the population mean μ is given by Equation (8.1) and calculated as follows:

$$\bar{y} = \frac{\sum_{i=1}^{n} y_i}{\sum_{i=1}^{n} m_i} = \frac{\$1,329,000}{151} = \frac{\$53,160}{6.04} = \$8801$$

Since M is not known, the $\bar{M}$ appearing in Equation (8.2) must be estimated by $\bar{m}$, where

$$\bar{m} = \frac{\sum_{i=1}^{n} m_i}{n} = \frac{151}{25} = 6.04$$

Example 8.1 gives $N = 415$. Then from Equation (8.2),

$$\hat{V}(\bar{y}) = \left(\frac{N - n}{Nn\bar{M}^2}\right) \frac{\sum_{i=1}^{n} (y_i - \bar{y}m_i)^2}{n - 1}$$

$$= \left[\frac{415 - 25}{(415)(25)(6.04)^2}\right](25,189)^2 = 653,785$$

Thus, the estimate of μ with a bound on the error of estimation is given by

$$\bar{y} \pm 2\sqrt{\hat{V}(\bar{y})} = 8801 \pm 2\sqrt{653,785} = 8801 \pm 1617$$

The best estimate of the average per-capita income is \$8801, and the error of estimation should be less than \$1617 with probability close to .95. This bound on the error of estimation is rather large; it could be reduced by sampling more clusters and, consequently, increasing the sample size.

Since the $\bar{y}$ given by Equation (8.1) is a ratio estimator, we can use alternative calculation formulas derived in Chapter 6. A summary of the data in Table 8.1 is as follows:

	N	Mean	SD
y	25	53.16	21.78
m	25	6.040	2.317

Correlation of y and m = .303.

Here, the measurements are in thousands (the last three zeros having been suppressed). Now, $\hat{V}(\bar{y})$ can be calculated as

$$\hat{V}(\bar{y}) = \left(\frac{N-n}{Nn}\right)\left(\frac{1}{\bar{m}^2}\right)[s_y^2 + r^2 s_m^2 - 2r\hat{\rho}s_y s_m]$$

where

$$s_m^2 = \frac{1}{m-1}\sum_{i=1}^{m}(m_i - \bar{m})^2$$

and the r used here is the $\bar{y}$ of Equation (8.1). This computation yields

$$\hat{V}(\bar{y}) = \left[\frac{390}{415(25)}\right]\frac{1}{(6.04)^2}[(21.78)^2 + (8.801)^2(2.371)^2$$
$$- 2(8.801)(.303)(21.78)(2.371)]$$
$$= .6524$$

and

$$2\sqrt{\hat{V}(\bar{y})} = 1.616$$

which is equivalent to \$1616. This is nearly the same bound as calculated above.

Recall that the ratio estimator is nearly unbiased when the plot of y versus m shows points falling close to a straight line through the origin. A plot of the data from Table 8.1 is shown in Figure 8.1.

FIGURE 8.1 Data from Table 8.1

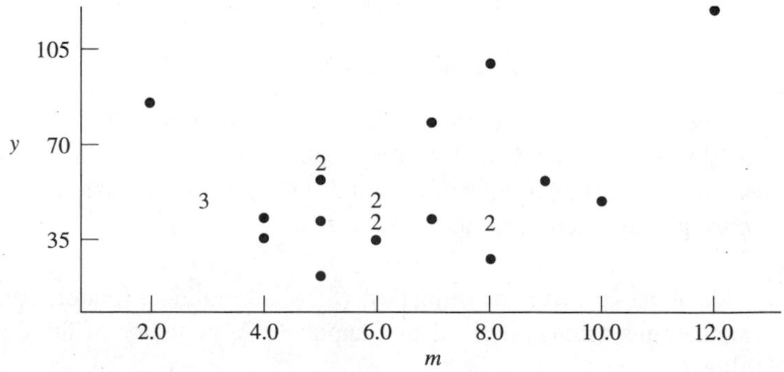

Although there is something of a linear trend here, it does not appear to be strong ($\hat{\rho} = .303$). Even so, the relative bias, approximated by

$$\left(\frac{N-n}{Nn}\right)\left[\left(\frac{s_m}{\bar{m}}\right)^2 - \hat{\rho}\left(\frac{s_m}{\bar{m}}\right)\left(\frac{s_y}{\bar{y}}\right)\right] = .004$$

is small, and the ratio estimate of μ should be reasonably good.

The population total τ is now $M\mu$ because M denotes the total number of elements in the population. Consequently, as in simple random sampling, $M\bar{y}$ provides an estimator of τ.

Estimator of the population total τ:

$$M\bar{y} = M\frac{\sum_{i=1}^{n} y_i}{\sum_{i=1}^{n} m_i} \tag{8.4}$$

Estimated variance of $M\bar{y}$:

$$\hat{V}(M\bar{y}) = M^2\hat{V}(\bar{y}) = N^2\left(\frac{N-n}{Nn}\right)s_r^2 \tag{8.5}$$

Note that the estimator $M\bar{y}$ is useful only if the number of elements in the population, M, is known.

Example 8.3 Use the data in Table 8.1 to estimate the total income of all residents of the city and place a bound on the error of estimation. There are 2500 residents of the city.

Solution

The sample mean $\bar{y}$ is calculated to be \$8801 in Example 8.2. Thus, the estimate of τ is

$$M\bar{y} = 2500(8801) = \$22{,}002{,}500$$

The quantity $\hat{V}(\bar{y})$ is calculated by the method used in Example 8.2, except that $\bar{M}$ can now be used in place of $\bar{m}$. The estimate of τ with a bound on the error of estimation is

$$M\bar{y} \pm 2\sqrt{\hat{V}(M\bar{y})} = M\bar{y} \pm 2\sqrt{M^2\hat{V}(\bar{y})}$$

$$22{,}002{,}500 \pm 2\sqrt{(2500)^2(653{,}785)}$$

$$22{,}002{,}500 \pm 4{,}042{,}848$$

Again, this bound on the error of estimation is large, and it could be reduced by increasing the sample size.

Often the number of elements in the population is not known in problems for which cluster sampling is appropriate. Thus, we cannot use the estimator $M\bar{y}$, but we can form another estimator of the population total that does not depend on M. The quantity $\bar{y}_t$, given by

$$\bar{y}_t = \frac{1}{n}\sum_{i=1}^{n} y_i \tag{8.6}$$

is the average of the cluster totals for the n sampled clusters. Hence, $\bar{y}_t$ is an unbiased estimator of the average of the N cluster totals in the population. By the same reasoning as employed in Chapter 4, $N\bar{y}_t$ is an unbiased estimator of the sum of the cluster totals or, equivalently, of the population total τ.

For example, it is highly unlikely that the number of adult males in a city would be known, and hence the estimator $N\bar{y}_t$, rather than $M\bar{y}$, would have to be used to estimate τ.

Estimator of the population total τ, which does not depend on M:

$$N\bar{y}_t = \frac{N}{n}\sum_{i=1}^{n} y_i \tag{8.7}$$

Estimated variance of $N\bar{y}$:

$$\hat{V}(N\bar{y}_t) = N^2\hat{V}(\bar{y}_t) = N^2\left(\frac{N-n}{Nn}\right)s_t^2 \tag{8.8}$$

where

$$s_t^2 = \frac{\sum_{i=1}^{n}(y_i - \bar{y}_t)^2}{n-1} \tag{8.9}$$

If there is a large amount of variation among the cluster sizes and if cluster sizes are highly correlated with cluster totals, the variance of $N\bar{y}_t$ [Equation (8.8)] is generally larger than the variance of $M\bar{y}$ [Equation (8.5)]. The estimator $N\bar{y}_t$ does not use the information provided by the cluster sizes $m_1, m_2, \ldots, m_n$ and hence may be less precise.

Example 8.4

Use the data of Table 8.1 to estimate the total income of all residents of the city if M is not known. Place a bound on the error of estimation.

Solution

Example 8.1 gives $N = 415$. From Equation (8.7) and Table 8.1, the estimate of the total income τ is

$$N\bar{y}_t = \frac{N}{n}\sum_{i=1}^{n} y_i = \frac{415}{25}(1{,}329{,}000) = \$22{,}061{,}400$$

This figure is fairly close to the estimate given in Example 8.3.

To place a bound on the error of estimation, we first note that

$$s_t^2 = \frac{\sum_{i=1}^{n}(y_i - \bar{y}_t)^2}{n-1} = (21{,}784)^2$$

from the data summary in Example 8.2. Then the estimate of the total income of all residents of the city, with a bound on the error of estimation, is $N\bar{y}_t \pm 2\sqrt{\hat{V}(N\bar{y}_t)}$. Substituting into Equation (8.8), we calculate

$$N\bar{y}_t \pm 2\sqrt{N^2\left(\frac{N-n}{Nn}\right)s_t^2}$$

$$22{,}061{,}400 \pm 2\sqrt{(415)^2\left[\frac{415-25}{(415)(25)}\right](21{,}784)^2}$$

$$22{,}061{,}400 \pm 3{,}505{,}532$$

The bound on the error of estimation is slightly smaller than the bound for the estimator $M\bar{y}$ (Example 8.3), partly because the cluster sizes are not highly correlated with the cluster total in this example. In other words, the cluster sizes are providing little information on cluster totals; hence, the unbiased estimator $N\bar{y}_t$ appears to be better than the estimator $M\bar{y}$.

8.4

EQUAL CLUSTER SIZES; COMPARISON TO SIMPLE RANDOM SAMPLING

For a more precise study of the relationships between cluster sampling and simple random sampling, we confine our discussion to the case in which all of the m_i's are equal to a common value—say, m. We assume this to be true for the entire population of clusters, as in the case of sampling cartons of canned foods where each carton contains exactly 24 cans. In this case, $M = Nm$, and the total sample size is nm elements (n clusters of m elements each).

The estimators of μ and τ possess special properties when all cluster sizes are equal (that is, $m_1 = m_2 = \cdots = m_N$). First, the estimator $\bar{y}$, given by Equation (8.1), is an unbiased estimator of the population mean μ. Second, $\hat{V}(\bar{y})$, given by Equation (8.2), is an unbiased estimator of the variance of $\bar{y}$. Finally, the two estimators, $M\bar{y}$ and $N\bar{y}_t$, of the population total τ are equivalent.

The estimator (8.1) of the population mean per element will be denoted in this equal cluster size case by $\bar{\bar{y}}_c$, and it becomes

$$\bar{\bar{y}}_c = \frac{1}{m}\left[\frac{1}{n}\sum_{i=1}^{n} y_i\right] = \frac{1}{mn}\sum_{i=1}^{n}\sum_{j=1}^{m} y_{ij}$$

where y_{ij} denotes the jth-sample observation from cluster i. Note that $\bar{\bar{y}}_c$ can be thought of as the overall average of all nm-sample measurements, or as the average of the sampled cluster totals divided by m. From the latter point of view, it is easy to see that

$$\hat{V}(\bar{\bar{y}}_c) = \left(\frac{N-n}{N}\right)\left(\frac{1}{nm^2}\right)\left(\frac{1}{n-1}\right)\sum_{i=1}^{n}(y_i - \bar{y}_t)^2$$

where

$$\bar{y}_t = \frac{1}{n}\sum_{i=1}^{n} y_i = m\bar{\bar{y}}_c$$

If we let the sample average for cluster i be denoted by $\bar{y}_i$, we have $\bar{y}_i = y_i/m$, or $y_i = m\bar{y}_i$. We can then write

$$\frac{1}{m^2 n(n-1)}\sum_{i=1}^{n}(y_i - \bar{y}_t)^2 = \frac{1}{m^2 n(n-1)}\sum_{i=1}^{n}(m\bar{y}_i - m\bar{\bar{y}}_c)^2$$
$$= \frac{1}{n(n-1)}\sum_{i=1}^{n}(\bar{y}_i - \bar{\bar{y}}_c)^2$$

To simplify the variance computations and to explore the relationship between cluster sampling and simple random sampling, we use a sum-of-squares identity similar to that developed in classical analysis of variance (ANOVA) arguments. It can be shown that

$$\sum_{i=1}^{n}\sum_{j=1}^{m}(y_{ij} - \bar{\bar{y}}_c)^2 = \sum_{i=1}^{n}\sum_{j=1}^{m}(y_{ij} - \bar{y}_i)^2 + \sum_{i=1}^{n}\sum_{j=1}^{m}(\bar{y}_i - \bar{\bar{y}}_c)^2$$
$$= \sum_{i=1}^{n}\sum_{j=1}^{m}(y_{ij} - \bar{y}_i)^2 + m\sum_{i=1}^{n}(\bar{y}_i - \bar{\bar{y}}_c)^2$$

The three terms, from the left, are named *total sum of squares* (SST), *within-cluster sum of squares* (SSW) and *between-cluster sum of squares* (SSB). The above equality is then

$$\text{SST} = \text{SSW} + \text{SSB}$$

With appropriate divisors, these sums of squares become the usual mean squares of ANOVA. Thus, the *between-cluster mean square* MSB is given by

$$\text{MSB} = \frac{\text{SSB}}{n-1} = \frac{m}{n-1}\sum_{i=1}^{n}(\bar{y}_i - \bar{\bar{y}}_c)^2$$

and the *within-cluster mean square* MSW is given by

$$\text{MSW} = \frac{\text{SSW}}{n(m-1)} = \frac{1}{n(m-1)} \sum_{i=1}^{n} \sum_{j=1}^{m} (y_{ij} - \bar{y}_i)^2$$

It now follows that

$$\hat{V}(\bar{\bar{y}}_c) = \left(\frac{N-n}{N}\right) \frac{1}{nm} \text{MSB}$$

Example 8.5 The circulation manager of a newspaper wishes to estimate the average number of newspapers purchased per household in a given community. Travel costs from household to household are substantial. Therefore, the 4000 households in the community are listed in 400 geographical clusters of 10 households each, and a simple random sample of 4 clusters is selected. Interviews are conducted, with the results as shown in the accompanying table. Estimate the average number of newspapers per household for the community and place a bound on the error of estimation.

Cluster	Number of newspapers										Total
1	1	2	1	3	3	2	1	4	1	1	19
2	1	3	2	2	3	1	4	1	1	2	20
3	2	1	1	1	1	3	2	1	3	1	16
4	1	1	3	2	1	5	1	2	3	1	20

Solution

From Equation (8.1),

$$\bar{y} = \frac{\sum_{i=1}^{n} y_i}{\sum_{i=1}^{n} m_i}$$

When $m_1 = m_2 = \cdots = m_n = m$, the equation becomes

$$\bar{\bar{y}}_c = \frac{\sum_{i=1}^{n} y_i}{nm} = \frac{19 + 20 + 16 + 20}{4(10)} = 1.875$$

Standard ANOVA computations were performed (using MINITAB) on the data with the following results:

ANOVA

Source	DF	SS	MS
Factor	3	1.07	.36
Error	36	43.30	1.20
Total	39	44.38	

In this output, "Factor" denotes the between-cluster calculations, and "Error" denotes the within-cluster calculations. Thus, MSB = .36, and MST = 1.20. It follows that

$$\hat{V}(\bar{\bar{y}}_c) = \left(\frac{N-n}{N}\right)\left(\frac{1}{nm}\right)\text{MSB}$$
$$\left(\frac{396}{400}\right)\frac{1}{4(10)}(.36)$$
$$= .0089$$

and

$$2\sqrt{\hat{V}(\bar{\bar{y}}_c)} = .19$$

Therefore, our best estimate of the number of newspapers per household is $1.88 \pm .19$.

How can we compare the precision of cluster sampling with that of simple random sampling? If we had taken the nm observations in a simple random sample and computed the mean $\bar{y}$ and variance s^2, then we would have

$$\hat{V}(\bar{y}) = \left(\frac{Nm - nm}{Nm}\right)\cdot\frac{s^2}{nm} = \left(\frac{N-n}{N}\right)\frac{s^2}{nm}$$

since there would be Nm total observations in the population. Thus, we can measure the relative efficiency of $\bar{\bar{y}}_c$ to $\bar{y}$ by comparing MSB to s^2. But we did *not* take a simple random sample; we took a cluster sample, so s^2 is not available. Fortunately, it turns out that we can approximate s^2 (the variance we would have obtained in a simple random sample) from quantities available in the cluster sample results. This approximation is

$$\hat{s}^2 = \frac{N(m-1)\text{MSW} + (N-1)\text{MSB}}{Nm-1}$$
$$\approx \frac{1}{m}[(m-1)\text{MSW} + \text{MSB}]$$

when N is large.

Using the calculations from Example 8.5, we see that

$$\hat{s}^2 \approx \frac{1}{10}[(9)(1.20) + .36] = 1.12$$

The estimated relative efficiency of $\bar{\bar{y}}_c$ to $\bar{y}$ is thus

$$\widehat{\text{RE}}(\bar{\bar{y}}_c/\bar{y}) = \frac{\hat{s}^2}{\text{MSB}} = \frac{1.12}{.36} = 3.11$$

In this case, cluster sampling is more efficient because there is so little variation between clusters (each cluster seems to be fairly representative of the entire population). This is somewhat unusual because, in most cases of naturally occurring clusters, cluster sampling will be less efficient than simple random sampling.

As another example, we sampled $n = 4$ clusters of $m = 20$ contiguous random digits from a random number table. If the goal is to estimate the mean of the random digits (known to be 4.5 in this case), how should our cluster sample compare to taking $80 = 4(20)$ random digits in a simple random sample? Since the clusters themselves contain randomly generated digits, we would expect the relative efficiency to be close to 1.

The ANOVA results for our sample were as follows:

ANOVA

Source	DF	SS	MS
Factor	3	40.54	13.51
Error	76	610.85	8.04
Total	79	651.39	

From this,

$$\hat{s}^2 = \frac{1}{20}[19(8.04) + 13.51] = 8.31$$

and

$$\widehat{\text{RE}}(\bar{\bar{y}}_c/\bar{y}) = \frac{8.31}{13.51} = .62$$

This is a little lower than expected, but not far from 1.0.

We will continue this discussion of comparisons between cluster sampling and simple random sampling in Chapter 9.

8.5

SELECTING THE SAMPLE SIZE FOR ESTIMATING POPULATION MEANS AND TOTALS

The quantity of information in a cluster sample is affected by two factors, the number of clusters and the relative cluster size. We have not encountered the latter factor in any of the sampling procedures discussed previously. In the problem of estimating the number of homes with inadequate fire insurance in a state, the clusters could be counties, voting districts, school districts, communities, or any other convenient grouping of homes. As we have already seen, the size of the bound on the error of estimation depends crucially upon the variation among the cluster *totals*. Thus, in attempting to achieve small bounds on the error of estimation, select clusters with as little variation as possible among these totals. We will now assume that the cluster size (sampling unit) has been chosen and will consider only the problem of choosing the number of clusters, n.

From Equation (8.2), the estimated variance of $\bar{y}$ is

$$\hat{V}(\bar{y}) = \frac{N-n}{Nn\bar{M}^2}(s_r^2)$$

where

$$s_r^2 = \frac{\sum_{i=1}^{n}(y_i - \bar{y}m_i)^2}{n-1} \tag{8.10}$$

The actual variance of $\bar{y}$ is approximately

$$V(\bar{y}) = \frac{N-n}{Nn\bar{M}^2}(\sigma_r^2) \tag{8.11}$$

where σ_r^2 is the population quantity estimated by s_r^2.

Because we do not know σ_r^2 or the average cluster size $\bar{M}$, choice of the sample size—that is, the number of clusters necessary to purchase a specified quantity of information concerning a population parameter—is difficult. We overcome this difficulty by using the same method we used for ratio estimation. That is, we use an estimate of σ_r^2 and $\bar{M}$ available from a prior survey, or we select a preliminary sample containing n' elements. Estimates of σ_r^2 and $\bar{M}$ can be computed from the preliminary sample and used to acquire an approximate total sample size n. Thus, as in all problems of selecting a sample size, we equate 2 SD of our estimator to a bound on the error of estimation, B. This bound is chosen by the experimenter and represents the maximum error that he or she is willing to tolerate. That is,

$$2\sqrt{V(\bar{y})} = B$$

Using Equation (8.11), we can solve for n.

We obtain similar results when using $M\bar{y}$ to estimate the population total τ because $V(M\bar{y}) = M^2V(\bar{y})$.

Approximate sample size required to estimate μ with a bound B on the error of estimation:

$$n = \frac{N\sigma_r^2}{ND + \sigma_r^2} \tag{8.12}$$

where σ_r^2 is estimated by s_r^2 and $D = (B^2\bar{M}^2)/4$.

Example 8.6 Suppose the data in Table 8.1 represent a preliminary sample of incomes in the city. How large a sample should be taken in a future survey in order to estimate the average per-capita income μ with a bound of \$500 on the error of estimation?

Solution

To use Equation (8.12), we must estimate σ_r^2; the best estimate available is s_r^2, which can be calculated by using the data in Table 8.1. Using the calculations in Example 8.2, we have

$$s_r^2 = \frac{\sum_{i=1}^{n} (y_i - \bar{y}m_i)^2}{n-1} = (25{,}189)^2$$

Quantity $\bar{M}$ can be estimated by $\bar{m} = 6.04$ calculated from Table 8.1. Then D is approximately

$$\frac{B^2\bar{m}^2}{4} = \frac{(500)^2(6.04)^2}{4} = (62{,}500)(6.04)^2$$

Using Equation (8.12) yields

$$n = \frac{N\sigma_r^2}{ND + \sigma_r^2} = \frac{415(25{,}189)^2}{415(6.04)^2(62{,}500) + (25{,}189)^2} = 166.58$$

Thus, 167 clusters should be sampled.

Approximate size required to estimate τ, using $M\bar{y}$, with a bound B on the error of estimation:

$$n = \frac{N\sigma_r^2}{ND + \sigma_r^2} \tag{8.13}$$

where σ_r^2 is estimated by s_r^2 and $D = B^2/4N^2$.

Example 8.7 Again using the data in Table 8.1 as a preliminary sample of incomes in the city, how large a sample is necessary to estimate the total income of all residents, τ, with a bound of \$1,000,000 on the error of estimation? There are 2500 residents of the city ($M = 2500$).

Solution

We use Equation (8.13) and estimate σ_r^2 by

$$s_r^2 = (25{,}189)^2$$

as in Example 8.6. When estimating τ, we use

$$D = \frac{B^2}{4N^2} = \frac{(1{,}000{,}000)^2}{4(415)^2}$$

$$ND = \frac{(1{,}000{,}000)^2}{4(415)} = 602{,}409{,}000$$

Then using Equation (8.13) gives

$$n = \frac{N\sigma_r^2}{ND + \sigma_r^2} = \frac{415(25{,}189)^2}{602{,}409{,}000 + (25{,}189)^2} = 212.88$$

Thus, 213 clusters should be sampled to estimate the total income with a bound of \$1,000,000 on the error of estimation.

The estimator $N\bar{y}_t$, shown in Equation (8.7), is used to estimate τ when M is unknown. The estimated variance of $N\bar{y}_t$, shown in Equation (8.8), is

$$\hat{V}(N\bar{y}_t) = N^2\left(\frac{N-n}{Nn}\right)s_t^2$$

where

$$s_t^2 = \frac{\sum_{i=1}^{n}(y_i - \bar{y}_t)^2}{n-1}$$

Thus, the population variance of $N\bar{y}_t$ is

$$V(N\bar{y}_t) = N^2 V(\bar{y}_t) = N^2\left(\frac{N-n}{Nn}\right)\sigma_t^2 \qquad (8.14)$$

where σ_t^2 is the population quantity estimated by s_t^2.

Estimation of τ with a bound of B units on the error of estimation leads to the following equation:

$$2\sqrt{V(N\bar{y}_t)} = B$$

Using Equation (8.14), we can solve for n.

Approximate sample size required to estimate τ, using $N\bar{y}_t$ with a bound B on the error of estimation:

$$n = \frac{N\sigma_t^2}{ND + \sigma_t^2} \qquad (8.15)$$

where σ_t^2 is estimated by s_t^2 and $D = B^2/4N^2$.

Example 8.8

Assume the data of Table 8.1 are from a preliminary study of incomes in the city and M is not known. How large a sample must be taken to estimate the total income of all residents, τ, with a bound of \$1,000,000 on the error of estimation?

Solution

The quantity σ_t^2 must be estimated by s_t^2, which is calculated from the data of Table 8.1. Using the calculations of Example 8.4 gives

$$s_t^2 = \frac{\sum_{i=1}^{n} (y_i - \bar{y}_t)^2}{n - 1} = (21{,}784)^2$$

The bound on the error of estimation is $B = \$1{,}000{,}000$. Hence,

$$D = \frac{B^2}{4N^2} = \frac{(1{,}000{,}000)^2}{4(415)^2}$$

From Equation (8.16),

$$n = \frac{N\sigma_t^2}{ND + \sigma_t^2} = \frac{415(21{,}784)^2}{415(1{,}000{,}000)^2/4(415)^2 + (21{,}784)^2} = 182.88$$

Thus, a sample of 183 clusters must be taken to have a bound of \$1,000,000 on the error of estimation.

8.6

ESTIMATION OF A POPULATION PROPORTION

Suppose an experimenter wishes to estimate a population proportion, or fraction, such as the proportion of houses in a state with inadequate plumbing or the proportion of corporation presidents who are college graduates. The best estimator of the population proportion p is the sample proportion $\hat{p}$. Let a_i denote the total number of elements in cluster i that possess the characteristic of interest. Then the proportion of elements in the sample of n clusters possessing the characteristic is given by

$$\hat{p} = \frac{\sum_{i=1}^{n} a_i}{\sum_{i=1}^{n} m_i}$$

where m_i is the number of elements in the ith cluster, $i = 1, 2, \ldots, n$. Note that $\hat{p}$ has the same form as $\bar{y}$ [see Equation (8.1)], except that y_i is replaced by a_i. The estimated variance of $\hat{p}$ is similar to that of $\bar{y}$.

Estimator of the population proportion p:

$$\hat{p} = \frac{\sum_{i=1}^{n} a_i}{\sum_{i=1}^{n} m_i} \tag{8.16}$$

Estimated variance of $\hat{p}$:

$$\hat{V}(\hat{p}) = \left(\frac{N - n}{Nn\bar{M}^2}\right) s_p^2 \tag{8.17}$$

where

$$s_p^2 = \frac{\sum_{i=1}^{n} (a_i - \hat{p}m_i)^2}{n - 1} \tag{8.18}$$

The variance formula, (8.17), is a good estimator only when the sample size n is large—say, $n \geq 20$. If $m_1 = m_2 = \cdots = m_N$, then $\hat{p}$ is an unbiased estimator of p, and $\hat{V}(\hat{p})$, shown in Equation (8.17), is an unbiased estimator of the actual variance of $\hat{p}$ for any sample size.

Example 8.9

In addition to being asked about their income, the residents of the sample survey of Example 8.2 are asked whether they rent or own their homes. The results are given in Table 8.2. Use the data in Table 8.2 to estimate the proportion of residents who live in rented housing. Place a bound on the error of estimation.

Solution

The best estimate of the population of renters is $\hat{p}$, shown in Equation (8.16), where

$$\hat{p} = \frac{\sum_{i=1}^{n} a_i}{\sum_{i=1}^{n} m_i} = \frac{72}{151} = \frac{2.88}{6.04} = .48$$

$$s_p^2 = (0.726)^2$$

TABLE 8.2 **Number of renters**

Cluster	Residents, m	Renters, a	$a_i - \hat{p}m_i$
1	8	4	0.16000
2	12	7	1.24000
3	4	1	− 0.92000
4	5	3	0.60000
5	6	3	0.12000
6	6	4	1.12000
7	7	4	0.64000
8	5	2	− 0.40000
9	8	3	− 0.84000
10	3	2	0.56000
11	2	1	0.04000
12	6	3	0.12000
13	5	2	− 0.40000
14	10	5	0.20000
15	9	4	− 0.32000
16	3	1	− 0.44000
17	6	4	1.12000
18	5	2	− 0.40000
19	5	3	0.60000
20	4	1	− 0.92000
21	6	3	0.12000
22	8	3	− 0.84000
23	7	4	0.64000
24	3	0	− 1.44000
25	8	3	− 0.84000

	N	Mean	Median	SD
Resident	25	6.040	6.000	2.371
Renters	25	2.880	3.000	1.509
$a_i - \hat{p}m_i$	25	− 0.019	0.120	0.726

The quantity $\bar{M}$ is estimated by $\bar{m}$, where

$$\bar{m} = \frac{\sum_{i=1}^{n} m_i}{n} = \frac{151}{25} = 6.04$$

Then from Equation (8.17),

$$\hat{V}(\hat{p}) = \left(\frac{N - n}{Nn\bar{M}^2}\right)s_p^2$$

$$= \frac{(415 - 25)}{415(25)(6.04)^2}(.726)^2 = 0.00054$$

The estimate of p with a bound on the error is

$$\hat{p} \pm 2\sqrt{\hat{V}(\hat{p})} = .48 \pm 2\sqrt{.00054} = .48 \pm .05$$

Thus, the best estimate of the proportion of people who rent homes is .48. The error of estimation should be less than .05 with probability of approximately .95. The fact that we are estimating a proportion here saves us no calculation effort as it did in the case of simple random sampling, and so other versions of $\hat{V}(\hat{p})$ may be considered. A summary of the data in Table 8.2 is given by

	N	Mean	SD
a	25	2.880	1.509
m	25	6.040	2.371

Correlation of *a* and *m* = .886.

From this summary, it is easy to calculate

$$\hat{V}(\hat{p}) = \left(\frac{N-n}{Nn}\right)\left(\frac{1}{\bar{m}}\right)^2 [s_a^2 + \hat{p}^2 s_m^2 - 2\hat{p}\hat{\rho} s_a s_m]$$

$$= \left(\frac{390}{415(25)}\right)\left(\frac{1}{6.04}\right)^2 [(1.509)^2 + (.477)^2(2.371)^2$$

$$- 2(.477)(.886)(1.509)(2.371)]$$

$$= .00055.$$

The result agrees with the previous calculation.

A plot of *a* versus *m* for the data in Table 8.2 is given in Figure 8.2. Note that the linearity is fairly strong here.

FIGURE 8.2 Data from Table 8.2

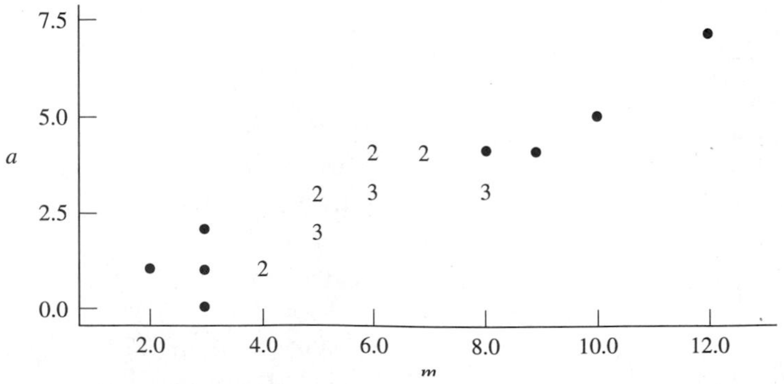

8.7

SELECTING THE SAMPLE SIZE FOR ESTIMATING PROPORTIONS

Estimation of the population proportion p with a bound of B units on the error of estimation implies that the experimenter wants

$$2\sqrt{V(\hat{p})} = B$$

This equation can be solved for n, and the solution is similar to Equation (8.12). That is,

$$n = \frac{N\sigma_p^2}{ND + \sigma_p^2}$$

where $D = B^2\bar{M}^2/4$, and σ_p^2 is estimated by

$$s_p^2 = \frac{\sum_{i=1}^{n}(a_i - \hat{p}m_i)^2}{n-1}$$

Example 8.10 The data in Table 8.2 are out of date. A new study will be conducted in the same city for the purpose of estimating the proportion p of residents who rent their homes. How large a sample should be taken to estimate p with a bound of .04 on the error of estimation?

Solution

The best estimate of σ_p^2 is s_p^2, which is calculated by using data from Table 8.2:

$$s_p^2 = \frac{\sum_{i=1}^{n}(a_i - \hat{p}m_i)^2}{n-1} = (.726)^2 = .527$$

Quantity $\bar{M}$ is estimated by $\bar{m} = 6.04$. Also, D is approximated by

$$\frac{B^2\bar{m}^2}{4} = \frac{(.04)^2(6.04)^2}{4} = .0146$$

Then

$$n = \frac{N\sigma_p^2}{ND + \sigma_p^2} = \frac{(415)(.527)}{(415)(.0146) + .527} = 33.20$$

Thus 34 clusters should be sampled to estimate p with a bound of .04 on the error of estimation.

8.8

CLUSTER SAMPLING COMBINED WITH STRATIFICATION

As is the case with all other sampling methods, cluster sampling can be combined with stratified sampling, in the sense that the population may be divided into L strata and a cluster sample can then be selected from each stratum.

Recall that Equation (8.1) has the form of a ratio estimator and can be thought of as the ratio of an estimator of the average cluster total to an estimator of the average cluster size. Thinking in terms of ratio estimators, then, we have two ways to form the estimator of a population mean across strata, the separate estimator and the combined estimator. A little investigation will show that if the separate estimator is employed, the total number of elements in each stratum must be known in order to assign proper stratum weights. Since these quantities are usually unknown, we will investigate only the combined form of the ratio estimator in the context of cluster sampling.

Instead of presenting formidable-looking general formulas, we will illustrate the technique with a numerical example.

Example 8.11 Let the data of Table 8.1 form the sample of stratum 1, with, as in Example 8.2, $N_1 = 415$ and $n_1 = 25$. A smaller neighboring city is taken to be stratum 2. For stratum 2, $n_2 = 10$ blocks are to be sampled from $N_2 = 168$. Estimate the average per-capita income in the two cities combined and place a bound on the error of estimation, given the additional data shown in the accompanying table.

Cluster i	Number of residents, m_i	Total income per cluster, y_i
1	2	\$18,000
2	5	52,000
3	7	68,000
4	4	36,000
5	3	45,000
6	8	96,000
7	6	64,000
8	10	115,000
9	3	41,000
10	1	12,000

Solution

The average cluster totals in the respective samples are $\bar{y}_{t1} = 53{,}160$ and $\bar{y}_{t2} = 54{,}700$. The average cluster sizes in the respective samples are $\bar{m}_1 = 6.04$ and $\bar{m}_2 = 4.90$. The estimate of the population average cluster total is then

$$\frac{1}{N}(N_1\bar{y}_{t1} + N_2\bar{y}_{t2})$$

while the estimate of the average cluster size is

$$\frac{1}{N}(N_1\bar{m}_1 + N_2\bar{m}_2)$$

An estimate of the population mean per element is then

$$\bar{y}_c = \frac{N_1\bar{y}_{t1} + N_2\bar{y}_{t2}}{N_1\bar{m}_1 + N_2\bar{m}_2}$$

and this equation does have the form of a combined ratio estimate. Analogous to the variance used in Section 6.6, the variance of $\bar{y}_c$ can be estimated by

$$\bar{V}(\bar{y}_c) = \frac{1}{M^2}\left\{\frac{N_1(N_1 - n_1)}{n_1}s_{c1}^2 + \frac{N_2(N_2 - n_2)}{n_2}s_{c2}^2\right\}$$

where M is the total number of elements in the population and can be estimated by $N_1\bar{m}_1 + N_2\bar{m}_2$ if it is not known. The first variance, s_{c1}^2, is the variance of terms $(y_i - \bar{y}_c m_i)$ from stratum 1. The second variance, s_{c2}^2, is the variance of terms $(y_i - \bar{y}_c m_i)$ from stratum 2.

For the data given in the table,

$$\bar{y}_c = \frac{415(53{,}160) + 168(54{,}700)}{415(6.04) + 168(4.90)} = 9385$$

For stratum 1,

$$s_{c1}^2 = \left(\frac{1}{n_1 - 1}\right)\sum_{i=1}^{n_1}[(y_i - \bar{y}_{t1}) - \bar{y}_c(m_i - \bar{m}_1)]^2 = (25{,}998)^2$$

and for stratum 2,

$$s_{c2}^2 = \left(\frac{1}{n_2 - 1}\right)\sum_{i=1}^{n_2}[(y_i - \bar{y}_{t2}) - \bar{y}_c(m_i - \bar{m}_2)]^2 = (8657)^2$$

Since

$$N_1\bar{m}_1 + N_2\bar{m}_2 = 3329.8$$

it follows that

$$\hat{V}(\bar{y}_c) = 412{,}563.8$$

and

$$2\sqrt{\hat{V}(\bar{y}_c)} = 1285$$

Thus, the average per-capita income for the two cities combined is $9385 ± $1285. We see that the bound on the error of estimation is slightly smaller than the bound for stratum 1 alone, as found in Example 8.2.

8.9

CLUSTER SAMPLING WITH PROBABILITIES PROPORTIONAL TO SIZE

Up to this point, all the sampling designs that have been discussed involve random samples of elements or clusters of elements. It turns out that this is not always the best way to conduct the sampling. Sometimes estimates can be improved by varying the probabilities with which units are sampled from the population. Suppose, for example, we want to estimate the number of job openings in a city by sampling industrial firms from within that city. Typically, many such firms will be small and employ few workers, whereas some firms will be large and employ many workers. In a simple random sample of firms, size of firms is not taken into account, and a typical sample will consist of mostly small firms. The number of job openings, however, is heavily influenced by the large firms. Thus, we should be able to improve the estimate of the number of job openings by giving the large firms a greater chance to appear in the sample. A method for accomplishing this is called *sampling with probabilities proportional to size*, or *pps sampling*.

Suppose the population consists of the set of elements $\{u_1, u_2, \ldots, u_N\}$ and a sample of n elements is to be selected *with replacement*, so that the sampled elements are selected independently from one another. Further, let $\{z_1, z_2, \ldots, z_N\}$ represent the respective probabilities of selection for the population elements. That is, z_i is the probability that u_i is selected on any one draw. For the case of random sampling with replacement, each z_i is simply $1/N$.

If n elements are sampled in this way and these elements are denoted by $\{y_1, y_2, \ldots, y_n\}$, then each y is really one of the u's. To simplify notation, we can think of rearranging the z_i's so that, after the sample is selected, z_i is the probability of selecting the value from the u's that y_i actually assumes.

An unbiased estimator of the population total is given by

$$\hat{\tau}_{\text{pps}} = \frac{1}{n}\sum_{i=1}^{n}\left(\frac{y_i}{z_i}\right)$$

To see that this is a reasonable estimator, think of random sampling with replacement and replace z_i with $1/N$. Then, the estimator becomes identical in form to the one introduced in Chapter 4.

The estimator $\hat{\tau}_{\text{pps}}$ is unbiased for any choices of z_i, but it is clearly in the best interest of the experimenter to choose these z_i's so that the variances of the estimators are as small as possible. How should this choice be made? Suppose, for the moment, that the value of y_i is known for each of the N units in the population. Thus, the population total τ is also known. Under these conditions, we can select each unit for the sample with probability proportional to its actual measured value y_i, assuming all measurements are positive. That is, we can make $z_i = y_i/\tau$.

With $z_i = y_i/\tau$ for each sampled item, $\hat{\tau}_{\text{pps}}$ becomes

$$\begin{aligned}\hat{\tau}_{\text{pps}} &= \frac{1}{n}\sum_{i=1}^{n}\frac{y_i}{z_i} = \frac{1}{n}\sum_{i=1}^{n}\frac{y_i}{(y_i/\tau)} \\ &= \frac{1}{n}\sum_{i=1}^{n}\tau = \frac{1}{n}(n\tau) = \tau\end{aligned}$$

Thus, every $\hat{\tau}_{\text{pps}}$ estimates τ exactly. In addition,

$$\begin{aligned}\hat{V}(\hat{\tau}_{\text{pps}}) &= \frac{1}{n(n-1)}\sum_{i=1}^{n}\left(\frac{y_i}{z_i} - \hat{\tau}_{\text{pps}}\right)^2 \\ &= \frac{1}{n(n-1)}\sum_{i=1}^{n}\left(\frac{y_i}{y_i/\tau} - \tau\right)^2 \\ &= \frac{1}{n(n-1)}\sum_{i=1}^{n}(\tau - \tau)^2 = 0\end{aligned}$$

which again shows that we have a precise estimator.

Now to know the values y_i for every unit in the population before sampling is impossible. (If they were known, no sampling would be necessary.) Hence, the choice $z_i = y_i/\tau$ is not possible, but it does provide a criterion for selecting z_i's that can be used in sampling. The best practical way to choose the z_i's is to choose them proportional to a known measurement that is highly correlated with y_i. In the problem of estimating total number of job openings, firms can be sampled with probabilities proportional to their total work force, which should be known fairly accurately before the sample is selected. The number of job openings per firm is not known before sampling, but it should be highly correlated with the total number of workers in the firm.

To show how pps sampling works numerically, assume a population of $N=4$ elements, $\{1, 2, 3\ 4\}$. For simple random samples of size $n=2$, $E(\bar{y})=2.5$ and $V(\bar{y})=\frac{5}{12}=.417$. Suppose we decide to sample $n=2$ elements with varying probabilities and choose $z_1=.1$, $z_2=.1$, $z_3=.4$, and $z_4=.4$. To accomplish this sampling, we can choose a random digit from the random number table and take our first sampled element to be

1 if the random digit is 0
2 if the random digit is 1
3 if the random digit is 2, 3, 4, or 5
4 if the random digit is 6, 7, 8, or 9

The process is then repeated for the second sampled element. (Note that the same element can be selected twice, which is not possible in simple random sampling.) This choice for z_1, z_2, z_3, and z_4 gives the smaller values in the population, (1, 2), the smaller chance of getting selected in the sample, and the larger values, (3, 4), have an appreciably larger chance of getting selected. Note that these probabilities are not exactly proportional to size, but they do tend in that direction.

Table 8.3 contains a listing of the ten possible samples, the probability of obtaining each sample, and the $\hat{\tau}_{\text{pps}}$ estimate produced from each sample. The sample {1, 2} will result if our first random digit is a 0, and our second random digit is a 1 or our first is a 1 and our second is a 0. The $\hat{\tau}_{\text{pps}}$ value then becomes

$$\hat{\tau}_{\text{pps}} = \frac{1}{n}\sum_{i=1}^{n}\left(\frac{y_i}{z_i}\right) = \frac{1}{2}\left(\frac{1}{.1}+\frac{2}{.1}\right)$$
$$= \frac{1}{2}(10+20) = 15$$

From Table 8.3, we see that

$$E(\hat{\tau}_{\text{pps}}) = 15(.02) + \frac{35}{4}(.80) + \cdots + 10(.16) = 10$$

And so for this example $\hat{\tau}_{\text{pps}}$ is demonstrated to be an unbiased estimator of τ. Also,

$$V(\hat{\tau}_{\text{pps}}) = (15-10)^2(.02) + \left(\frac{35}{4} - 10\right)^2(.08) + \cdots + (10-10)^2(.16)$$
$$= 6.250$$

TABLE 8.3 **Sampling with varying probabilities (samples of size $n = 2$ from {1, 2, 3, 4} with $z_1 = z_2 = .1$, $z_3 = z_4 = .4$)**

Sample	Probability of obtaining sample	$\hat{\tau}_{pps}$
{1, 2}	.02	15
{1, 3}	.08	$\frac{35}{4}$
{1, 4}	.08	10
{2, 3}	.08	$\frac{55}{4}$
{2, 4}	.08	15
{3, 4}	.32	$\frac{35}{4}$
{1, 1}	.01	10
{2, 2}	.01	20
{3, 3}	.16	$\frac{15}{2}$
{4, 4}	.16	10
	1.00	

If we use simple random sampling with $n = 2$, our estimate $N\bar{y}$ of τ will have a variance computed as follows:

$$V(N\bar{y}) = N^2 V(\bar{y}) = (4)^2(.417) = 6.672$$

which is larger than $V(\hat{\tau}_{\text{pps}})$. The proportional reduction in variance by using a pps estimator would be greater if the population measurements had more variability and if the z_i's were even closer to being truly proportional to the size of the population measurements.

Sampling with varying probabilities will lower the variance of an estimator, thus allowing for more precise estimates, if the probabilities are proportional, or approximately proportional, to the size of the sampled measurements. If, however, the probabilities are improperly chosen, then $\hat{\tau}_{\text{pps}}$ can have *larger* variance than the corresponding $\hat{\tau}$ from simple random sampling. Suppose, for example, in samples of size $n = 2$ from $\{1, 2, 3, 4\}$, we choose $z_1 = z_2 = .4$ and $z_3 = z_4 = .1$. Then $\hat{\tau}_{\text{pps}}$ is still an unbiased estimator of τ, but $V(\hat{\tau}_{\text{pps}}) = 81.25$. In this case, $\hat{\tau}_{\text{pps}}$ is a much poorer estimator than $\hat{\tau}$ from simple random sampling.

In summary, pps sampling involves sampling *with* replacement, which means that a sampled item is not removed from the population after it is selected for the sample. Thus, a particular sampling unit can be selected more than once. A repeated selection is usually undesirable, but it will not happen often if n/N is small. Moreover, this undesirable feature is often more than offset by the reduction in variance that can occur. The pps estimator of τ only produces smaller variance if the probabilities z_i are proportional, or approximately proportional, to the size of the y_i's under investigation.

Cluster sampling often provides an ideal situation in which to use pps sampling since the number of elements in a cluster, m_i, forms a natural measure of the size of the cluster. Sampling with probabilities proportional to m_i pays big dividends in terms of reducing the bound on the error of estimation when the cluster total y_i is highly correlated with the number of elements in the cluster, which is often the case.

Let z_i be given by

$$z_i = \frac{m_i}{M}$$

Thus, the estimator of a population total $\hat{\tau}_{\text{pps}}$ becomes

$$\begin{aligned}\hat{\tau}_{\text{pps}} &= \frac{1}{n}\sum_{i=1}^{n}\frac{y_i}{z_i} = \frac{1}{n}\sum_{i=1}^{n}\frac{y_i}{(m_i/M)} \\ &= \frac{M}{n}\sum_{i=1}^{n}\frac{y_i}{m_i} = \frac{M}{n}\sum_{i=1}^{n}\bar{y}_i\end{aligned}$$

where $\bar{y}_i$ is the average of the observations in the ith cluster. The estimated variance of $\hat{\tau}_{\text{pps}}$ has a particularly simple form, as given later.

Since there are now M elements in the population, the estimator of the population mean, $\hat{\mu}_{pps}$, is simply

$$\hat{\mu}_{pps} = \frac{1}{M}\hat{\tau}_{pps} = \frac{1}{n}\sum_{i=1}^{n}\bar{y}_i$$

The estimated variance of $\hat{\mu}_{pps}$ is also easy to calculate.

Estimator of the population mean μ:

$$\hat{\mu}_{pps} = \frac{1}{n}\sum_{i=1}^{n}\bar{y}_i \qquad (8.19)$$

where $\bar{y}_i$ is the mean for the ith cluster.

Estimated variance of $\hat{\mu}_{pps}$:

$$\hat{V}(\hat{\mu}_{pps}) = \frac{1}{n(n-1)}\sum_{i=1}^{n}(\bar{y}_i - \hat{\mu}_{pps})^2 \qquad (8.20)$$

Estimator of the population total τ:

$$\hat{\tau}_{pps} = \frac{M}{n}\sum_{i=1}^{n}\bar{y}_i \qquad (8.21)$$

Estimated variance of $\hat{\tau}_{pps}$:

$$\hat{V}(\hat{\tau}_{pps}) = \frac{M^2}{n(n-1)}\sum_{i=1}^{n}(\bar{y}_i - \hat{\mu}_{pps})^2 \qquad (8.22)$$

We illustrate the technique of sampling with probabilities proportional to cluster sizes and the use of the formulas just given in the next two examples.

Example 8.12

An auditor wishes to sample sick-leave records of a large firm in order to estimate the average number of days of sick leave per employee over the past quarter. The firm has eight divisions, with varying numbers of employees per division. Since number of days of sick leave used within each division should be highly correlated with the number of employees, the auditor decides to sample $n = 3$ divisions with probabilities proportional to number of employees. Show how to select the sample if the respective numbers of employees are 1200, 450, 2100, 860, 2840, 1910, 290, 3200.

Solution

We first list the number of employees and the cumulative range for each division, as follows:

Division	Number of employees	Cumulative range
1	1,200	1–1200
2	450	1201–1650
3	2,100	1651–3750
4	860	3751–4610
5	2,840	4611–7450
6	1,910	7451–9360
7	390	9361–9750
8	3,200	9751–12,950
	12,950	

Since $n = 3$ divisions are to be sampled, we must select three random numbers between 00001 and 12,950. We can make this selection by starting anywhere on a random number table and selecting five-digit numbers, but we chose to start on line 1, column 4 of Appendix Table 2. The first three numbers between 00001 and 12,950, as we proceed down the column, are 02011, 07972, and 10281. The first appears in the cumulative range of division 3, the second appears in the range of division 6, and the third appears in the range of division 8. Thus divisions 3, 6, and 8 constitute the sample. (Note that one division can be selected more than once. In that event, we treat the resulting data as two separate but equal sample values.)

Example 8.13 Suppose the total number of sick days used by the three sampled divisions during the past quarter are, respectively,

$$y_1 = 4320 \qquad y_2 = 4160 \qquad y_3 = 5790$$

Estimate the average number of sick days used per person for the entire firm and place a bound on the error of estimation.

Solution

We must first compute the cluster means for the sampled clusters, which are

$$\bar{y}_1 = \frac{4320}{2100} = 2.06 \qquad \bar{y}_2 = \frac{4160}{1910} = 2.18 \qquad \bar{y}_3 = \frac{5790}{3200} = 1.81$$

(Note that the numbers of employees per sampled firm come from the data in Example 8.12.)

Now by Equation (8.19),

$$\hat{\mu}_{\text{pps}} = \frac{1}{n}\sum_{i=1}^{n} \bar{y}_i = \frac{1}{3}(2.06 + 2.18 + 1.81) = 2.02$$

Also, by Equation (8.20),

$$\hat{V}(\hat{\mu}_{\text{pps}}) = \frac{1}{n(n-1)} \sum_{i=1}^{n} (\bar{y}_i - \hat{\mu}_{\text{pps}})^2$$

$$= \frac{1}{3(2)} [(2.06 - 2.02)^2 + (2.18 - 2.02)^2 + (1.81 - 2.02)^2]$$

$$= .0119$$

Thus, the bound on the error of estimation is

$$2\sqrt{.0119} = .22$$

Our estimate of the average number of sick days used by employees of the firm is $2.02 \pm .22$.

We now have three estimators of the population total in cluster sampling, the ratio estimator (8.4), the unbiased estimator (8.7), and the pps estimator (8.21). How do we know which is best? Here are some guidelines about how to answer this question: If y_i is uncorrelated with m_i, then the unbiased estimator is better than either of the other two. If y_i is correlated with m_i, then the ratio and pps estimators are more precise than the unbiased estimator. The pps estimator is better than the ratio estimator if the within-cluster variation does not change with changing m_i. The ratio estimator is better than the pps estimator if the within-cluster variation increases with increasing m_i.

In Examples 8.12 and 8.13, the number of sick-leave days used should increase as the number of employees increases. Thus, the unbiased estimator is a poor choice here. The variation of sick-leave days within divisions, however, may remain relatively constant across divisions. In that case the pps estimator is the best choice.

8.10

SUMMARY

This chapter introduces a third sample survey design, cluster sampling. In this design, each sampling unit is a group, or cluster, of elements. Cluster sampling may provide maximum information at minimum cost when a frame listing population elements is not available or when the cost of obtaining observations increases with increasing distance between elements.

The estimator of the population mean μ is the sample mean $\bar{y}$, given by Equation (8.1). The estimated variance of $\bar{y}$ is given by Equation (8.2). Two estimators of the population total τ were given with their estimated variances. The estimator $M\bar{y}$ is presented in Equation (8.4); it is used when the number of elements M in the population is known. Estimator $N\bar{y}_t$ [see Equation (8.7)] is used when M is unknown.

In Section 8.4, we discussed an appropriate sample size for estimating μ or τ with a specified bound on the error of estimation.

In cluster sampling, the estimator of a population proportion p is the sample proportion $\hat{p}$, given by Equation (8.16). The estimated variance of $\hat{p}$ is given by Equation (8.17). The problem of selecting a sample size for estimating a proportion is similar to the problem for estimating a mean.

Cluster sampling can also be used within strata in a stratified population, and an example was given in Section 8.8.

CASE STUDY REVISITED

THE NEIGHBORHOOD CHARACTERISTICS PROBLEM

At the beginning of this chapter we suggested using U.S. census data on block statistics to estimate the proportion of residents aged 65 and over in a 40-block area. The $n = 5$ blocks were randomly sampled from the 40 and the following data obtained:

Number of residents, m_i	Number aged 65 and over, a_i	$\hat{p}m_i$	$a_i - \hat{p}m_i$	$(a_i - \hat{p}m_i)^2$
90	15	21.60	− 6.60	43.5600
32	8	7.68	0.32	0.1024
47	14	11.28	2.72	7.3984
25	9	6.00	3.00	9.0000
16	4	3.84	0.16	0.0256
210	50			60.0864

$$\hat{p} = \frac{\sum_{i=1}^{n} a_i}{\sum_{i=1}^{n} m_i} = \frac{50}{210} = .24$$

So the best estimate of the proportion of people aged 65 or over is .24.

The bound on the error of estimation is

$$2\sqrt{\hat{V}(\hat{p})} = 2\sqrt{\left(\frac{N-n}{Nn\bar{m}^2}\right)\left(\frac{1}{n-1}\right)\sum_{i=1}^{n}(a_i - pm_i)^2}$$

$$= 2\sqrt{\left[\frac{35}{(40)(5)(42)^2}\right]\left(\frac{1}{4}\right)(60.0864)}$$

$$= .08$$

Thus, the estimate of the true proportion for the 40-block area is $.24 \pm .08$, or .16 to .32. We are confident that over 16% of the residents have ages of 65 or more.

EXERCISES

8.1 An experimenter working in an urban area desires to estimate the average value of a variable highly correlated with race. She thinks she should use cluster sampling, with city blocks as clusters and adults within blocks as elements. Explain why you would, or would not, use cluster sampling in each of the following situations.
(a) Most of the adults in certain blocks are white, and most in other blocks are nonwhite.
(b) The proportion of nonwhites is the same in every block and is not close to 0 or 1.
(c) The proportion of nonwhites differs from block to block in the manner that would be expected if the clusters were made up by randomly assigning adults in the population to clusters.

8.2 A manufacturer of band saws wants to estimate the average repair cost per month for the saws he has sold to certain industries. He cannot obtain a repair cost for each saw, but he can obtain the total amount spent for saw repairs and the number of saws owned by each industry. Thus, he decides to use cluster sampling, with each industry as a cluster. The manufacturer selects a simple random sample of $n = 20$ from the $N = 96$ industries he services. The data on total cost of repairs per industry and number of saws per industry are as given in the accompanying table. Estimate the average repair cost per saw for the past month, and place a bound on the error of estimation.

Industry	Number of saws	Total repair cost for past month (in dollars)
1	3	50
2	7	110
3	11	230
4	9	140
5	2	60
6	12	280
7	14	240
8	3	45
9	5	60
10	9	230
11	8	140
12	6	130
13	3	70
14	2	50
15	1	10
16	4	60
17	12	280
18	6	150
19	5	110
20	8	120

8.3 For the data in Exercise 8.2, estimate the total amount spent by the 96 industries on band saw repairs. Place a bound on the error of estimation.

8.4 After checking his sales records, the manufacturer of Exercise 8.2 finds that he sold a total of 710 band saws to these industries. Using this additional information, estimate the total amount spent on saw repairs by these industries and place a bound on the error of estimation.

8.5 The same manufacturer (Exercise 8.2) wants to estimate the average repair cost per saw for next month. How many clusters should he select for his sample if he wants the bound on the error of estimation to be less than $2?

8.6 A political scientist developed a test designed to measure the degree of awareness of current events. She wants to estimate the average score that would be achieved on this test by all students in a certain high school. The administration at the school will not allow the experimenter to randomly select students out of classes in session, but it will allow her to interrupt a small number of classes for the purpose of giving the test to every member of the class. Thus, the experimenter selects 25 classes at random from the 108 classes in session at a particular hour. The test is given to each member of the sampled classes, with results as shown in the accompanying table. Estimate the average score that would be achieved on this test by all students in the school. Place a bound on the error of estimation.

Class	Number of students	Total score
1	31	1590
2	29	1510
3	25	1490
4	35	1610
5	15	800
6	31	1720
7	22	1310
8	27	1427
9	25	1290
10	19	860
11	30	1620
12	18	710
13	21	1140
14	40	1980
15	38	1990
16	28	1420
17	17	900
18	22	1080
19	41	2010
20	32	1740
21	35	1750
22	19	890
23	29	1470
24	18	910
25	31	1740

8.7 The political scientist of Exercise 8.6 wants to estimate the average test score for a similar high school. She wants the bound on the error of estimation to be less than 2 points. How many classes should she sample? Assume the school has 100 classes in session during each hour.

8.8 An industry is considering revision of its retirement policy and wants to estimate the proportion of employees that favor the new policy. The industry consists of 87 separate plants located throughout the United States. Since results must be obtained quickly and with little cost, the industry decides to use cluster sampling with each plant as a cluster. A simple random sample of 15 plants is selected, and the opinions of the employees in these plants are obtained by questionnaire. The results are as shown in the accompanying table. Estimate the proportion of employees in the industry who favor the new retirement policy and place a bound on the error of estimation.

Plant	Numbers of employees	Number favoring new policy
1	51	42
2	62	53
3	49	40
4	73	45
5	101	63

(continued)

(continued)

Plant	Numbers of employees	Number favoring new policy
6	48	31
7	65	38
8	49	30
9	73	54
10	61	45
11	58	51
12	52	29
13	65	46
14	49	37
15	55	42

8.9 The industry of Exercise 8.8 modified its retirement policy after obtaining the results of the survey. It now wants to estimate the proportion of employees in favor of the modified policy. How many plants should be sampled to have a bound of .08 on the error of estimation? Use the data from Exercise 8.8 to approximate the results of the new survey.

8.10 An economic survey is designed to estimate the average amount spent on utilities for households in a city. Since no list of households is available, cluster sampling is used, with divisions (wards) forming the clusters. A simple random sample of 20 wards is selected from the 60 wards of the city. Interviewers then obtain the cost of utilities from each household within the sampled wards; the total costs are shown in the accompanying table. Estimate the average amount a household in the city spends on utilities and place a bound on the error of estimation.

Sampled ward	Number of households	Total amount spent on utilities	Sampled ward	Number of households	Total amount spent on utilities
1	55	\$2210	11	73	\$2930
2	60	2390	12	64	2470
3	63	2430	13	69	2830
4	58	2380	14	58	2370
5	71	2760	15	63	2390
6	78	3110	16	75	2870
7	69	2780	17	78	3210
8	58	2370	18	51	2430
9	52	1990	19	67	2730
10	71	2810	20	70	2880

8.11 In the survey of Exercise 8.10, the number of households in the city is not known. Estimate the total amount spent on utilities for all households in the city and place a bound on the error of estimation.

8.12 The economic survey of Exercise 8.10 is to be performed in a neighboring city of similar structure. The objective is to estimate the total amount spent on utilities by households in the city, with a bound of $5000 on the error of estimation. Use the data in Exercise 8.10 to find the appropriate number of clusters needed to achieve this bound.

8.13 An inspector wants to estimate the average weight of fill for cereal boxes packaged in a certain factory. The cereal is available to him in cartons containing 12 boxes each. The inspector randomly selects 5 cartons and measures the weight of fill for every box in the sampled cartons, with the results (in ounces) as shown in the accompanying table. Estimate the average weight of fill for boxes packaged by this factory and place a bound on the error of estimation. Assume that the total number of cartons packaged by the factory is large enough for the finite population correction to be ignored.

Carton	Ounces of fill											
1	16.1	15.9	16.1	16.2	15.9	15.8	16.1	16.2	16.0	15.9	15.8	16.0
2	15.9	16.2	15.8	16.0	16.3	16.1	15.8	15.9	16.0	16.1	16.1	15.9
3	16.2	16.0	15.7	16.3	15.8	16.0	15.9	16.0	16.1	16.0	15.9	16.1
4	15.9	16.1	16.2	16.1	16.1	16.3	15.9	16.1	15.9	15.9	16.0	16.0
5	16.0	15.8	16.3	15.7	16.1	15.9	16.0	16.1	15.8	16.0	16.1	15.9

8.14 A newspaper wants to estimate the proportion of voters favoring a certain candidate, candidate A, in a statewide election. Since selecting and interviewing a simple random sample of registered voters is very expensive, cluster sampling is used, with precincts as clusters. A simple random sample of 50 precincts is selected from the 497 precincts in the state. The newspaper wants to make the estimation on election day but before final returns are tallied. Therefore reporters are sent to the polls of each sample precinct to obtain the pertinent information directly from the voters. The results are shown in the accompanying table. Estimate the proportion of voters favoring candidate A and place a bound on the error of estimation.

Number of voters	Number favoring A	Number of voters	Number favoring A	Number of voters	Number favoring A
1290	680	1893	1143	843	321
1170	631	1942	1187	1066	487
840	475	971	542	1171	596
1620	935	1143	973	1213	782
1381	472	2041	1541	1741	980
1492	820	2530	1679	983	693
1785	933	1567	982	1865	1033
2010	1171	1493	863	1888	987
974	542	1271	742	1947	872
832	457	1873	1010	2021	1093
1247	983	2142	1092	2001	1461
1896	1462	2380	1242	1493	1301
1943	873	1693	973	1783	1167
798	372	1661	652	1461	932
1020	621	1555	523	1237	481
1141	642	1492	831	1843	999
1820	975	1957	932		

8.15 The newspaper of Exercise 8.14 wants to conduct a similar survey during the next election. How large a sample size will be needed to estimate the proportion of voters favoring a similar candidate with a bound of .05 on the error of estimation? Use the data in Exercise 8.14.

8.16 A forester wishes to estimate the average height of trees on a plantation. The plantation is divided into quarter-acre plots. A simple random sample of 20 plots is selected from the 386 plots on the plantation. All trees on the sampled plots are measured, with the results as shown in the accompanying table. Estimate the average height of trees on the plantation and place a bound on the error of estimation. [*Hint:* The total for cluster i can be found by taking m_i times the cluster average.]

Number of trees	Average height (in feet)
42	6.2
51	5.8
49	6.7
55	4.9
47	5.2
58	6.9
60	6.3
52	6.7
61	5.9
49	6.1
57	6.0
63	4.9
43	4.3
59	5.2
48	5.7
41	6.1
45	5.3
46	6.7
62	6.1
58	7.0

8.17 To emphasize safety, a taxicab company wants to estimate the proportion of unsafe tires on their 175 cabs. (Ignore spare tires.) Selecting a simple random sample of tires is impractical, so cluster sampling is used, with each cab as a cluster. A random sample of 25 cabs gives the following number of unsafe tires per cab:

$$2, 4, 0, 1, 2, 0, 4, 1, 3, 1, 2, 0, 1, 1, 2, 2, 4, 1, 0, 0, 3, 1, 2, 2, 1$$

Estimate the proportion of unsafe tires being used on the company's cabs and place a bound on the error of estimation.

8.18 Accountants frequently require their business clients to provide cost inventories. Since a complete inventory is costly, quarterly inventories can conveniently be accomplished by sampling. Suppose a plumbing supply firm desires a cost inventory for many small items in stock. To obtain a simple random sample of items is difficult. However, the items are arranged on shelves, and selecting a simple random sample of shelves is relatively easy, treating each shelf as a cluster of items.

Sampling 10 of the 48 shelves gave the results shown in the accompanying table. Estimate the total dollar amount of the items on the shelves and place a bound on the error of estimation.

Cluster	Number of items, m_i	Total dollar amount, y_i
1	42	83
2	27	62
3	38	45
4	63	112
5	72	96
6	12	58
7	24	75
8	14	58
9	32	67
10	41	80

8.19 A certain firm specializing in the manufacture and sale of leisure clothing has 80 retail stores in Florida and 140 in California. With each state as a stratum, the firm wishes to estimate average sick-leave time per employee for the past year. Each outlet can be viewed as a cluster of employees, and total sick leave time for each store can be determined from records. Simple random samples of 8 stores from Florida and 10 stores from California gave the results shown in the accompanying table (m_i denotes the number of employees, and y_i denotes total days sick leave for the ith store). Estimate the average amount of sick leave per employee and calculate an estimate of the variance of your estimator.

Florida		California	
m_i	y_i	m_i	y_i
12	40	16	51
20	52	8	32
8	30	4	11
14	36	3	10
24	71	12	33
15	48	17	39
10	39	24	61
6	21	30	37
		21	40
		9	41

8.20 Block statistics report the number of housing units, the number of residents, and the total number of rooms within housing units for a random sample of eight blocks selected from a large city. (Assume the number of blocks in the city is very large.) The data are given in the accompanying table.

Block	Number of housing units	Number of residents	Number of rooms
1	12	40	58
2	14	39	72
3	3	12	26
4	20	52	98
5	12	37	74
6	8	33	57
7	10	41	76
8	6	14	48

(a) Estimate the average number of residents per housing unit and place a bound on the error of estimation.

(b) Estimate the average number of rooms per resident and place a bound on the error of estimation.

8.21 A certain type of circuit board manufactured for installation in computers has 12 microchips per board. During the quality control inspection of 10 of these boards, the numbers of defective microchips per board were as follows:

$$2, 0, 1, 3, 2, 0, 0, 1, 3, 4$$

Estimate the proportion of defective microchips in the population from which this sample was drawn and place a bound on the error of estimation.

8.22 Refer to the setting of Exercise 8.21. Suppose the sample of 10 boards used there came from a shipment of 50 such boards. Estimate the total number of defective microchips in the shipment and place a bound on the error of estimation.

8.23 A large firm has its equipment inventories listed separately by department. From the 15 departments in the firm, 5 are to be randomly sampled by an auditor, who will then check to make sure that all equipment is properly identified and located. The proportion of equipment items not properly identified is of interest to the auditor. The data are given in the accompanying table. Estimate the proportion of equipment items in the firm not properly identified and place a bound on the error of estimation.

Department	Number of equipment items	Number of items not properly identified
1	15	2
2	27	3
3	9	1
4	31	1
5	16	2

8.24 Suppose that for the firm of Exercise 8.23 the 15 departments have the number of equipment items given in the accompanying table. Select a sample of 3 departments with probabilities proportional to number of equipment items.

Department	Number of items	Department	Number of items
1	12	9	31
2	9	10	26
3	27	11	22
4	40	12	19
5	35	13	16
6	15	14	33
7	18	15	6
8	10		

8.25 Suppose the three departments selected in Exercise 8.24 each have two improperly identified equipment items. Estimate the total number of improperly identified items in the firm and place a bound on the error of estimation.

8.26 An investigator wishes to estimate the average number of defects per board on boards of electronic components manufactured for installation in computers. The boards contain varying numbers of components, and the investigator feels that the number of defects should be positively correlated with the number of components on a board. Thus, pps sampling is used, with the probability of selecting any one board for the sample being proportional to the number of components on that board. A sample of $n=4$ boards is to be selected from the $N=10$ boards of one day's production. The number of components on the 10 boards are, respectively,

$$10, 12, 22, 8, 16, 24, 9, 10, 8, 31$$

Show how to select $n=4$ boards with probabilities proportional to size.

8.27 After the sampling of Exercise 8.26 was completed, the respective number of defects found on boards 2, 3, 5, and 7 were 1, 3, 2, and 1. Estimate the average number of defects per board and place a bound on the error of estimation.

8.28 A state agriculture department wants to measure total yield of tomatoes for a sample of fields, with the goal of estimating total tomato yields for the state. Discuss the merits of simple random sampling as compared with sampling with probabilities proportional to size.

8.29 Refer to the U.S. population figures given in USPOP in the appendix. From the nine northeastern states, select a sample of four states with probabilities proportional to their total population sizes in 1980. Is this procedure an appropriate sampling scheme for estimating total unemployment in the Northeast? Is this procedure an appropriate sampling scheme for estimating acres of forest land in the Northeast?

8.30 For the $n = 4$ states selected in Exercise 8.29, record their 1990 population sizes from the appendix. Use these data to estimate the total 1990 population in the northeastern states and place a bound on the error of estimation. Is the total given in the table included in your interval estimate? Do you think this method of sampling is better than selecting a simple random sample of four states for purposes of estimating total population? Why?

8.31 A large shipment of frozen seafood is packaged in cartons, each containing twenty-four 50-pound packages. There are 100 cartons in the shipment. The seafood thawed and then refroze. The total weight (in pounds) of spoiled seafood is determined by a government inspector for each of a sample of 5 cartons. These data are as follows:

$$9, 6, 3, 10, 2$$

Estimate the total weight of spoiled seafood in the shipment and place a bound on the error of estimation.

8.32 Using the data of Exercise 8.31, estimate the average amount of spoiled seafood per 5-pound package and place a bound on the error of estimation.

8.33 A political scientist wishes to sample resident students on a large-university campus. Individual housing units can be conveniently used as clusters of students, or collections of housing units (freshmen dormitories, fraternity houses, and so on) can be used as strata. Discuss the merits of cluster versus stratified random sampling if the goal is to estimate the proportion of students favoring a certain candidate in the following types of elections:
(a) A student government election
(b) A national presidential election

8.34 Under what conditions does cluster sampling produce a smaller bound on the error of estimation for a mean than simple random sampling?

8.35 Disregarding costs of sampling, what criteria would you use for selecting appropriate clusters in a cluster sampling problem?

8.36 A school district has $N = 30$ schools, with the ith school containing m_i students. A simple random sample of $n = 5$ schools is selected and asked whether they provide a nurse whose services are available to all students. The results of the survey are as follows:

School	m_i	Provides nurse
1	1200	Yes
2	800	No
3	1000	No
4	600	Yes
5	1000	No

(a) Estimate the proportion of *students* in the district who have nursing services available to them.
(b) Show how to calculate a variance for the estimator in part (a). You need not carry out the calculations.

8.37 An industry produces customized truck bodies in 20 plants scattered throughout the country. Only a small number of truck bodies are completed at each plant on any given day. For a particular day, it is desired to estimate the average man-hours of work for items completed that day in the industry. Discuss the relative merits of cluster sampling (plants as clusters) versus stratified random sampling (plants as strata) for estimating the desired average. Which design would you recommend?

EXPERIENCES WITH REAL DATA

8.1 The data set USPOP in the appendix and on the data disk gives the percentage of people over age 65 for each state. Using these percentages from a random sample of ten states, estimate the percentage over age 65 for the United States as a whole for 1990, with a bound on the error of estimation, using
(a) A ratio estimator based on each state being a cluster of people.
(b) A simple average of the percentages for the sampled states.
Compare the two techniques and comment on which one you prefer.

8.2 Using the same sample of states as in 8.1, estimate the population density (people per square mile) of the United States in 2010, with a bound on the error.

8.3 From the WORLDPOP data on the data disk, select a random sample of 12 countries from Africa. Based on this sample,
(a) Estimate the birth rate for Africa, with a bound on the error of estimation.
(b) Estimate the total gross national product (GNP) for Africa, with a bound on the error of estimation.

8.4 From the SCHOOLS data in the appendix or on the data disk, estimate the average teacher salary for the United States in 1992 based on a random sample of 15 states. Place a bound on the error of estimation.

8.5 Try an economic study, perhaps by treating households in a certain fixed geographic area (perhaps a few city blocks) as clusters of people. Sample n households and, upon gaining permission for an interview, record the total weekly amount spent on food by all individuals in the household and the number of individuals. Then estimate the average amount spent on food per person among the households in this population. Even if all the money is actually spent by one person (say, the mother), that total amount is the same as would have been recorded if each individual had purchased his or her own food. Thus, the cluster total is available even though the observations per element may not be.

C A S E   S T U D Y

HOW MUCH DO STUDENTS SPEND FOR ENTERTAINMENT?

The entertainment dollars are important to the businesses in a town containing a university. How can we estimate the average monthly amount spent on entertainment per student? Locating students who may be randomly selected from a directory is difficult, but locating randomly selected classrooms, all of which should contain students at a prime class hour, like 10:00 A.M. on Monday, is relatively easy. Since classes may be large, sufficient information can be obtained by sampling a subset of those students in each sampled class. The result is a two-stage cluster sample.

A certain midsized university has 12,000 students divided into 150 classes at 10:00 A.M. on Mondays. Almost all the students should be in class at this hour. For the purpose of estimating the average monthly amount spent on entertainment, 4 classes are randomly selected, and approximately 10% of the students in each class are interviewed. The methods of two-stage cluster sampling are used in the analysis.

CHAPTER 9

TWO-STAGE CLUSTER SAMPLING

9.1 INTRODUCTION

Two-stage cluster sampling is an extension of the concept of cluster sampling. Recall from the discussion of cluster sampling in Chapter 8 that a cluster is usually a convenient or natural collection of elements, such as blocks of households or cartons of flashbulbs. A cluster often contains too many elements to obtain a measurement on each, or it contains elements so nearly alike that measurement of only a few elements provides information on an entire cluster. When either situation occurs, the experimenter can select a probability sample of clusters and then take a probability sample of elements within each cluster. The result is a two-stage cluster sample.

Definition 9.1

A *two-stage cluster sample* is obtained by first selecting a probability sample of clusters and then selecting a probability sample of elements from each sampled cluster.

Our discussion will be limited to simple random samples at each stage. For example, a national survey of university students' opinions can be conducted by selecting a simple random sample of universities from all those in the country and then selecting a simple random sample of students from each university. Thus, a university corresponds to a cluster of students. Similarly, the total amount of accounts receivable for a chain store can be estimated by first taking a simple random sample of stores and then selecting a simple random sample of accounts from each. Thus, each chain store provides a cluster of accounts.

Two-stage cluster sampling is commonly used in large surveys involving the sampling of housing units. We mentioned in Chapter 4 that the Gallup poll samples approximately 300 election districts from around the United States. At the second stage, this poll randomly (or systematically) selects approximately 5 households per district, for a total sample size of about 1500 households. In other polls, block statistics from the U.S. Census Bureau form clusters of households, as discussed in Chapter 8, which are then subsampled before interviews are conducted.

Sampling for quality control purposes often involves two (or more) stages of sampling. For example, when an inspector samples packaged products, such as frozen food, he or she commonly samples cartons and then samples packages from within cartons. When one is sampling products turned out at various workstations, one might sample workstations and then sample items produced at each sampled station. When sampling requires detailed investigation of components of products, such as measuring plate thicknesses in automobile batteries, a quite natural procedure is to sample some of the products (batteries) and then sample components (plates) within these products.

There is a certain similarity between cluster sampling and stratified random sampling. Think of a population being divided into nonoverlapped groups of elements. If these groups are considered to be strata, then a simple random sample is selected from *each* group. If these groups are considered to be clusters, then a simple random sample of *groups* is selected, and the sampled groups are then subsampled. Stratified random sampling provides estimators with small variance when there is little variation among elements within each group. Cluster sampling does well when the elements within each group are highly variable, and all groups are quite similar to one another.

The advantages of two-stage cluster sampling over other designs are the same as those listed in Chapter 8 for cluster sampling. First, a frame listing all elements in the population may be impossible or costly to obtain, whereas obtaining a list of all clusters may be easy. For example, compiling a list of all university students in the country would be expensive and time-consuming, but a list of universities could be readily acquired. Second, the cost of obtaining data may be inflated by travel costs if the sampled elements are spread over a large geographic area. Thus, sampling clusters of elements that are physically close together is often economical.

9.2

HOW TO DRAW A TWO-STAGE CLUSTER SAMPLE

The first problem in selecting a two-stage cluster sample is the choice of appropriate clusters. Two conditions are desirable: (1) geographic proximity of the elements within a cluster and (2) cluster sizes that are convenient to administer.

The selection of appropriate clusters also depends on whether we want to sample a few clusters and many elements from each or many clusters and a few elements from each. Ultimately, the choice is based on costs. Large clusters tend to possess heterogeneous elements, and hence a large sample is required from each in order to acquire accurate estimates of population parameters. In contrast, small clusters frequently contain relatively homogeneous elements, in which case accurate information on the characteristics of a cluster can be obtained by selecting a small sample from each cluster.

Consider the problem of sampling personal incomes in a large city. The city can be divided into large clusters—for example, precincts, which contain a heterogeneous assortment of incomes. Thus, a small number of precincts may yield a representative cross-section of incomes within the city, but a fairly large sample of elements from each cluster will be required in order to accurately estimate its mean (because of the heterogeneity of incomes within the cluster). In contrast, the city can be divided into small, relatively homogeneous clusters—say, city blocks. Then a small sample of people from each block will give adequate information on each cluster's mean, but obtaining accurate information on the mean income for the entire city will require many blocks.

As another example, consider the university student opinion poll. If students within a university hold similar opinions on the question of interest but opinions differ widely from university to university, then the sample should contain a few representatives from many different universities. If the opinions vary greatly within each university, then the survey should include many representatives from each of a few universities.

To select the sample, we first obtain a frame listing all clusters in the population. We then draw a simple random sample of clusters, using the random sampling procedures presented in Chapter 4. Third, we obtain frames that list all elements in each of the sampled clusters. Finally, we select a simple random sample of elements from each of these frames.

9.3

UNBIASED ESTIMATION OF A POPULATION MEAN AND TOTAL

As in previous chapters, we are interested in estimating a population mean μ or a population total τ and placing a bound on the error of estimation. The following notation is used:

$N =$ The number of clusters in the population

$n =$ The number of clusters selected in a simple random sample

$M_i =$ The number of elements in cluster i

$m_i =$ The number of elements selected in a simple random sample from cluster i

$$M = \sum_{i=1}^{N} M_i = \text{the number of elements in the population}$$

$$\bar{M} = \frac{M}{N} = \text{the average cluster size for the population}$$

$y_{ij} =$ the jth observation in the sample from the ith cluster

$$\bar{y}_i = \frac{1}{m_i} \sum_{j=1}^{m_i} y_{ij} = \text{the sample mean for the } i\text{th cluster}$$

In constructing an estimator of the population mean μ, we might try to parallel what was done in Chapter 8 on single-stage cluster sampling. Equation (8.7) gives

$$\frac{N}{n} \sum_{i=1}^{n} y_i$$

as an unbiased estimator of τ. Thus, if we divide by M,

$$\frac{N}{Mn} \sum_{i=1}^{n} y_i$$

becomes an unbiased estimator of μ, but we cannot evaluate this estimator now since we no longer know the cluster totals, y_i. We can, however, estimate y_i by $M_i \bar{y}_i$, and, on substituting $M_i \bar{y}_i$ for y_i, we have an unbiased estimator of μ, which we can calculate from our sample data.

Unbiased estimator of the population mean μ:

$$\hat{\mu} = \left(\frac{N}{M}\right) \frac{\sum_{i=1}^{n} M_i \bar{y}_i}{n} = \frac{1}{\bar{M}} \frac{\sum_{i=1}^{n} M_i \bar{y}_i}{n} \tag{9.1}$$

assuming simple random sampling at each stage.

Estimated variance of $\hat{\mu}$:

$$\hat{V}(\hat{\mu}) = \left(\frac{N-n}{N}\right)\left(\frac{1}{n\bar{M}^2}\right) s_b^2 + \frac{1}{nN\bar{M}^2} \sum_{i=1}^{n} M_i^2 \left(\frac{M_i - m_i}{M_i}\right)\left(\frac{s_i^2}{m_i}\right) \tag{9.2}$$

where

$$s_b^2 = \frac{\sum_{i=1}^{n} (M_i \bar{y}_i - \bar{M}\hat{\mu})^2}{n-1} \qquad (9.3)$$

and

$$s_i^2 = \frac{\sum_{j=1}^{m_i} (y_{ij} - \bar{y}_i)^2}{m_i - 1} \qquad i = 1, 2, \ldots, n \qquad (9.4)$$

Notice that s_b^2 is simply the sample variance among the terms $M_i \bar{y}_i$.

The estimator $\hat{\mu}$ shown in Equation (9.1) depends on M, the number of elements in the population. A method of estimating μ when M is unknown is given in the next section.

Note that s_i^2 is the sample variance for the sample selected from cluster i.

Example 9.1

A garment manufacturer has 90 plants located throughout the United States and wants to estimate the average number of hours that the sewing machines were down for repairs in the past months. Because the plants are widely scattered, she decides to use cluster sampling, specifying each plant as a cluster of machines. Each plant contains many machines, and checking the repair record for each machine would be time-consuming. Therefore, she uses two-stage sampling. Enough time and money are available to sample $n = 10$ plants and approximately 20% of the machines in each plant.

Using the data in Table 9.1, estimate the average downtime per machine and place a bound on the error of estimation. The manufacturer knows she has a combined total of 4500 machines in all plants.

TABLE 9.1 **Downtime for sewing machines**

Plant	M_i	m_i	Downtime (in hours)	$\bar{y}_i$	s_i^2
1	50	10	5, 7, 9, 0, 11, 2, 8, 4, 3, 5	5.40	11.38
2	65	13	4, 3, 7, 2, 11, 0, 1, 9, 4, 3, 2, 1, 5	4.00	10.67
3	45	9	5, 6, 4, 11, 12, 0, 1, 8, 4	5.67	16.75
4	48	10	6, 4, 0, 1, 0, 9, 8, 4, 6, 10	4.80	13.29
5	52	10	11, 4, 3, 1, 0, 2, 8, 6, 5, 3	4.30	11.12
6	58	12	12, 11, 3, 4, 2, 0, 0, 1, 4, 3, 2, 4	3.83	14.88
7	42	8	3, 7, 6, 7, 8, 4, 3, 2	5.00	5.14
8	66	13	3, 6, 4, 3, 2, 2, 8, 4, 0, 4, 5, 6, 3	3.85	4.31
9	40	8	6, 4, 7, 3, 9, 1, 4, 5	4.88	6.13
10	56	11	6, 7, 5, 10, 11, 2, 1, 4, 0, 5, 4	5.00	11.80

Solution

The following display shows a convenient spreadsheet formulation of the data and summary calculations for completing the analysis in two-stage cluster sampling. Data from each cluster form a row of the display, with 13 columns allowed for data even though not all samples are this large. The estimator of the mean and the between-cluster component of the variance are both based on the $M_i y_u$ values, $\hat{\mu}$ being based on their average and s_b being their standard deviation. The s_i terms are the within-sample standard deviations for the respective samples, and the "within" term for each sample is its component of the within-sample variance, namely,

$$M_i(M_i - m_i)s_i^2/m_i$$

All of these computations, including the statistical summary at the bottom, are easily made on a computer or graphing calculator.

Row	Plant	Mi	mi	C4	C5	C6	C7	C8	C9	C10	C11	C12
1	1	50	10	5	7	9	0	11	2	8	4	3
2	2	65	13	4	3	7	2	11	0	1	9	4
3	3	45	9	5	6	4	11	12	0	1	8	4
4	4	48	10	6	4	0	1	0	9	8	4	6
5	5	52	10	11	4	3	1	0	2	8	6	5
6	6	58	12	12	11	3	4	2	0	0	1	4
7	7	42	8	3	7	6	7	8	4	3	2	*
8	8	66	13	3	6	4	3	2	2	8	4	0
9	9	40	8	6	4	7	3	9	1	4	5	*
10	10	56	11	6	7	5	10	11	2	1	4	0

Row	C13	C14	C15	C16	ȳi	Miȳi	si	Within
1	5	*	*	*	5.40000	270.000	3.37310	2275.56
2	3	2	1	5	4.00000	260.000	3.26599	2773.33
3	*	*	*	*	5.66667	255.000	4.09268	3015.00
4	10	*	*	*	4.80000	230.400	3.64539	2423.89
5	3	*	*	*	4.30000	223.600	3.33500	2429.09
6	3	2	4	*	3.83333	222.333	3.85730	3308.05
7	*	*	*	*	5.00000	210.000	2.26779	918.00
8	4	5	6	3	3.84615	253.846	2.07550	1159.10
9	*	*	*	*	4.87500	195.000	2.47487	980.00
10	5	4	*	*	5.00000	280.000	3.43511	2703.27

	N	Mean	Median	SD
mi	10	10.400	10.000	1.838
ȳi	10	4.672	4.838	0.647
Miȳi	10	240.02	242.12	27.72

The best estimate of μ is $\hat{\mu}$, shown in Equation (9.1), which yields

$$\hat{\mu} = \frac{N}{Mn}\sum_{i=1}^{n} M_i\bar{y}_i = \frac{1}{\bar{M}}\cdot\frac{1}{n}\cdot\sum_{c=1}^{n} M_i\bar{y}_i = \frac{1}{50}(240.02) = 4.80$$

To estimate the variance of $\hat{\mu}$, we must calculate

$$s_b^2 = \frac{1}{n-1}\sum_{i=1}^{n}(M_i\hat{y}_i - \bar{M}\hat{\mu})^2 = (27.72)^2$$

and

$$\sum_{i=1}^{n} M_i^2\left(\frac{M_i - m_i}{M_i}\right)\left(\frac{s_i^2}{m_i}\right) = 21{,}985$$

the sum of the last column of the spreadsheet. Then from Equation (9.2),

$$\begin{aligned}\hat{V}(\hat{\mu}) &= \left(\frac{N-n}{N}\right)\left(\frac{1}{n\bar{M}^2}\right)s_b^2 + \frac{1}{nN\bar{M}^2}\sum_{i=1}^{n} M_i^2\left(\frac{M_i - m_i}{M_i}\right)\left(\frac{s_i^2}{m_i}\right)\\ &= \left(\frac{90-10}{90}\right)\left[\frac{1}{(10)(50)^2}\right](27.72)^2 + \frac{1}{(10)(90)(50)^2}(21{,}985)\\ &= .0371\end{aligned}$$

The estimate of μ with a bound on the error of estimation is given by

$$\hat{\mu} \pm 2\sqrt{\hat{V}(\hat{\mu})}$$

$$4.80 \pm 2\sqrt{.0371}$$

$$4.80 \pm .39$$

Thus, the average downtime is estimated to be 4.80 hours. The error of estimation should be less than .39 hour with a probability of approximately .95.

An unbiased estimator of a population total can be found by taking an unbiased estimator of the population mean and multiplying by the number of elements in the population in a manner similar to that used in simple random sampling. Thus, $M\hat{\mu}$ is an unbiased estimator of τ for two-stage cluster sampling.

Estimation of the population total τ:

$$\hat{\tau} = M\hat{\mu} = N\frac{\sum_{i=1}^{n} M_i\bar{y}_i}{n} \qquad (9.5)$$

assuming simple random sampling at each stage.

Estimated variance of $\hat{\tau}$:

$$\hat{V}(\hat{\tau}) = M^2\hat{V}(\hat{\mu})$$

$$= \left(\frac{N-n}{N}\right)\left(\frac{N^2}{n}\right)s_b^2 + \frac{N}{n}\sum_{i=1}^{n} M_i^2\left(\frac{M_i - m_i}{M_i}\right)\left(\frac{s_i^2}{m_i}\right) \quad (9.6)$$

where s_b^2 is given by Equation (9.3) and s_i^2 is given by Equation (9.4).

Example 9.2 Estimate the total amount of downtime during the past month for all machines owned by the manufacturer in Example 9.1. Place a bound on the error of estimation.

Solution

The best estimate of τ is

$$\hat{\tau} = M\hat{\mu} = \frac{N}{n}\sum_{i=1}^{n} M_i\bar{y}_i = 90(240.02) = 21{,}602$$

The estimated variance of $\hat{\tau}$ is found by using the value of $\hat{V}(\hat{\mu})$ calculated in Example 9.1 and substituting as follows:

$$\hat{V}(\hat{\tau}) = M^2\hat{V}(\hat{\mu}) = (4500)^2(.0371)$$

The estimate of τ with a bound on the error of estimation is

$$\hat{\tau} \pm \sqrt{\hat{V}(\hat{\tau})}$$

$$21{,}605.31 \pm 2\sqrt{(4500)^2(.0371)}$$

$$21{,}602 \pm 1733$$

Thus, the estimate of total downtime is 21,602 hours. We are fairly confident that the error of estimation is less than 1733 hours.

9.4

RATIO ESTIMATION OF A POPULATION MEAN

The estimator $\hat{\mu}$, given by Equation (9.1), depends on the total number of elements in the population, M. When M is unknown, as is frequently the case, it must be estimated from the sample data. We obtain an estimator of M by multiplying the average cluster size, $\sum_{i=1}^{n} M_i/n$, by the number of clusters in the population, N. If we replace M by its estimator, we obtain a ratio estimator, denoted by $\hat{\mu}_r$, because the numerator and denominator are both random variables.

Ratio estimator of the population mean μ:

$$\hat{\mu}_r \frac{\sum_{i=1}^{n} M_i \bar{y}_i}{\sum_{i=1}^{n} M_i} \tag{9.7}$$

Estimated variance of $\hat{\mu}_r$:

$$\hat{V}(\hat{\mu}_r) = \left(\frac{N-n}{N}\right)\left(\frac{1}{n\bar{M}^2}\right)s_r^2 + \frac{1}{nN\bar{M}^2}\sum_{i=1}^{n} M_i^2 \left(\frac{M_i - m_i}{M_i}\right)\left(\frac{s_i^2}{m_i}\right) \tag{9.8}$$

where

$$s_r^2 = \frac{\sum_{i=1}^{n} M_i^2(\bar{y}_i - \hat{\mu}_r)^2}{n-1} = \frac{\sum_{i=1}^{n} (M_i\bar{y}_i - \hat{\mu}_r M_i)^2}{n-1} \tag{9.9}$$

and

$$s_i^2 = \frac{\sum_{i=1}^{m_i} (y_{ij} - \bar{y}_i)^2}{m_i - 1} \qquad i = 1, 2, \ldots, n \tag{9.10}$$

The estimator $\hat{\mu}_r$ is biased, but the bias is negligible when n is large.

Example 9.3 Using the data in Table 9.1, estimate the average downtime per machine and place a bound on the error of estimation. Assume the manufacturer does not know how many machines there are in all plants combined.

Solution

The following display shows a convenient spreadsheet formulation of the data and summary calculations for completing the analysis in two-stage cluster sampling using a ratio estimator. The estimator of the mean and the between-cluster component of the variance are still based on the $M_i\bar{y}_i$ values, $\hat{\mu}_r$ being based on their average and s_r being the standard deviation of terms of the form $(M_i\bar{y}_i - \hat{\mu}_r M_i)$. The s_i terms are the within-sample standard deviations for the respective samples, and the "within" term for each sample is its component of the within-sample variance, namely,

$$M_i(M_i - m_i)s_i^2/m_i$$

just as in the case of unbiased estimation. All of these computations, including the statistical summary at the bottom, are easily made on a computer or graphing calculator.

Row	Plant	Mi	mi	ȳi	Miȳi	Miȳi-$\hat{\mu}$rMi	si	Within
1	1	50	10	5.40000	270.000	40.0000	3.37310	2275.56
2	2	65	13	4.00000	260.000	−39.0000	3.26599	2773.33
3	3	45	9	5.66667	255.000	48.0000	4.09268	3015.00
4	4	48	10	4.80000	230.400	9.6000	3.64539	2423.89
5	5	52	10	4.30000	223.600	−15.6000	3.33500	2429.09
6	6	58	12	3.83333	222.333	−44.4667	3.85730	3308.05
7	7	42	8	5.00000	210.000	16.8000	2.26779	0918.00
8	8	66	13	3.84615	253.846	−49.7539	2.07550	1159.10
9	9	40	8	4.87500	195.000	11.0000	2.47487	980.00
10	10	56	11	5.00000	280.000	22.4000	3.43511	2703.27

	N	Mean	Median	SD
Mi	10	52.20	51.00	9.00
mi	10	10.400	10.000	1.84
Miȳi	10	240.02	242.12	27.72
Miȳi-$\hat{\mu}$rMi	10	−0.1	10.3	35.1

Solution

Because M is unknown, we must use $\hat{\mu}_r$, given by Equation (9.7), to estimate μ. Our calculations yield

$$\hat{\mu}_r = \frac{\sum_{i=1}^{n} M_i \bar{y}_i}{\sum_{i=1}^{n} M_i} = \frac{\frac{1}{n}\sum_{i=1}^{n} M_i \bar{y}_i}{\frac{1}{n}\sum_{i=1}^{n} M_i} = \frac{240.02}{52.2}$$
$$= 4.60$$

To find the estimated variance of $\hat{\mu}_r$, we must calculate

$$s_r^2 = \frac{1}{n-1}\sum_{i=1}^{n} M_i^2(\bar{y}_i - \hat{\mu}_r)^2$$
$$= (35.1)^2$$

from the standard deviation of the values for $(M_i\bar{y}_i - \hat{\mu}_r M_i)$.

Note that as in Example 9.1,

$$\sum_{i=1}^{n} M_i^2 \left(\frac{M_i - m_i}{M_i}\right)\left(\frac{s_i^2}{m_i}\right) = 21{,}985$$

We can estimate $\bar{M}$ by using the average cluster size for the sample:

$$\frac{\sum_{i=1}^{n} M_i}{n} = \frac{522}{10} = 52.2$$

Substituting into Equation (9.8) yields the estimated variance of $\hat{\mu}_r$:

$$\begin{aligned}\hat{V}(\hat{\mu}_r) &= \left(\frac{N-n}{N}\right)\left(\frac{1}{n\bar{M}^2}\right)s_r^2 + \frac{1}{nN\bar{M}^2}\sum_{i=1}^{n} M_i^2\left(\frac{M_i - m_i}{M_i}\right)\left(\frac{s_i^2}{m_i}\right) \\ &= \left(\frac{90-10}{90}\right)\left[\frac{1}{(10)(52.2)^2}\right](35.1)^2 \\ &\quad + \frac{1}{(10)(90)(52.2)^2}(21{,}985) \\ &= .0492\end{aligned}$$

The estimate of the average downtime with a bound on the error of estimation is

$$\hat{\mu}_r \pm 2\sqrt{\hat{V}(\hat{\mu})}$$
$$4.60 \pm 2\sqrt{.0492}$$
$$4.60 \pm .44$$

Thus, the estimated mean downtime per machine is 4.60 hours with a bound on the error of estimation of .44 hour.

9.5

ESTIMATION OF A POPULATION PROPORTION

Consider the problem of estimating a population proportion p such as the proportion of university students in favor of a certain law or the proportion of machines that have had no downtime for the past month. An estimate of p can be obtained by using $\hat{\mu}$, given in Equation (9.1), or $\hat{\mu}_r$, given in Equation (9.7), and letting $y_{ij} = 1$ or 0 depending on whether or not the jth element in the ith cluster falls into the category of interest.

Because M is usually unknown, we present the formula for estimating p with a ratio estimator analogous to $\hat{\mu}_r$, given in Equation (9.7). Let $\hat{p}_i$ denote the proportion of sampled elements from cluster i that fall into the category of interest.

Estimator of a population proportion p:

$$\hat{p} = \frac{\sum_{i=1}^{n} M_i \hat{p}_i}{\sum_{i=1}^{n} M_i} \tag{9.11}$$

Estimated variance of $\hat{p}$:

$$\hat{V}(\hat{p}) = \left(\frac{N-n}{N}\right)\left(\frac{1}{n\bar{M}^2}\right)s_r^2 + \frac{1}{nN\bar{M}^2}\sum_{i=1}^{n} M_i^2 \left(\frac{M_i - m_i}{M_i}\right)\left(\frac{\hat{p}_i\hat{q}_i}{m_i - 1}\right) \tag{9.12}$$

where

$$s_r^2 = \frac{\sum_{i=1}^{n} M_i^2(\hat{p}_i - \hat{p})^2}{n-1} = \frac{\sum_{i=1}^{n} (M_i\hat{p}_i - \hat{p}M_i)^2}{n-1} \tag{9.13}$$

and $\hat{q}_i = 1 - \hat{p}_i$.

Example 9.4

The manufacturer in Example 9.1 wants to estimate the proportion of machines that have been shut down for major repairs (those requiring parts from stock outside the factory). The sample proportions of machines requiring major repairs are given in Table 9.2. The data are for the machines sampled in Example 9.1. Estimate p, the proportion of machines involved in major repairs for all plants combined, and place a bound on the error of estimation.

TABLE 9.2 **Proportion of sewing machines requiring major repairs**

Plant	M_i	m_i	Proportion of machines requiring major repairs, $\hat{p}_i$
1	50	10	.40
2	65	13	.38
3	45	9	.22
4	48	10	.30
5	52	10	.50
6	58	12	.25
7	42	8	.38
8	66	13	.31
9	40	8	.25
10	56	11	.36

Solution

The following display shows a convenient spreadsheet formulation of the data and summary calculations for completing the analysis in two-stage cluster sampling using a ratio estimator of a proportion. The estimator of the proportion and the between-cluster component of the variance are based on the $M_i \hat{p}_i$ values, $\hat{p}$ being based on their average and s_r being the standard deviation of terms of the form $(M_i \hat{p}_i - \hat{p} M_i)$. The "within" term for each sample is its component of the within-sample variance, namely,

$$M_i(M_i - m_i)\hat{p}_i(1 - \hat{p}_i)/(m_i - 1)$$

All of these computations, including the statistical summary at the bottom, are easily made on a computer or graphing calculator.

Row	Plant	Mi	mi	p̂i	Mip̂i	Mip̂i−p̂Mi	Within
1	1	50	10	0.40	20.00	3.00	53.3333
2	2	65	13	0.38	24.70	2.60	66.3607
3	3	45	9	0.22	9.90	−5.40	34.7490
4	4	48	10	0.30	14.40	−1.92	42.5600
5	5	52	10	0.50	26.00	8.32	60.6667
6	6	58	12	0.25	14.50	−5.22	45.4773
7	7	42	8	0.38	15.96	1.68	48.0624
8	8	66	13	0.31	20.46	−1.98	62.3519
9	9	40	8	0.25	10.00	−3.60	34.2857
10	10	56	11	0.36	20.16	1.12	58.0608

	N	Mean	Median	SD
Mi	10	52.20	51.00	9.00
mi	10	10.400	10.000	1.84
Mip̂i	10	17.61	17.98	5.59
Mip̂i-p̂Mi	10	−0.14	−0.40	4.29

The best estimate of p is given by

$$\hat{p} = \frac{\sum_{i=1}^{n} M_i \hat{p}_i}{\sum_{i=1}^{n} M_i} = \frac{\frac{1}{n}\sum_{i=1}^{n} M_i \hat{p}_i}{\frac{1}{n}\sum_{i=1}^{n} M_i} = \frac{17.61}{52.20} = .34$$

To estimate the variance of $\hat{p}$, we calculate

$$s_r^2 = \frac{1}{n-1}\sum_{i=1}^{n} M_i^2(\hat{p}_i - \hat{p})^2 = (4.29)^2$$

the square of the standard deviation among terms $(M_i\hat{p}_i - \hat{p}M_i)$ and

$$\sum_{i=1}^{n} M_i^2\left(\frac{M_i - m_i}{M_i}\right)\left(\frac{\hat{p}_i\hat{q}_i}{m_i - 1}\right) = 505.91$$

the sum of the last column on the spreadsheet. Then the estimated variance of $\hat{p}$ when $\bar{M}$ is estimated by the sample average, 52.2, is

$$\begin{aligned}\hat{V}(\hat{p}) &= \left(\frac{N-n}{N}\right)\left(\frac{1}{n\bar{M}^2}\right)s_r^2 + \frac{1}{nN\bar{M}^2}\sum_{i=1}^{n} M_i^2\left(\frac{M_i - m_i}{M_i}\right)\left(\frac{\hat{p}_i\hat{q}_i}{m_i - 1}\right)\\ &= \left(\frac{90-10}{90}\right)\left[\frac{1}{(10)(52.2)^2}\right](4.29)^2 + \frac{1}{(10)(90)(52.2)^2}(505.91)\\ &= .00081\end{aligned}$$

The best estimate of the proportion of machines that have undergone major repairs is

$$\hat{p} \pm 2\sqrt{\hat{V}(\hat{p})}$$

$$.34 \pm 2\sqrt{.00081}$$

$$.34 \pm .057$$

We estimate the proportion of machines involved in major repairs to be .34, with a bound of .056 on the error of estimation.

9.6

SAMPLING EQUAL-SIZED CLUSTERS

Suppose that each cluster contains $\bar{M}$ elements; that is

$$M_1 = M_2 = \cdots = M_N = \bar{M}$$

In this case, it is common to take samples of equal size from each cluster, so that

$$m_1 = m_2 = \cdots = m_n = m$$

Under these conditions, Equation (9.1) gives

$$\hat{\mu} = \frac{1}{n}\sum_{i=1}^{n}\bar{y}_i \tag{9.14}$$

which is equivalent to the overall sample average

$$\hat{\mu} = \frac{1}{nm}\sum_{i=1}^{n}\sum_{j=1}^{m} y_{ij}$$

where y_{ij} denotes the jth measurement within the ith cluster. This situation might occur in sampling packaged products (each cluster being, say, a carton of 24 cans of vegetables) or in sampling manufactured items (each cluster perhaps being the soldered connections inside a microcomputer).

Equation (9.2) now becomes

$$\hat{V}(\hat{\mu}) = (1 - f_1)\frac{\text{MSB}}{nm} + (1 - f_2)\left(\frac{1}{N}\right)\frac{\text{MSW}}{m} \qquad (9.15)$$

where $f_1 = n/N$, $f_2 = m/\bar{M}$,

$$\text{MSB} = \frac{m}{n-1}\sum_{i=1}^{n}(\bar{y}_i - \hat{\mu})^2$$

and

$$\text{MSW} = \frac{1}{n(m-1)}\sum_{i=1}^{n}\sum_{j=1}^{m}(y_{ij} - \bar{y}_i)^2$$

$$= \frac{1}{n}\sum_{i=1}^{n} s_i^2$$

Remember that MSB is the between-cluster mean square and MSW is the within-cluster mean square from Chapter 8.

Given Equation (9.15), we can make a series of important observations on the behavior of two-stage cluster sampling:

1. If N is large, $\hat{V}(\hat{\mu}) = \text{MSB}/nm$ and depends only on the cluster means. Thus, we can produce a good estimator of the variance of $\hat{\mu}$ even if the s_i^2 terms are poor estimators of the within-cluster variance. This may happen, for example, if systematic sampling is used within the clusters.
2. If $m = \bar{M}$ (or $f_2 = 1$), then two-stage cluster sampling reduces to one-stage cluster sampling, as in Chapter 8.
3. If $n = N$, then

$$\hat{V}(\hat{\mu}) = (1 - f_2)\frac{\text{MSW}}{nm}$$

which is the variance estimate one would obtain in a stratified random sample with $n = N$ strata and m observations from each.

Thus, we see that when m is close to $\bar{M}$, two-stage cluster sampling behaves like single-stage cluster sampling. When n is close to N, two-stage cluster sampling behaves like stratified random sampling. If clusters are heterogeneous within, we should choose to sample as in the former case. If clusters are homogeneous within, we should choose to sample as in the latter case. More specific information on sample size selection is provided below.

When N is large, the variance estimate

$$\hat{V}(\hat{\mu}) = \frac{1}{nm}\text{MSB} \tag{9.16}$$

actually estimates the true variance

$$V(\hat{\mu}) = \frac{1}{n}\left(\sigma_b^2 + \frac{\sigma_w^2}{m}\right) \tag{9.17}$$

where

σ_b^2 = Variance among the cluster means

σ_w^2 = Variance among the elements within clusters.

Since MSB/m estimates $\sigma_b^2 + \sigma_w^2/m$, and MSW estimates σ_w^2, it follows that

$$\frac{1}{m}(\text{MSB} - \text{MSW}) \tag{9.18}$$

estimates σ_b^2. Separate estimators of σ_b^2 and σ_w^2 will be needed for sample-size calculations.

As in the case of stratified random sampling, we now want to find sample sizes m and n that either minimize $V(\hat{\mu})$ for a fixed cost or minimize total cost of sampling for a fixed $V(\hat{\mu})$. Suppose the cost associated with sampling each cluster is c_1 and the cost associated with sampling each element within a cluster is c_2. Then the total cost c is

$$c = nc_1 + nmc_2 \tag{9.19}$$

The value of m that minimizes $V(\hat{\mu})$ for fixed cost, or minimizes c for fixed variance, is given by

$$m = \sqrt{\frac{\sigma_w^2 c_1}{\sigma_b^2 c_2}} \tag{9.20}$$

After m is determined, n is found from Equation (9.17) if $V(\hat{\mu})$ is fixed or Equation (9.19) if c is fixed. Note that m increases as σ_w^2 increases and decreases as σ_b^2 increases. Thus, more elements are sampled from within clusters (and hence fewer clusters are sampled) as σ_w^2 gets large compared to σ_b^2.

Example 9.5

One of the key quality assurance measurements in the manufacturing of automobile batteries is the thickness of the lead plates. Positive plates are manufactured to be thicker than negative plates, so the two must be treated separately. It is desired to set up a sampling plant to sample n batteries per day and make m negative plate thickness measurements per battery, so that the variance of the estimate mean plate thickness is

0.1. (Measurements of thickness are in thousandths of an inch.) The cost of cutting a battery open is six times the cost of measuring a plate.

A preliminary study of four batteries, with nine plate thickness measurements per battery gave the following data:

B_1: 97, 101, 97, 97, 99, 100, 96, 100, 100

B_2: 95, 96, 96, 99, 96, 97, 95, 96, 100

B_3: 99, 96, 97, 97, 96, 98, 99, 98, 100

B_4: 94, 95, 97, 98, 97, 97, 97, 95, 96

Solution

The computations for MSB and MSW are quickly found by analysis of variance (the following are done on MINITAB):

ANOVA

Source	DF	SS	MS
Factor	3	30.31	10.10
Error	32	79.33	2.48
Total	35	109.64	

Note that MSW = 2.48 and MSB = 10.10, from which the estimate of σ_b^2, Equation (9.18), becomes

$$\frac{1}{9}(\text{MSB} - \text{MSW}) = .85$$

Now, the optimum value of m is

$$m = \sqrt{\frac{\sigma_w^2 c_1}{\sigma_b^2 c_2}} \approx \sqrt{\frac{2.48}{.85}\left(\frac{6}{1}\right)} = 4.18$$

We would take $m = 4$ thickness measurements per battery. Setting $V(\hat{\mu}) = .1$ and using Equation (9.17) yields

$$.1 = \frac{1}{n}\left(.85 + \frac{2.48}{4}\right)$$

$$n = \frac{1}{.1}(1.47) = 14.7 \approx 15$$

Thus, the quality assurance plan should call for sampling $n = 15$ batteries and making $m = 4$ negative plate thickness measurements per battery. This should result in an estimated mean thickness with variance of around 0.1, for the day, at minimum cost. The fact that $n > m$ should not be surprising in view of the fact that the data show little variability within batteries.

9.7

TWO-STAGE CLUSTER SAMPLING WITH PROBABILITIES PROPORTIONAL TO SIZE

Since the number of elements in a cluster may vary greatly from cluster to cluster, a technique often advantageous is to sample clusters with probabilities proportional to their sizes, as was discussed in Section 8.8. Generally, pps sampling is used only at the first stage of a two-stage sampling procedure because the elements within clusters tend to be somewhat similar in size. Hence, we will present estimators of μ and τ for two-stage cluster sampling in which the first-stage sampling is carried out with probabilities proportional to size.

Equation (8.19) provides an estimator of μ, in the case of single-stage cluster sampling, of the form

$$\frac{1}{n}\sum_{i=1}^{n} \bar{y}_i \tag{9.21}$$

In Chapter 8, $\bar{y}_i$ was calculated from all the elements in cluster i and was *exactly* the cluster mean. In this chapter, $\bar{y}_i$ is calculated from a sample of elements from cluster i and is only an *estimate* of the cluster mean. Nevertheless, Equation (9.21) still forms an unbiased estimate of μ, with an estimated variance as given in Equation (8.20).

To form an unbiased estimate of τ, one merely multiplies Equation (9.21) by M, the number of elements in the population.

Estimator of the population mean μ:

$$\hat{\mu}_{\text{pps}} = \frac{1}{n}\sum_{i=1}^{n} \bar{y}_i \tag{9.22}$$

Estimated variance of $\hat{\mu}_{\text{pps}}$:

$$\hat{V}(\hat{\mu}_{\text{pps}}) = \frac{1}{n(n-1)}\sum_{i=1}^{n} (\bar{y}_i - \hat{\mu}_{\text{pps}})^2 \tag{9.23}$$

Estimator of the population total τ:

$$\hat{\tau}_{\text{pps}} = \frac{M}{n}\sum_{i=1}^{n} \bar{y}_i \tag{9.24}$$

Estimated variance of τ_{pps}:

$$\hat{V}(\hat{\tau}_{\text{pps}}) = \frac{M^2}{n(n-1)}\sum_{i=1}^{n} (\bar{y}_i - \hat{\mu}_{\text{pps}})^2 \tag{9.25}$$

We illustrate this pps procedure with the following examples.

Example 9.6

From the six hospitals in a city, a researcher desires to sample three hospitals for the purpose of estimating the proportion of current patients who have been (or will be) in the hospital for more than two consecutive days. Since the hospitals vary in size, they will be sampled with probabilities proportional to their numbers of patients. For the three sampled hospitals, 10% of the records of current patients will be examined to determine how many patients will stay in the hospital for more than two days. Given the information on hospital sizes in the accompanying table, select a sample of three hospitals with probabilities proportional to size.

Hospital	Number of patients	Cumulative range	Hospital	Number of patients	Cumulative range
1	328	1–328	4	220	870–1089
2	109	329–437	5	280	1090–1369
3	432	438–869	6	190	1370–1559

Solution

Since three hospitals are to be selected, three random numbers between 0001 and 1559 must be chosen from the random number table. Our numbers turned out to be 1505, 1256, and 0827. Locating these numbers in the cumulative range column leads to the selection of hospitals 3, 5, and 6.

Example 9.7

Suppose the sampled hospitals in Example 9.6 yielded the following data on number of patients staying more than two days:

Hospital	Number of patients sampled	Number staying more than two days
3	43	25
5	28	15
6	19	8

Estimate the proportion of patients staying more than two days, for all six hospitals, and place a bound on the error of estimation.

Solution

The proportion of interest for each hospital is simply the sample mean and, by Equation (9.22), the best estimate of the population proportion is the average of the three sample means. Thus,

$$\hat{\mu}_{\text{pps}} = \frac{1}{3}\left(\frac{25}{43} + \frac{15}{28} + \frac{8}{19}\right) = \frac{1}{3}(.58 + .54 + .42) = .51$$

By Equation (9.23),

$$\hat{V}(\hat{\mu}_{\text{pps}}) = \frac{1}{3(2)}[(.58 - .51)^2 + (.54 - .51)^2 + (.42 - .51)^2]$$
$$= .0025$$

Thus, the bound on the error of estimation is

$$2\sqrt{\hat{V}(\hat{\mu}_{\text{pps}})} = 2\sqrt{.0025} = .10$$

and our estimate of the true population proportion is $.51 \pm .10$.

One further comment about when one uses pps sampling is in order. If the variation as measured by s_b^2 is small in comparison with that measured by s_i^2 [that is, if the second term dominates the variance expression (9.2)], then we would want to select few clusters and many elements from within each sampled cluster. In that case, any sampling plan for clusters will work well.

If, however, the s_i^2 terms are small compared with s_b^2 [Equation (9.2) is dominated by the first term], then great care should be taken in planning the selection of clusters. In this case, the comments made at the end of Section 8.8 still hold; the pps method works well if the cluster sizes vary appreciably.

9.8

SUMMARY

The concept of cluster sampling can be extended to two-stage sampling by taking a simple random sample of elements from each sampled cluster. Two-stage cluster sampling is advantageous when one wishes to have sample elements in geographic proximity because of travel costs.

Two-stage cluster sampling eliminates the need to sample all elements in each sampled cluster. Thus, the cost of sampling can often be reduced with little loss of information.

An unbiased estimator of μ was presented for the case when M, the total number of elements in the population, is known. When M is unknown, a ratio estimator is employed. Estimators were also given for a population total τ and for a population proportion p.

CASE STUDY REVISITED

THE ESTIMATION OF ENTERTAINMENT EXPENSES

The sampling plan for estimating average monthly entertainment expenses among students, outlined at the beginning of this chapter, involved the selection of $n = 4$ classrooms from $N = 150$ and subsampling students from each. The data are as follows (averages in dollars):

No. of students per class, M_i	No. of students sampled, m_i	$\bar{y}_i$	s_i^2	$M_i\bar{y}_i$	$[M_i(\bar{y}_i - \hat{\mu}_r)]^2$
80	8	16	9	1280	518,400
47	5	30	15	1410	55,225
62	6	21	22	1302	61,504
39	4	45	18	1755	608,400
228				5747	1,243,529

$$\hat{\mu}_r = \frac{\sum_{i=1}^{n} M_i \bar{y}_i}{\sum_{i=1}^{n} M_i} = \frac{5747}{228} = 25 \qquad \bar{M} = 57$$

$$s_r^2 = \left(\frac{1}{n-1}\right) \sum_{i=1}^{n} [M_i(\bar{y}_i - \hat{\mu}_r)]^2 = 414{,}510$$

$$s_w^2 = \sum_{i=1}^{n} M_i(M_i - m_i)\left(\frac{s_i^2}{m_i}\right) = 31{,}275$$

The best estimate of the average monthly amount spent on entertainment among all students at the university is $\hat{\mu}_r = \$25$. The bound on the error of estimation is

$$2\sqrt{\hat{V}(\hat{\mu}_r)} = 2\sqrt{\left(\frac{N-n}{N}\right)\left(\frac{1}{n\bar{M}^2}\right)(s_r^2) + \left(\frac{1}{nN\bar{M}^2}\right)(s_w^2)}$$

$$= 2\sqrt{\left(\frac{146}{150}\right)\left[\frac{1}{4(57)^2}\right](414{,}510) + \left[\frac{1}{4(150)(57)^2}\right](31{,}275)} = 11$$

Hence, we are confident that the true average monthly entertainment amount is between $25 - 11 = \$14$ and $25 + 11 = \$36$. (This interval could be reduced in size by sampling more classrooms or more students per classroom. Which procedure would you recommend?)

EXERCISES

9.1 Suppose a large retail store has its accounts receivable listed by department. The firm wishes to estimate the total accounts receivable on a given day by sampling. Discuss the relative merits of stratified random sampling, single-stage cluster sampling, systematic sampling, and two-stage cluster sampling. What extra information would you like to have on these accounts before selecting the sampling design?

9.2 A nurseryman wants to estimate the average height of seedlings in a large field that is divided into 50 plots that vary slightly in size. He believes the heights are fairly constant throughout each plot but may vary considerably from plot to plot. Therefore, he decides to sample 10% of the trees within each of 10 plots using a two-stage cluster sample. The data are as given in the accompanying table. Estimate the average height of seedlings in the field and place a bound on the error of estimation.

Plot	Number of seedlings	Number of seedlings sampled	Heights of seedlings (in inches)
1	52	5	12, 11, 12, 10, 13
2	56	6	10, 9, 7, 9, 8, 10
3	60	6	6, 5, 7, 5, 6, 4
4	46	5	7, 8, 7, 7, 6
5	49	5	10, 11, 13, 12, 12
6	51	5	14, 15, 13, 12, 13
7	50	5	6, 7, 6, 8, 7
8	61	6	9, 10, 8, 9, 9, 10
9	60	6	7, 10, 8, 9, 9, 10
10	45	6	12, 11, 12, 13, 12, 12

9.3 In Exercise 9.2, assume that the nurseryman knows there are approximately 2600 seedlings in the field. Use this additional information to estimate the average height and place a bound on the error of estimation.

9.4 A supermarket chain has stores in 32 cities. A company official wants to estimate the proportion of stores in the chain that do not meet a specified cleanliness criterion. Stores within each city appear to possess similar characteristics; therefore, she decides to select a two-stage cluster sample containing one-half of the stores within each of 4 cities. Cluster sampling is desirable in this situation because of travel costs. The data collected are given in the accompanying table. Estimate the proportion of stores not meeting the cleanliness criterion and place a bound on the error of estimation.

City	Number of stores in city	Number of stores sampled	Number of stores not meeting criterion
1	25	13	3
2	10	5	1
3	18	9	4
4	16	8	2

9.5 Repeat Exercise 9.4, given that the chain contains 450 stores. [*Hint*: Use the unbiased estimator of Equation (9.1) and adapt it to proportions.]

9.6 To improve telephone service, an executive of a certain company wants to estimate the total number of phone calls placed by secretaries in the company during one day. The company contains 12 departments, each making approximately the same number of calls per day. Each department employs approximately 20 secretaries, and the number of calls made varies considerably from secretary to secretary. The executive decides to employ two-stage cluster sampling, using a small number of departments (clusters) and selecting a fairly large number of secretaries (elements) from each. Ten secretaries are sampled from each of 4 departments. The data are summarized in the accompanying table. Estimate the total number of calls placed by the secretaries in this company and place a bound on the error of estimation.

Department	Number of secretaries	Number of secretaries sampled	Mean, $\bar{y}_i$	Variance, s_i^2
1	21	10	15.5	2.8
2	23	10	15.8	3.1
3	20	10	17.0	3.5
4	20	10	14.9	3.4

9.7 A city-zoning commission wants to estimate the proportion of property owners in a certain section of a city who favor a proposed zoning change. The section is divided into seven distinct residential areas, each containing similar residents. Because the results must be obtained in a short period of time, two-stage cluster sampling is used. Three of the seven areas are selected at random, and 20% of the property owners in each area selected are sampled. The figure of 20% seems reasonable because the people living within each area seem to be in the same socioeconomic class and hence tend to hold similar opinions on the zoning question. The results are given in the accompanying table. Estimate the proportion of property owners who favor the proposed zoning change and place a bound on the error of estimation.

Area	Number of property owners	Number of property owners sampled	Number in favor of zoning change
1	46	9	1
2	67	13	2
3	93	20	2

9.8 A forester wants to estimate the total number of trees in a certain county that are infected with a particular disease. There are ten well-defined forest areas in the county; these areas can be subdivided into plots of approximately the same size. Four crews are available to conduct the survey, which must be completed in one day. Hence, two-stage cluster sampling is used. Four areas (clusters) are chosen with six plots (elements) randomly selected from each. (Each crew can survey six plots in one day.) The data are given in the accompanying table. Estimate the total number of infected trees in the county and place a bound on the error of estimation.

Area	Number of plots	Number of plots sampled	Number of infected trees per plot
1	12	6	15, 14, 21, 13, 9, 10
2	15	6	4, 6, 10, 9, 8, 5
3	14	6	10, 11, 14, 10, 9, 15
4	21	6	8, 3, 4, 1, 2, 5

9.9 A new bottling machine is being tested by a company. During a test run, the machine fills 24 cases, each containing 12 bottles. The company wishes to estimate the average number of ounces of fill per bottle. A two-stage cluster sample is employed using 6 cases (clusters), with 4 bottles (elements) randomly selected from each. The results are given in the accompanying table. Estimate the average number of ounces per bottle and place a bound on the error of estimation.

Case	Average ounces of fill for sample, $\bar{y}_i$	Sample variance, s_i^2
1	7.9	.15
2	8.0	.12
3	7.8	.09
4	7.9	.11
5	8.1	.10
6	7.9	.12

9.10 A certain manufacturing plant contains 40 machines, all producing the same product (say, boxes of cereal). An estimate of the proportion of defective products (say, boxes underfilled) for a given day is desired. Discuss the relative merits of two-stage cluster sampling (machines as clusters of boxes) and stratified random sampling (machines as strata) as possible designs for this study.

9.11 A market research firm constructed a sampling plan to estimate the weekly sales of brand A cereal in a certain geographic area. The firm decided to sample cities within the area and then to sample supermarkets within cities. The number of boxes of brand A cereal sold in a specified week is the measurement of interest. Five cities are sampled from the 20 in the area. Using the data given in the accompanying table, estimate the average sales for the week for all supermarkets in the area. Place a bound on the error of the estimation. Is the estimator you used unbiased?

City	Number of supermarkets	Number of supermarkets sampled	$\bar{y}_i$	s_i^2
1	45	9	102	20
2	36	7	90	16
3	20	4	76	22
4	18	4	94	26
5	28	6	120	12

9.12 In Exercise 9.11, do you have enough information to estimate the total number of boxes of cereal sold by all supermarkets in the area during the week? If so, explain how you would estimate this total, and place a bound on the error of estimation.

9.13 If a study like the one outlined in Exercise 9.11 is to be done again, would you recommend that cities be sampled with probabilities proportional to their numbers of supermarkets? Why?

9.14 Suppose a sociologist wants to estimate the total number of retired people residing in a certain city. She decides to sample blocks and then sample households within blocks. (Block statistics from the Census Bureau aid in determining the number of households in each block.) Four blocks are randomly selected from the 300 of the city. From the data in the accompanying table, estimate the total number of retired residents in the city and place a bound on the error of estimation.

Block	Number of households	Number of households sampled	Number of retired residents per household
1	18	3	1, 0, 2
2	14	3	0, 3, 0
3	9	3	1, 1, 2
4	12	3	0, 1, 1

9.15 Using the data in Exercise 9.14, estimate the average number of retired residents per household and place a bound on the error of estimation.

9.16 From the data in Exercise 9.14, can you estimate the average number of retired residents per block? How can you construct this estimate and place a bound on the error of estimation?

9.17 In the estimation of the amount of impurities in a bulk product like sugar, the sampling procedure may select bags of sugar from a warehouse and then select small test samples from each bag. The test samples are analyzed for amount of impurities. Discuss how one might choose the number of bags to sample and the number and size of the test samples taken from each bag.

9.18 A quality assurance program requires a sampling of manufactured products as they come off assembly lines in a production facility. One could treat the assembly lines as clusters or as strata. Discuss the relative merits of these two options and design a sampling plan for each case.

9.19 It is desired to estimate the *total* amount of money spent during a quarter for entertainment by students housed on a particular campus. There are N students housed on campus, of which a simple random sample of n could be selected. There are also M housing units on campus, of which a simple random sample of m could be selected. The housing units range from small fraternity houses to large dormitories. Discuss the relative merits of using individual students or housing units as sampling units. Which would you recommend, and why?

EXPERIENCES WITH REAL DATA

9.1 Refer to the U.S. population data in the appendix. Construct a two-stage cluster sampling estimate of the total population in the United States for 2010 by first sampling divisions and then sampling states within divisions. Sample four of the nine divisions and at least two states within each sampled division. Compute an estimate of the variance attached to your estimator. Would you recommend this procedure over stratified random sampling? Why?

9.2 When one is sampling people, the naturally occurring frames typically involve people grouped in clusters. Hence, two-stage cluster sampling is commonly employed as a matter of economic convenience. For example, try estimating the total number of library books currently checked out by students on your campus. (Any other numerical variable of interest to you can be substituted for the number of library books.) Some naturally occurring clusters of students are those in residence halls, classrooms, fraternities and sororities, and pages of a student directory. (Can you think of others?) Estimate the total of interest, and place a bound on the error of estimation, by using the following procedures:

(a) Sampling residences and students within residences
(b) Sampling classrooms in use and students within classrooms
(c) Sampling pages of the student directory and students' names within pages

Whichever method you choose, think carefully about the relative sample sizes for the first and second stages. If the experiment with students is not applicable to your situation, a simpler problem to carry out is to estimate the number of words in this (or any other) book by randomly sampling pages and then sampling lines within a page. Should the two-stage sampling scheme for a statistics book with formulas and tables differ from the scheme for a novel?

CASE STUDY

HOW MANY PEOPLE ATTENDED THE CONCERT?

A newspaper reporter wants to estimate the number of people in attendance at a free rock concert, and she goes to a statistician for advice. The statistician observes that the newspaper is handing out free orange hats to some attendees, and he asks how many are to be given out. Armed with this knowledge, he suggests the following estimation scheme.

Suppose t hats are being worn by a subgroup of the N people in attendance. After the concert begins, n people are randomly selected, and the number s wearing the orange hats is observed. Now the sample proportion of those wearing the hats, $\hat{p} = s/n$, is an estimate of the population proportion t/N. In other words,

$$\hat{p} = \frac{s}{n} \approx \frac{t}{N}$$

or

$$N \approx \frac{nt}{s} = \hat{N}$$

The details of this estimator are developed in Chapter 10.

CHAPTER 10

ESTIMATING THE POPULATION SIZE

10.1

INTRODUCTION

In the preceding chapters, we estimated means, totals, and proportions, assuming that the population size was either known or so large that it could be ignored if not expressly needed to calculate an estimate. Frequently, however, the population size is not known and is important to the goals of the study. In fact, in some studies, estimation of the population size is the main goal. The study of the growth, evolution, and maintenance of wildlife populations depends crucially on accurate estimates of population sizes, and estimating the size of such populations will motivate much of our discussion in this chapter. The techniques can also be used for estimating the number of people at a concert or a sporting event, the number of defects in a bolt of material, and many similar quantities. We will present and discuss four methods for estimating population sizes.

The first method is *direct sampling*. This procedure entails drawing a random sample from a wildlife population of interest, tagging each animal sampled, and returning the tagged animals to the population. At a later date, another random sample (of a fixed size) is drawn from the same population, and the number of tagged animals is observed. If N represents the total population size, t represents the number of animals tagged in the initial sample, and p represents the proportion of tagged animals in the population, then

$$\frac{t}{N} = p$$

Consequently, $N = t/p$. We can obtain an estimate of N because t is known and p can be estimated by $\hat{p}$, the proportion of tagged animals in the second sample. Thus,

$$\hat{N} = \frac{\text{the number of animals tagged}}{\text{the proportion of tagged animals in the second sample}}$$

or, equivalently,

$$\hat{N} = \frac{t}{\hat{p}}$$

The second technique is *inverse sampling*. It is similar to direct sampling, but the second sample size is not fixed. That is, we sample until a fixed number of tagged animals is observed. Using this procedure, we can also obtain an estimate of N, the total population size, using

$$\hat{N} = \frac{t}{\hat{p}}$$

The third technique depends on first estimating the density of elements in the population and then multiplying by an appropriate measure of area. If we estimate that there are $\hat{\lambda}$ animals per acre and the area of interest contains A acres, then $A\hat{\lambda}$ provides an estimate of the population size.

The fourth method is similar to the third but depends only on being able to identify the presence or absence of animals on the sampled plots. Then under certain conditions the density and the total number of animals can still be estimated.

10.2

ESTIMATION OF A POPULATION SIZE USING DIRECT SAMPLING

Direct sampling can be used to estimate the size of a mobile population. First, a random sample of size t is drawn from the population. At a later date a second sample of size n is drawn. For example, suppose a conservationist is concerned about the apparent decline in the number of seals in Alaska. Estimates of the population size are available from previous years. For determination of whether or not there has been a decline, a random sample of $t = 200$ seals is caught, tagged, and then released. A month later a second sample of size $n = 100$ is obtained. Using these data (often called recapture data), we can estimate N, the population size. This method assumes that tagging does not affect the likelihood of recapture.

Let s be the number of tagged individuals observed in the second sample. The proportion of tagged individuals in the sample is

$$\hat{p} = \frac{s}{n}$$

An estimate of N is given by

$$\hat{N} = \frac{t}{\hat{p}} = \frac{nt}{s}$$

Estimator of N:

$$\hat{N} = \frac{nt}{s} \tag{10.1}$$

Estimated variance of $\hat{N}$:

$$\hat{V}(\hat{N}) = \frac{t^2 n(n-s)}{s^3} \tag{10.2}$$

Note that s must be greater than zero for Equations (10.1) and (10.2) to hold. We will assume that n is large enough so that s is greater than zero with high probability.

You should also realize that $\hat{N}$, which is presented in Equation (10.1), is not an unbiased estimator of N. For $s > 0$,

$$E(\hat{N}) \approx N + \frac{N(N-t)}{nt}$$

Hence, for fairly large sample sizes—that is, t and n large—the term

$$\frac{N(N-t)}{nt}$$

is small, and the bias of the estimator $\hat{N}$ approaches zero. Estimator $\hat{N}$ tends to overestimate the true value of N. Chapman (1952) gives another estimator of N, along with its approximate variance, which is nearly unbiased for most direct sampling situations.

Example 10.1

Before posting a schedule for the upcoming hunting season, the game commission for a particular county wishes to estimate the size of the deer population. A random sample of 300 deer is captured ($t = 300$). The deer are tagged and released. A second sample of 200 is taken two weeks later ($n = 200$). If 62 tagged deer are recaptured in the second sample ($s = 62$), estimate N and place a bound on the error of estimation.

Solution

Using Equation (10.1), we have

$$\hat{N} = \frac{nt}{s} = \frac{200(300)}{62} = 967.74$$

or $\hat{N} = 968$.

A bound on the error of estimation is given by

$$2\sqrt{\hat{V}(\hat{N})} = 2\sqrt{\frac{t^2 n(n-s)}{s^3}} = 2\sqrt{\frac{(300)^2(200)(138)}{(62)^3}} = 204.18$$

Thus, the game commission estimates the total number of deer is 968, with a bound on the error of estimation of approximately 205 deer.

You may be concerned about the magnitude of the bound on the error of estimation in this example. As might be expected, we can obtain a more accurate estimator of N by increasing the two sample sizes (n and t). Further information on the choice of t and n is given in Section 10.4.

10.3

ESTIMATION OF A POPULATION SIZE USING INVERSE SAMPLING

Inverse sampling is the second method for estimating N, the total size of a population. We again assume that an initial sample of t individuals is drawn, tagged, and released. Later, random sampling is conducted until exactly s tagged animals are recaptured. If the sample contains n individuals, the proportion of tagged individuals in the sample is given by $\hat{p} = s/n$. We use this sample proportion to estimate the proportion of tagged individuals in the population.

Again, the estimator of N is given by

$$\hat{N} = \frac{t}{\hat{p}} = \frac{nt}{s}$$

but note that s is fixed and n is random.

Estimator of N:

$$\hat{N} = \frac{nt}{s} \tag{10.3}$$

Estimated variance of $\hat{N}$:

$$\hat{V}(\hat{N}) = \frac{t^2 n(n-s)}{s^2(s+1)} \tag{10.4}$$

Note that Equations (10.3) and (10.4) hold only for $s > 0$. This restriction offers no difficulty; we simply specify that s must be greater than zero, and we sample until s tagged individuals are recaptured. The estimator $\hat{N} = nt/s$, obtained by using inverse sampling, provides an unbiased estimator of N, and the variance given by Equation (10.4) is an unbiased estimator of the true variance of $\hat{N}$.

Variance (10.4) for the inverse case looks very much like variance (10.2) for the direct case, and the estimators $\hat{N}$ appear to be identical. However, the inverse method offers the advantages that s can be fixed in advance, $\hat{N}$ is unbiased, and an unbiased estimator of the true variance of $\hat{N}$ is available.

Example 10.2

Authorities of a large wildlife preserve are interested in the total number of birds of a particular species that inhabit the preserve. A random sample of $t = 150$ birds is trapped, tagged, and then released. In the same month a second sample is drawn until 35 tagged birds are recaptured ($s = 35$). In total, 100 birds are recaptured in order to find 35 tagged ones ($n = 100$). Estimate N and place a bound on the error of estimation.

Solution

Using Equation (10.3), we estimate N by

$$\hat{N} = \frac{nt}{s} = \frac{100(150)}{35} = 428.57$$

A bound on the error of estimation is found by using Equation (10.4) as follows:

$$2\sqrt{\hat{V}(\hat{N})} = 2\sqrt{\frac{t^2 n(n-s)}{s^2(s+1)}} = 2\sqrt{\frac{(150)^2(100)(65)}{(35)^2(36)}} = 115.173$$

Hence, we estimate that 429 birds of the particular species inhabit the preserve. We are quite confident that our estimate is within approximately 116 birds of the true population size.

10.4

CHOOSING SAMPLE SIZES FOR DIRECT AND INVERSE SAMPLING

We have been discussing direct sampling and inverse sampling techniques. You probably wonder which is the better one to use. Either method can be used. Inverse sampling yields more precise information than does direct sampling, provided the second sample size n required to recapture s tagged individuals is small relative to the population size N. However, if nothing is known about the size of N, a poor choice of t could make n quite large when inverse sampling is used. For example, if $N = 10{,}000$

TABLE 10.1 **Values of $V(\hat{N})/N$ for direct sampling**

	$p_1 = t/N$					
$p_2 = n/N$	.001	.01	.1	.25	.50	1.0
.001	999,000	99,000	9000	3000	1000	0
.01	99,000	9,900	900	300	100	0
.1	9,990	990	90	30	10	0
.25	3,996	396	36	12	4	0
.50	1,998	198	18	6	2	0
1.0	999	99	9	3	1	0

and a first sample of $t = 50$ individuals is drawn, a large second sample would be needed to obtain exactly $s = 10$ tagged animals.

Table 10.1 is useful in determining the sample sizes (t and n) required to estimate N with a fixed bound on the error of estimation. However, to use these data requires some prior knowledge concerning the magnitude of N. Entries in Table 10.1 are $V(\hat{N})/N$ for direct sampling. If you know the approximate size of N, you can determine the variance of the estimator for fixed values of the sample sizes t and n. In Table 10.1, these sample sizes are expressed as fractions of N. These fractions, given by

$$p_1 = \frac{t}{N} \qquad \text{and} \qquad p_2 = \frac{n}{N}$$

are called sampling fractions.

Having a graph of the entries in this table would be convenient. However, the numbers are so large that we can only display a portion of Table 10.1. In Figure 10.1, we display the values of $V(\hat{N})/N$ for various values of the sampling fractions $p_1 = t/N$ and $p_2 = n/N$. Note that as either p_1 or p_2 increases, the variance of $\hat{N}$ divided by N decreases; consequently, $V(\hat{N})$ decreases for a fixed value of N. Intuitively, this result makes sense, since we should obtain a more accurate estimate of N by taking large sample sizes.

Example 10.3 The game commission in Example 10.1 believes that the size of the deer population this year is approximately the same as in the preceding year, when there were between 800 and 1000 deer. Determine the bound on the error of estimation associated with the sampling fractions of $p_1 = .25$ and $p_2 = .25$.

Solution

We take the larger of the two figures (N approximately 1000) to obtain a conservative estimate of $V(\hat{N})$ (one that is larger than would be expected). We see from Figure 10.1

FIGURE 10.1 Graph of values of Table 10.1

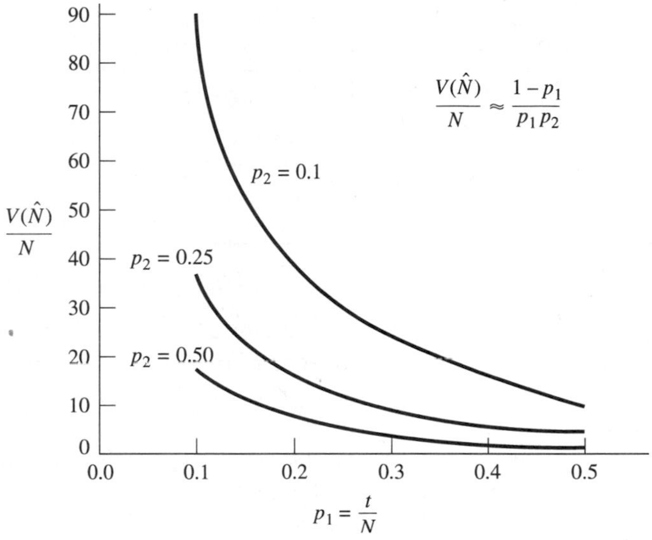

(or Table 10.1) that the sampling fractions of $p_1 = t/N = .25$ and $p_2 = s/N = .25$ yield

$$\frac{V(\hat{N})}{N} = 12$$

Taking $N = 1000$, we have

$$V(\hat{N}) = 1000(12) = 12{,}000$$

$$\sqrt{V(\hat{N})} = \sqrt{12{,}000} = 109.541$$

The corresponding bound on the error of estimation is

$$2\sqrt{V(\hat{N})} = 2(109.541) = 219.082$$

An investigator could use this information to plan his survey. If this bound on the error of estimation is acceptable, he could run a survey using $p_1 = .25$ and $p_2 = .25$; that is, he could draw an initial sample of

$$t = p_1 N = (.25)(1000) = 250$$

and a second sample of

$$n = p_2 N = (.25)(1000) = 250$$

He could then estimate N by using the data from the survey. The bound on the error of estimation should be approximately equal to 220, provided the original range for N is accurate.

TABLE 10.2 **Values of $V(\hat{N})/N$ for inverse sampling**

	$p_1 = t/N$					
$p_2 = s/N$	.001	.01	.1	.25	.5	1.0
.001	999	990	900	750	500	0
.01		99	90	75	50	0
.1			9	7.5	5	0
.25				3	2	0
.5					1	0
1.0						0

If the bound on the error for $\hat{N}$ is not acceptable for the sampling fractions of $p_1 = p_2 = .25$, the investigator can work with Table 10.1 (or Figure 10.1) to determine the sampling fractions required to achieve an acceptable bound on the error of estimation.

We can examine $V(\hat{N})$ for inverse sampling in the same manner as for direct sampling. Entries in Table 10.2 are the values of $V(\hat{N})/N$ for various sampling fractions $p_1 = t/N$ and $p_2 = s/N$ when inverse sampling is used. Recall that in inverse sampling we fix s rather than n; hence, the second sampling fraction is in terms of s. A graphical representation of these data would be helpful, but again the numbers are too large to plot conveniently. A portion of Table 10.2 is presented in Figure 10.2.

FIGURE 10.2 Graph of values of Table 10.2

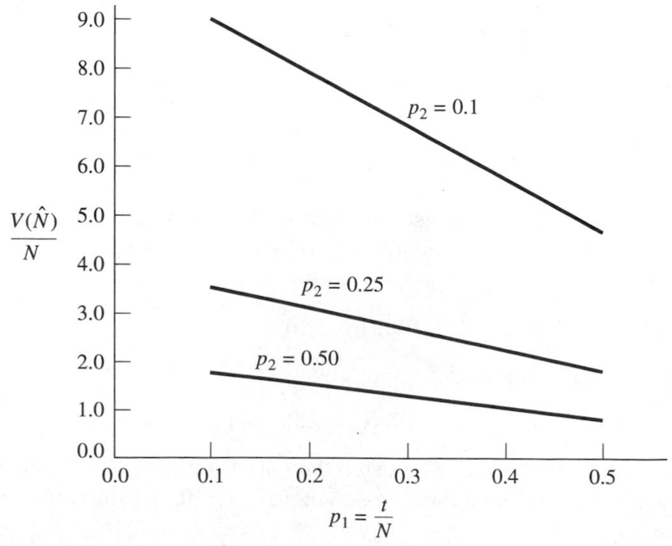

Note that $V(\hat{N})/N$ [or, equivalently, $V(\hat{N})$ for a given value of N] decreases as p_1 and p_2 increase. If the experimenter has an approximate range for N, he or she could use either Figure 10.2 or Table 10.2 to determine the sampling fractions ($p_1 = t/N$, $p_2 = s/N$) necessary to achieve a reasonable bound. Then the experimenter can conduct a survey with an initial sample of

$$t = p_1 N$$

The experimenter would begin a second sample at a future time and continue until

$$s = p_2 N$$

tagged animals are recaptured. The corresponding bound on the error of estimation for N will be acceptable provided the original estimate for N was reasonable.

The preceding tag–recapture techniques can be extended to more than two stages. At the second stage, the $(n - s)$ untagged animals can be tagged, and all n returned to the population. At a later time, a third sample can be taken, and the counting and tagging operation repeated. This multistage approach will result in an improved estimator of N and is especially useful in ongoing studies where samples might be taken every week or so.

10.5

ESTIMATING POPULATION DENSITY AND SIZE FROM QUADRAT SAMPLES

Estimation of the number of elements in a defined area can be accomplished by first estimating the number of elements per unit area (that is, the *density* of the elements) and then multiplying the estimated density by the size of the area under study. For example, if a loom produces 2 defects per square yard of material, on the average, then a bolt containing 40 square yards should contain approximately 80 defects. We will discuss estimates of both the density and the total number of elements. Our discussion, however, need not be confined to areas because the same methods work for estimating the total number of bacteria in a fixed volume of liquid or the total number of telephone calls coming into a switchboard over a fixed interval of time. Talking in terms of areas is convenient for illustrative purposes.

Suppose a region of total area A is to be sampled by randomly selecting n plots, each of area a. For convenience, we assume $A = Na$. Each plot will be called a *quadrat* (even though they may not be square). In the terminology of earlier chapters, a quadrat can be thought of as a cluster of elements. We let m_i denote the number of elements in quadrat i, and we let M, given by

$$M = \sum_{i=1}^{N} m_i$$

denote the total number of elements in the population (having area A). Also, we let

$$\lambda = \frac{M}{A}$$

denote the density of elements, or the number of elements per unit area. Our goal is to estimate λ and then $M = \lambda A$. In this discussion, note that the m_i's are random variables, since they are the numbers of elements that happen to be located in a randomly located quadrat of fixed area.

Suppose the element counts m_i are obtained from n independently and randomly selected quadrats, each of area a. Then

$$\bar{m} = \frac{1}{n}\sum_{i=1}^{n} m_i$$

is an estimator of the number of elements *per quadrat*, and $\hat{\lambda} = \bar{m}/a$ is an estimator of the number of elements *per unit area*. From principles discussed in Chapter 4, it follows that the variance of $\hat{\lambda}$ can be estimated by

$$\frac{1}{a^2}\hat{V}(\bar{m}) = \frac{1}{a^2}\frac{s_m^2}{n} \tag{10.5}$$

where

$$s_m^2 = \frac{\sum_{i=1}^{n}(m_i - \bar{m})^2}{n - 1}$$

From here it is easy to form $\hat{M} = \hat{\lambda}A$ and develop an estimator for the variance of $\hat{M}$.

It seems that there is nothing new here. However, it is often the case that the elements being counted (diseased trees, bacteria colonies, traffic accidents, . . .) are themselves randomly distributed over area, volume, or time. Then we can simplify the above results by taking into account the randomness of the elements. (The reader who has studied probability theory may recognize that we are assuming m_i to have a Poisson distribution.) Under the assumption of randomly dispersed elements, we have the following estimators of λ and M.

Estimator of the density λ:

$$\hat{\lambda} = \frac{\bar{m}}{a} \tag{10.6}$$

Estimated variance of $\hat{\lambda}$:

$$\hat{V}(\hat{\lambda}) = \frac{\hat{\lambda}}{an} \tag{10.7}$$

Estimator of the total M:

$$\hat{M} = \hat{\lambda} A \tag{10.8}$$

Estimated variance of $\hat{M}$:

$$\hat{V}(\hat{M}) = A^2 \hat{V}(\hat{\lambda}) = A^2 \left(\frac{\hat{\lambda}}{an} \right) \tag{10.9}$$

We illustrate the use of these estimators in the following examples.

Example 10.4 Florida has a serious problem with fire ants, and estimating the number of ant hills per unit area (at some key locations) is an important consideration in keeping track of population increases. Fifty sample guadrats of approximately 16 m^2 each gave the following results:

Number of hills	Frequency
0	13
1	8
2	12
3	10
4	5
5	2
	50

From these data, estimate the density of ant hills per unit area.

Solution

For the data given above,

$$\sum_{i=1}^{n} m_i = 0(13) + 1(8) + 2(12) + \cdots + 5(2) = 92$$

and

$$\sum_{i=1}^{n} m_i^2 = 0(13) + 1(8) + 4(12) + \cdots + 25(2) = 276$$

then

$$\hat{\lambda} = \frac{\bar{m}}{a} = \frac{1}{16}\left(\frac{92}{50}\right) = 0.11 \text{ hills}/m^2$$

The variance of $\hat{\lambda}$ can be estimated by

$$\frac{1}{a^2}\frac{s_m^2}{n} = \frac{1}{(16)^2(50)}\frac{1}{(49)}[276 - \frac{(92)^2}{50}]$$
$$= .00017$$

and the bound on the error of estimation is $2\sqrt{.00017} = .026$. Thus, our estimated density of .11 should be within .026 units of the true density.

If we think the ant hills are randomly dispersed in the region under study, we can estimate the variance of $\hat{\lambda}$ by

$$\hat{V}(\hat{\lambda}) = \frac{\hat{\lambda}}{an} = \frac{.11}{(16)(50)} = .00014$$

and

$$2\sqrt{\hat{V}(\hat{\lambda})} = .023$$

about the same as found above. The closeness of the two variance estimators suggests that the randomness assumption may be reasonable.

Note that, under the randomness assumption, both the estimator $\hat{\lambda}$ and $\hat{V}(\hat{\lambda})$ depend on the data only through $\bar{m}$.

Example 10.5

The density of trees having fusiform rust on a southern-pine plantation of 200 acres is to be estimated from a sample of $n = 10$ quadrats of .5 acre each. The ten sampled plots had an average $\bar{m}$ of 2.8 infected trees per quadrat. Estimate the density of infected trees and place a bound on the error of estimation.

Solution

Using Equation (10.6) with $a = .5$, we determine the estimated density as

$$\hat{\lambda} = \frac{\bar{m}}{a} = \frac{2.8}{.5} = 5.6 \text{ trees per acre}$$

The bound on the error is

$$2\sqrt{\frac{\hat{\lambda}}{an}} = 2\sqrt{\frac{5.6}{(.5)(10)}} = 2.1$$

Thus, we estimate the density as 5.6 ± 2.1, or from 3.5 to 7.7 infected trees per acre. This interval is a fairly large interval since the sample size is relatively small.

Example 10.6 For the situation and data in Example 10.5, estimate the total number of infected trees in the 200-acre plantation. Place a bound on the error of estimation.

Solution

Using Equation (10.8), we see that the estimated total is

$$\hat{M} = \hat{\lambda}A = (5.6)(200) = 1120 \text{ trees}$$

The bound on the error, from Equation (10.9), is

$$2\sqrt{\hat{V}(\hat{M})} = 2A\sqrt{\frac{\hat{\lambda}}{an}} = 2(200)\sqrt{\frac{5.6}{(.5)(10)}} = 420$$

Thus, we estimate the total number of infected trees as 1120 ± 420, or 700 to 1540.

Notice that the bound on the error of estimation, for both λ and M, contains both a and n in the denominator. Hence, this bound will decrease as a is increased or as n is increased. A useful sample size can be determined by fixing a at some desirable level for convenient fieldwork and then choosing n to produce the desired bound, assuming some preliminary knowledge of λ. A rough rule for optimally determining a will be given in the next section for a slightly different estimator of λ.

Quadrat sampling, as described above, assumes that counts are made within each sampled quadrat without error. In counting diseased trees, for example, each sampled quadrat is thoroughly searched so that every diseased tree is discovered. In contrast, *line transect sampling* generally allows for missed items and adjusts the estimate of a population total accordingly. In line transect sampling, the investigator walks along a predetermined line and counts all the items in the category of interest that can be seen on either side of the line, no matter how far the items might be from the line. An adjustment is made for the fact that items have a decreasing probability of being spotted as their distance from the line transect increases. Such adjustments can become quite complicated and are beyond the scope of this book. A good reference for line transect sampling is Thompson (1992).

10.6

ESTIMATING POPULATION DENSITY AND SIZE FROM STOCKED QUADRATS

In quadrat sampling of plants or animals, counting the exact number of the species under investigation is often difficult. In contrast, detecting the presence or absence of the species of interest is often easy. We will now show that just knowing whether or not a species is present in a sample quadrat can lead to an estimate of density and of population size.

Foresters refer to a quadrat that contains the species of interest as being *stocked*. We will adopt that terminology. For a sample of n quadrats, each of area a, from a population of area A, let y denote the number of sampled quadrats that are *not* stocked. Under the assumption of randomness of elements, introduced in Section 10.5, the proportion of unstocked quadrats in the population is approximately $e^{-\lambda a}$. We know from our discussions of estimating proportions in Chapter 4 that the sample proportion of unstocked quadrats is a good estimator of the population proportion. Thus, (y/n) is an estimator of $e^{-\lambda a}$. This result leads to the following estimators of λ and M.

Estimator of the density λ:

$$\hat{\lambda} = -\left(\frac{1}{a}\right)\ln\left(\frac{y}{n}\right) \tag{10.10}$$

(ln denotes natural logarithm.)

Estimated variance of $\hat{\lambda}$:

$$\hat{V}(\hat{\lambda}) = \frac{1}{na^2}(e^{\hat{\lambda}a} - 1) \tag{10.11}$$

Estimator of the total M:

$$\hat{M} = \hat{\lambda}A \tag{10.12}$$

Estimated variance of $\hat{M}$:

$$\hat{V}(\hat{M}) = A^2\hat{V}(\hat{\lambda}) = \frac{A^2}{na^2}(e^{\hat{\lambda}a} - 1) \tag{10.13}$$

The following example illustrates the use of these estimators.

Example 10.7

Again, refer to the 200-acre tree plantation of Example 10.5. Now for estimation of the density of trees infected by fusiform rust, $n = 20$ quadrats of .5 acre each will be sampled, but only the presence or absence of infected trees will be noted for each sampled quadrat. (Since this task is easier than counting trees, the sample size can be increased.) Suppose $y = 4$ of the 20 quadrats show no signs of fusiform rust. Estimate the density and number of infected trees, placing bounds on the error of estimation in both cases.

Solution

From Equation (10.10), we see that the density is estimated by

$$\hat{\lambda} = -\left(\frac{1}{a}\right)\ln\left(\frac{y}{n}\right)$$
$$= -\frac{1}{(.5)}\ln\left(\frac{4}{20}\right) = 3.2 \text{ trees per acre}$$

The bound on the error is, according to Equation (10.11),

$$2\sqrt{\frac{1}{na^2}(e^{\hat{\lambda}a}-1)} = 2\sqrt{\frac{1}{(10)(.5)^2}(e^{(3.2)(.5)}-1)} = 1.8$$

We then estimate the density as 3.2 ± 1.8, or 1.4 to 5.0 infected trees per acre.

From Equation (10.12), we have

$$\hat{M} = \hat{\lambda}A = (3.2)(200) = 640$$

and the bound on the error, from Equation (10.13), is

$$2\sqrt{\hat{V}(\hat{M})} = 2(200)\sqrt{\frac{1}{(10)(.5)}(e^{(3.2)(.5)}-1)} = 360$$

Our estimate of the total number of infected trees is 640 ± 360, or 280 to 1000.

Generally, the estimator based on stocked quadrats alone is less precise than the one based on actual count data. However, since making the measurements is easier when one is only looking for stocked quadrats, the sample size can usually be quite large. The stocked-quadrat estimator does not work if $y=0$ or $y=n$. Thus choice of the quadrat size a is very important. Swindel (1983) gives a rough rule for choosing a as

$$a = \frac{1.6}{\lambda}$$

when some preliminary knowledge of λ is available. If, for example, we expect to see approximately 4 infected trees per acre, then each sampled quadrat should be $1.6/4 = .4$ acre.

The stocked-quadrat technique can also be used with volume or time samples. Cochran (1950) discusses the use of this technique, and modifications of it, for estimating bacterial densities in liquids.

Many other techniques for estimating population sizes are available. An excellent reference is the manual of wildlife investigational techniques (Mosby, 1969) listed in the references.

10.7 SUMMARY

Estimation of the size of a population is often very important, especially when one is studying plant and animal populations. This chapter presents four procedures for estimating the total population size N.

The first technique is direct sampling. A random sample of t individuals is drawn from a population and tagged. At a later date, a fixed random sample of size n is drawn, and the number of tagged individuals is observed. Using these data, we can estimate N and place a bound on the error of estimation.

The second technique, inverse sampling, is similar to direct sampling, with the exception that we continue sampling until a fixed number s of tagged individuals is recaptured in the second sample. The sample data are then used to estimate N and to place a bound on the error of estimation.

When a choice is available between inverse and direct sampling procedures, the inverse procedure appears to provide more accurate results. However, in some instances, particularly when little or nothing is known concerning the relative size of N, the direct sampling procedure is the better choice.

The third and fourth methods both involve sampling quadrats, volumes, or intervals of time and then counting elements of interest within these relatively small units. This procedure leads to estimates of both the density of elements and the total number of elements in the population.

CASE STUDY REVISITED

THE ESTIMATION OF CONCERT ATTENDANCE

In the example used to introduce this chapter, the newspaper reporter tells the statistician that $t = 500$ orange hats are given out. From their seat high in a balcony the statistician and the reporter locate $n = 200$ seats in a random fashion. In those 200 seats, they observe $s = 40$ of the orange hats. Now N, the size of the crowd, is estimated to be

$$\hat{N} = \frac{nt}{s} = \frac{200(500)}{40} = 2500$$

The bound on the error of estimation is given by

$$2\sqrt{\hat{V}(\hat{N})} = 2\sqrt{\frac{t^2 n(n-s)}{s^3}}$$

$$= 2\sqrt{\frac{(500)^2(200)(160)}{(40)^3}} = 707$$

We confidently estimate the size of the crowd to be between $2500 - 707 = 1793$ and $2500 + 707 = 3207$. This interval is large and could be reduced by increasing n. (Note that this estimate assumes that the people wearing the orange hats are somewhat randomly dispersed through the crowd.)

EXERCISES

10.1 Discuss the differences between direct and inverse sampling.

10.2 Name the restriction implicit in the use of (a) direct sampling or (b) inverse sampling. How can this restriction be satisfied in each case?

10.3 Assuming the cost of sampling is not significant, how can you improve the bound on the error of estimation, using either direct or inverse sampling?

10.4 A particular sportsmen's club is concerned about the number of brook trout in a certain stream. During a period of several days, $t = 100$ trout are caught, tagged, and then returned to the stream. Note that the sample represents 100 different fish; hence, any fish caught during these dates that had already been tagged was immediately released. Several weeks later a second sample of $n = 120$ trout is caught and observed. Suppose 27 in the second sample were tagged ($s = 27$). Estimate N, the total size of the population, and place a bound on the error of estimation.

10.5 Wildlife biologists wish to estimate the total size of the bobwhite quail population in a section of southern Florida. A series of 50 traps is used. In the first sample, $t = 320$ quail are caught. After being captured, each bird is removed from the trap and tagged with a metal band on its left leg. All birds are then released. Several months later a second sample of $n = 515$ quail is obtained. Suppose $s = 91$ of these birds have tags. Estimate N and place a bound on the error of estimation.

10.6 A game commission is interested in estimating the number of large-mouth bass in a reservoir. A random sample of $t = 2876$ bass is caught. Each bass is marked and released. One month later, a second sample of $n = 2562$ is caught. Suppose $s = 678$ have tags in the second sample. Estimate the total population size and place a bound on the error of estimation.

10.7 A team of conservationists is interested in estimating the size of the pheasant population in a particular area prior to the hunting season. The team believes that the true population size is between 2000 and 3000. Assuming $N \approx 3000$, the sampling fractions of p_1 and p_2 equal to .25 should give a bound on the error of estimation approximately equal to $2(189.74) = 379.48$ (Figure 10.1). The conservationists feel that this bound on the error is reasonable and so decide to choose $t = 750$ and $n = 750$. By using traps, they catch 750 pheasants for the first sample. Each of these pheasants is tagged and released. Several weeks later the second sample of $n = 750$ is obtained. Suppose 168 of these pheasants have tags ($s = 168$). Estimate the population size and place a bound on the error of estimation.

10.8 City officials are concerned about the nuisance caused by pigeons around city hall. To emphasize the problem, they hire a team of investigators to estimate the number of pigeons occupying the building. With several different traps a sample of $t = 60$ pigeons is captured, tagged, and released. One month later, the process is repeated, using $n = 60$. Suppose $s = 18$ tagged pigeons are observed in the second sample. Estimate N and place a bound on the error of estimation.

10.9 Animal resource experts on a particular game preserve are concerned about an apparent decline in the rabbit population. In a study conducted two years ago, the population size was estimated to be $N = 2500$. Assume the population size is still of this magnitude, and use Figure 10.1 to determine the approximate sample sizes (t and n) required to estimate N with a bound equal to 356.

10.10 A zoologist wishes to estimate the size of the turtle population in a given geographical area. She believes that the turtle population size is between 500 and 1000; hence, an initial sample of 100 (10%) appears to be sufficient. The $t = 100$ turtles are caught, tagged, and released. A second sampling is begun one month later, and she decides to continue sampling until $s = 15$ tagged turtles are recaptured. She catches 160 turtles before obtaining 15 tagged turtles ($n = 160$, $s = 15$). Estimate N and place a bound on the error of estimation.

10.11 Because of a particularly harsh winter, state park officials are concerned about the number of squirrels inhabiting their parks. An initial sample of $t = 100$ squirrels is trapped, tagged, and released. As soon as the first sample is completed, the officials begin working on a second sample of $n = 75$. They trap 10 squirrels that were tagged previously. Estimate N and place a bound on the error of estimation.

10.12 Assume the costs of taking an observation in the first sample and in the second sample are the same. Determine which is the most desirable: Having $t > n$, $t = n$, or $t < n$ for a fixed cost of conducting the two samples. [*Hint*: Consult Figures 10.1 and 10.2.]

10.13 A team of wildlife ecologists is interested in the effectiveness of an antifertility drug in controlling the growth of pigeon populations. To measure effectiveness, they will estimate the size of the population this year and compare it with the estimated size for a previous year. A large trap was constructed for the experiment. The trap was then baited with a corn feed containing a fixed amount of the drug. An initial sample of $t = 120$ pigeons is trapped and allowed to eat the medicated feed. Each bird is then tagged on its leg and released. At a later date, a second sample of $n = 100$ pigeons is trapped. Suppose 48 of these birds have tags ($s = 48$): Estimate the size of the pigeon population and place a bound on the error of estimation.

10.14 Air samples of 100 cubic centimeters each are taken periodically from an industrial section of a city. The density of a certain type of harmful particle is the parameter of interest. Suppose 15 samples gave an average particle count of 210 per sample. Estimate the particle density, per cubic centimeter, and place a bound on the error of estimation.

10.15 Suppose in the air sampling of Exercise 10.14 detecting the presence or absence of particles is easy but counting the particles is difficult. Among 500 such samples, 410 showed the particles to be present. Estimate the particle density and place a bound on the error of estimation.

10.16 Cars passing through an intersection are counted during randomly selected ten-minute intervals throughout the working day. Twenty such samples showed an average of 40 cars per interval. Estimate, with a bound on the error, the number of cars that you expect to go through the intersection in an eight-hour period.

10.17 Would you recommend use of the stocked-quadrat method in counting cars, as in Exercise 10.16?

10.18 Discuss the problem of estimating highly mobile animal populations by using quadrat sampling.

10.19 The data in the accompanying table show the number of bacteria colonies observed in 240 microscopic fields. Estimate, with a bound on the error of estimation, the density of colonies per field. What assumptions are necessary for this procedure?

Colonies per field	Number of fields
0	11
1	37
2	64
3	55
4	37
5	24
6	12

Source: C. I. Bliss and R. A. Fisher, Fitting the Negative Binomial Distribution to Biological Data," *Biometrics* 9 (1953): 176–200. With permission from The Biometric Society.

10.20 Outline how you can estimate the number of cars in a city during the working day. Compare four different methods for making this estimate. Which of the four do you think will work best? Why?

10.21 Telephone calls coming into a switchboard were monitored for 10 different time periods of 15 minutes each. The frequency distribution of the sample data is as follows:

No. of calls	No. of time periods
0	1
1	3
2	6
	10

(a) Estimate the average number of calls per hour coming into this switchboard.
(b) Show how to calculate the variance of the estimate in part (a).
(c) What assumptions are necessary for the answers in parts (a) and (b) to be valid?

EXPERIENCES WITH REAL DATA

10.1 Simulate the tag–recapture technique for animal populations by conducting the following experiment. Put a known number N of beads in a jar. Mark t of them in some distinguishable manner and thoroughly mix the beads. Then sample n beads, record the number of "tags," and estimate N by the direct method, placing a bound on the error of estimation. Does the resulting interval include your known N?

Replace the n beads and repeat the sampling, using the inverse method (sample until you have s tagged beads), and place a bound on the error. Does this interval include N?

You might try various sample sizes and various degrees of mixing. How will you choose an appropriate sample size? What do you think will happen if the marked beads are not thoroughly mixed with the others? Does this question suggest a realistic difficulty with the tag–recapture method?

10.2 The structure of the problems discussed in this chapter require that there be t marked objects (tags) randomly distributed among the N objects in a population. If t is known, a random sample of size n will supply information to estimate N, provided that some marked objects show up in the sample. The marked objects can be entered into the population without taking an initial sample.

Try the following technique for estimating the size of a crowd at a sporting event, lecture, movie, or other similar event. Obtain the names and descriptions of t people that you know will be attending the event. Ask them to distribute themselves somewhat randomly in the crowd. Then sample n people at random, perhaps as people leave the building. Count the number of original t "tagged" individuals who appear in your sample and estimate N. (You can use the inverse sample method here also.)

10.3 Estimate the number of three-lettered words in this book by first estimating the density of three-lettered words per page. Place a bound on the error of estimation. Try two different techniques for making this estimate. Which works better in your opinion? What assumptions are necessary for either method to work?

10.4 Figures 10.3–10.5 show points distributed in planar regions. The objective is to estimate the density of points on the page, which is the same for all three figures. The estimation is to be accomplished through quadrat sampling.

(a) Construct a quadrat for sampling by cutting a 2-cm × 2-cm square opening in a small note card. A single quadrat sample of 4 cm^2 is then obtained by dropping the note card on the planar region. If the opening falls across the boundary of the planar region, slide it toward the center until the open square is entirely inside the planar region of the figure.

(b) Randomly select six quadrats from Figure 10.3, counting the number of points in each. Use these data to estimate the density of points in Figure 10.3, with a bound on the error of estimation.

(c) Repeat the instructions from part (b) for Figures 10.4 and 10.5. Compare the results. Which figure produces the largest bound on the error for the estimate of point density?

(d) Repeat the sampling of six quadrat samples from each figure but use systematic sampling this time. Estimate the density and approximate the bound on the error in each case. Discuss how systematic sampling compares with simple random sampling for each of the three figures.

10.5 For Figure 10.3, estimate the density of points in the plane by the stocked-quadrat method. Begin by constructing a quadrat of 1 cm^2. Then, select 20 samples of 1 cm^2 each and record simply the presence or absence of points for each sample. Estimate the density of points on the plane and find an approximate bound on the error of estimation. How does this estimate compare with the one from 10.4(b)? Could the precision of the estimate be improved by taking more samples of smaller quadrat size?

FIGURE 10.3 Planar region I

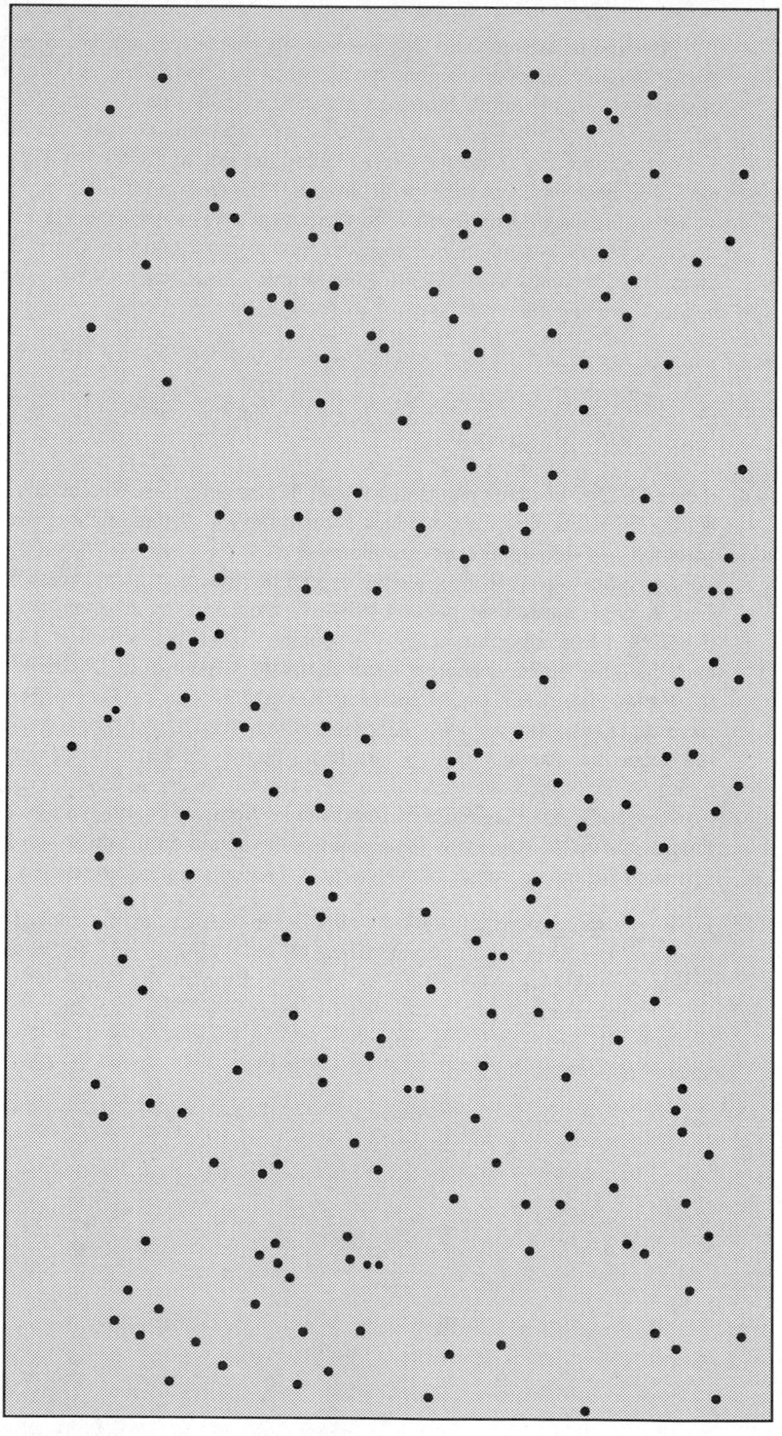

FIGURE 10.4 Planar region II

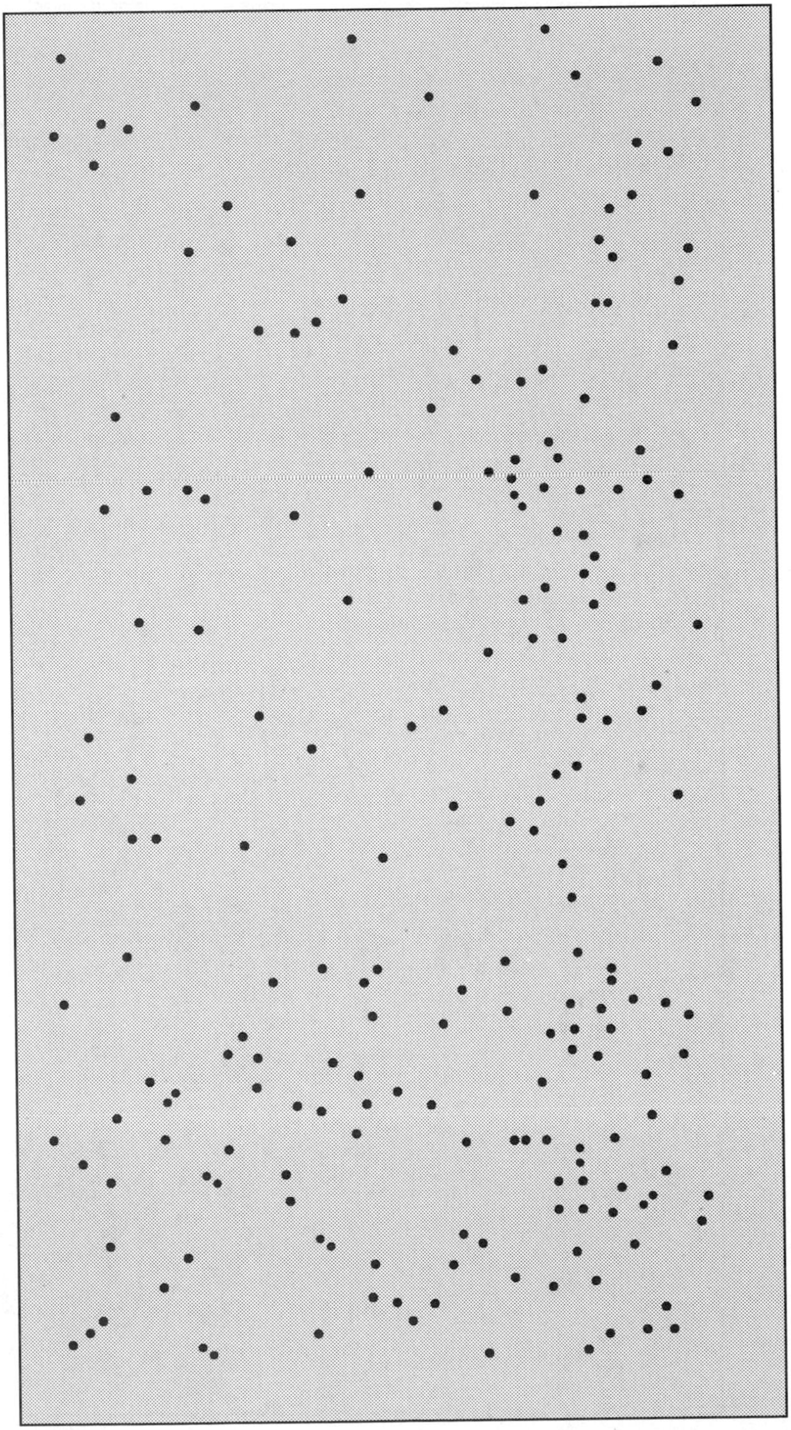

FIGURE 10.5 Planar region III

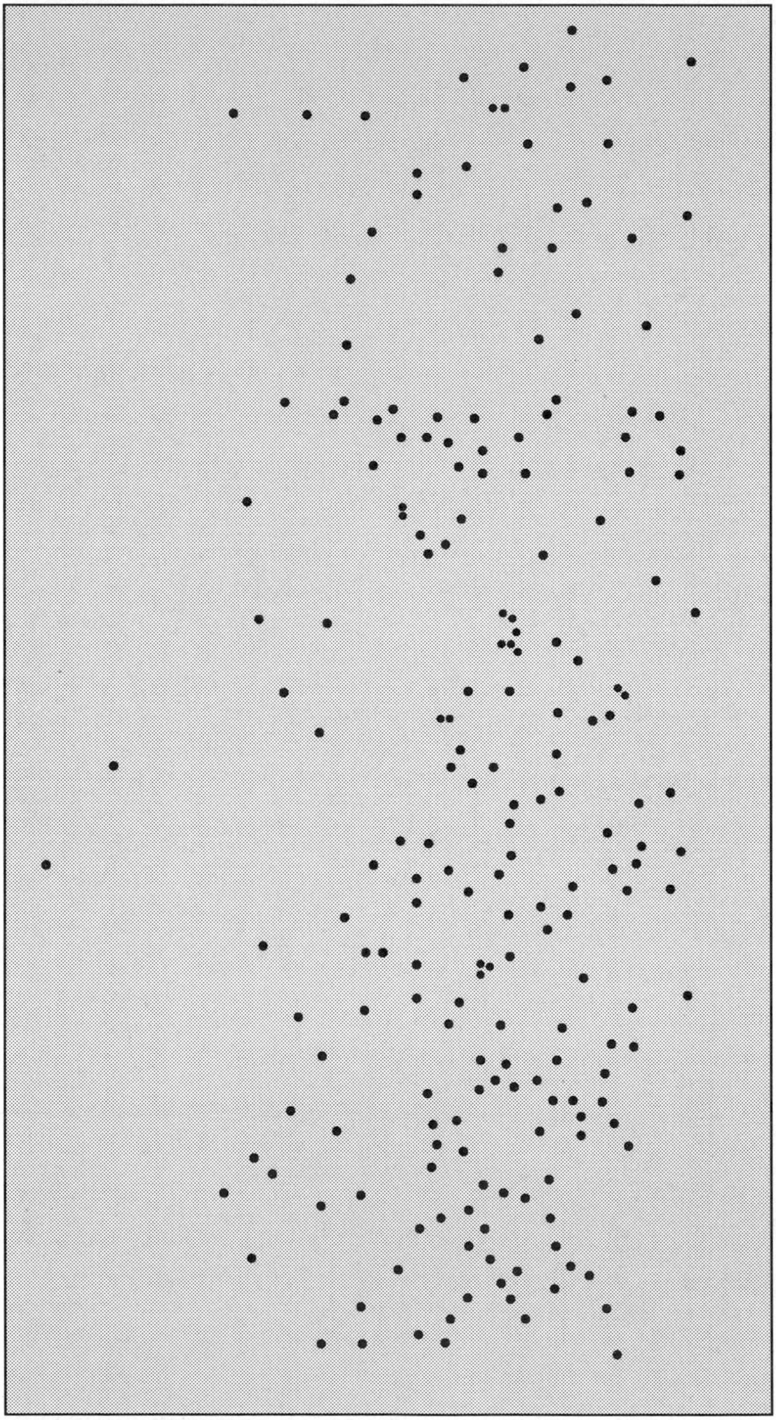

CHAPTER 11

SUPPLEMENTAL TOPICS

11.1 INTRODUCTION

Four sample survey designs—simple and stratified random sampling, cluster sampling, and systematic sampling—have been discussed in preceding chapters. For each design, we assumed that the data were correctly recorded and provided an accurate representation of the n elements sampled from the population. Under these assumptions, we were able to estimate certain population parameters and place a bound on the error of estimation.

There are many situations in which the assumptions underlying these designs are not fulfilled. First, the recorded measurements are not always accurate representations of the desired data because of biases of the interviewers or measuring equipment. Second, the frame is not always adequate, and hence the sample might not have been selected from the complete population. Third, obtaining accurate sample data might be impossible because of the sensitive nature of the questions.

In this chapter, we give some methods for analyzing data when measurement errors are present or an inadequate frame is used.

11.2

INTERPENETRATING SUBSAMPLES

An experimenter is interested in obtaining information from a simple random sample of n persons selected from a population of size N. She has k interviewers available to do the fieldwork, but the interviewers differ in their manner of interviewing and hence obtain slightly different responses from identical subjects. For example, suppose the interviewer is to rate the health of a respondent on a scale from 0 to 5, with 0 denoting poor health. Obtaining this type of data requires skill in interviewing and subjective judgment by the interviewer. One interviewer might not obtain enough information and might tend to rate the health of an individual too high, whereas another might obtain detailed information and tend to rate the health too low.

A good estimate of the population mean can be obtained by using the following technique. Randomly divide the n-sampled elements into k subsamples of m elements each and assign one interviewer to each of the k subsamples. Note that $m = n/k$ and n can always be chosen so that m is an integer. We consider the first subsample to be a simple random sample of size m selected from the n elements in the total sample. The second subsample is then a simple random sample of size m selected from the $(n - m)$ remaining elements. This process is continued until the n elements have been randomly divided into k subsamples. The k subsamples are sometimes called *interpenetrating subsamples*.

We expect some interviewers to give measurements that are too small and some too large, but the average of all sample measurements should be close to the population mean. That is, we expect the biases of the investigators to possess an average that is very near zero. Thus, the sample mean $\bar{y}$ is the best estimator of the population mean μ, even though the measurements are biased.

We use the following notation. Let y_{ij} denote the jth observation in the ith subsample, $j = 1, 2, \ldots, m$, $i = 1, 2, \ldots, k$. Then $\bar{y}_i$, given by

$$\bar{y}_i = \frac{1}{m} \sum_{i=1}^{k} y_{ij} \qquad (11.1)$$

is the average of all observations in the ith subsample. The sample mean $\bar{y}$ is the average of the k subsample means.

Estimator of the population mean μ:

$$\bar{y} = \frac{1}{k} \sum_{i=1}^{k} \bar{y}_i \qquad (11.2)$$

Estimated variance of $\bar{y}$:

$$\hat{V}(\bar{y}) = \left(\frac{N-n}{N}\right)\frac{s_k^2}{k} \tag{11.3}$$

where

$$s_k^2 = \frac{\sum_{i=1}^{k}(\bar{y}_i - \bar{y})^2}{(k-1)} \tag{11.4}$$

The technique of interpenetrating subsamples gives an estimate of the variance of $\bar{y}$, given in Equation (11.3), that accounts for interviewer biases. That is, the estimated variance given in Equation (11.3) is usually larger than the standard estimate of the variance of a sample mean obtained in simple random sampling because of the biases present in the measurements.

Example 11.1 A sociologist wants to estimate the average height of adult males in a community containing 800 men. He has 10 assistants, each with his or her own equipment, to acquire the measurements. Since the experimenter believes the assistants will produce slightly biased measurements, he decides to take a simple random sample of $n = 80$ males from the population and randomly divide the sample into 10 subsamples of 8 persons each. Each assistant is then assigned to one subsample. The measurements produce the following subsample means (measurements in feet):

$$\bar{y}_1 = 5.9 \qquad \bar{y}_6 = 5.7$$
$$\bar{y}_2 = 5.8 \qquad \bar{y}_7 = 5.8$$
$$\bar{y}_3 = 6.1 \qquad \bar{y}_8 = 5.6$$
$$\bar{y}_4 = 6.0 \qquad \bar{y}_9 = 5.9$$
$$\bar{y}_5 = 6.1 \qquad \bar{y}_{10} = 6.0$$

Estimate the mean height of adult males in the community and place a bound on the error of estimation.

Solution

The best estimate of the population mean is the sample mean $\bar{y}$. Thus, from Equation (11.2),

$$\bar{y} = \frac{1}{k}\sum_{i=1}^{k} y_i = \frac{1}{10}(5.9 + 5.8 + \cdots + 6.0) = 5.89$$

We must now estimate the variance of $\bar{y}$ by using Equation (11.3). The term s_k^2 is simply the sample variance among the $\bar{y}_i$'s. Or, the following identity can be established:

$$\sum_{i=1}^{k}(\bar{y}_i - \bar{y})^2 = \sum_{i=1}^{k}\bar{y}_i^2 - \frac{\left(\sum_{i=1}^{k}\bar{y}_i\right)^2}{k}$$

Substituting, we obtain

$$\sum_{i=1}^{k}(\bar{y}_i - \bar{y})^2 = 347.17 - \frac{(58.9)^2}{10} = .25$$

Then

$$\hat{V}(\bar{y}) = \left(\frac{N-n}{N}\right)\frac{\sum_{i=1}^{k}(\bar{y}_i - \bar{y})^2}{k(k-1)} = \left(\frac{800-80}{800}\right)\left[\frac{.25}{10(9)}\right] = .0025$$

The estimate of the mean height of adult males, with a bound on the error of estimation, is given by

$$\bar{y} \pm 2\sqrt{\hat{V}(\bar{y})}$$
$$5.89 \pm 2\sqrt{.0025}$$
$$5.89 \pm .10$$

To summarize, the best estimate of the mean height is 5.89 feet, and we are reasonably confident that our error of estimation is less than .10 foot.

11.3

ESTIMATION OF MEANS AND TOTALS OVER SUBPOPULATIONS

Obtaining a frame that lists only those elements in the population is often impossible. For example, we may wish to sample households containing children, but the best frame available may be a list of all households in a city. We may be interested in a firm's overdue accounts, but the only frame available may list all the firm's accounts receivable. In situations of this type, we wish to estimate parameters of a subpopulation of the population represented in the frame. Sampling is complicated because we do not know whether an element belongs to the subpopulation until after it has been sampled.

The problem of estimating a subpopulation mean is solved essentially in the same manner as in Chapter 4. Let N denote the number of elements in the population and N_1

the number of elements in the subpopulation. A simple random sample of n elements is selected from the population of N elements. Let n_1 denote the number of sampled elements that are from the subpopulation. Let y_{1j} denote the jth-sampled observation that falls in the subpopulation. Then the sample mean for elements from the subpopulation, denoted by $\bar{y}_1$, is given by

$$\bar{y}_1 = \frac{1}{n_1} \sum_{j=1}^{n_1} y_{1j}$$

The sample mean $\bar{y}_1$ is an unbiased estimate of the subpopulation mean μ_1.

Estimator of the subpopulation mean μ_1:

$$\bar{y}_1 = \frac{1}{n_1} \sum_{j=1}^{n_1} y_{1j} \tag{11.5}$$

Estimated variance of $\bar{y}_1$:

$$\hat{V}(\bar{y}_1) = \left(\frac{N_1 - n_1}{N_1}\right) \frac{s_1^2}{n_1} \tag{11.6}$$

where

$$s_1^2 = \frac{\sum_{j=1}^{n_1} (y_{1j} - \bar{y}_1)^2}{(n_1 - 1)}$$

The quantity $(N_1 - n_1)/N_1$ can be estimated by $(N - n)/N$ if N_1 is unknown.

Example 11.2

An economist wants to estimate the average weekly amount spent on food by families with children in a certain county known to be a poverty area. A complete list of all the 250 families in the county is available, but identifying those families with children is impossible. The economist selects a simple random sample of $n = 50$ families and finds that $n_1 = 42$ families have at least one child. The 42 families with children are interviewed and give the following information:

$$\sum_{j=1}^{42} y_{1j} = \$1720 \qquad \sum_{j=1}^{42} y_{1j}^2 = 72{,}200$$

Estimate the average weekly amount spent on food by all families with children and place a bound on the error of estimation.

Solution

The estimator of the population mean is $\bar{y}_1$, given by Equation (11.5). Calculations yield

$$\bar{y}_1 = \frac{1}{n_1}\sum_{j=1}^{n_1} y_{1j} = \frac{1}{42}(1720) = 40.95$$

We have the equality

$$\sum_{j=1}^{n_1}(y_{1j} - \bar{y}_1)^2 = \sum_{j=1}^{n_1} y_{1j}^2 - \frac{1}{n_1}\left(\sum_{j=1}^{n_1} y_{1j}\right)^2$$

and substituting gives

$$\sum_{j=1}^{n_1}(y_{1j} - \bar{y}_1)^2 = 72{,}200 - \frac{1}{42}(1720)^2 = 1762$$

The quantity $(N_1 - n_1)/N_1$ must be estimated by $(N - n)/N$, since N_1 is unknown. The estimated variance of $\bar{y}_1$, given in Equation (11.6), then becomes

$$\hat{V}(\bar{y}_1) = \left(\frac{N-n}{N}\right)\frac{\sum_{j=1}^{n_1}(y_{1j} - \bar{y}_1)^2}{n_1(n_1 - 1)} = \left(\frac{250-50}{250}\right)\left[\frac{1762}{42(41)}\right]$$
$$= .819$$

Thus, the estimate of the population average, with a bound on the error of estimation, is given by

$$\bar{y}_1 \pm 2\sqrt{\hat{V}(\bar{y}_1)}$$
$$40.95 \pm 2\sqrt{.819}$$
$$40.95 \pm 1.81$$

Our best estimate of the average weekly amount spent on food by families with children is \$40.95. The error of estimation should be less than \$1.81 with probability approximately .95.

If the number of elements in the subpopulation N_1 is known, the subpopulation total τ_1 can be estimated by $N_1\bar{y}_1$.

Estimator of the subpopulation total τ_1:

$$N_1\bar{y}_1 = \frac{N_1}{n_1}\sum_{j=1}^{n_1} y_{1j} \tag{11.7}$$

Estimated variance of $N_1\bar{y}_1$:

$$\hat{V}(N_1\bar{y}_1) = N_1^2\hat{V}(\bar{y}_1) = N_1^2\left(\frac{N_1 - n_1}{N_1}\right)\frac{s_1^2}{n_1} \tag{11.8}$$

Example 11.3

A preliminary study of the county in Example 11.2 reveals $N_1 = 205$ families with children. Using this information and data given in Example 11.2, estimate the total weekly amount spent on food by families with children. (*Note:* N_1 will vary over time. We assume that the value of N_1 used in this analysis is correct.)

Solution

The best estimator of the total is $N_1\bar{y}_1$, given in Equation (11.7), which yields an estimate of

$$N_1\bar{y}_1 = 205(40.95) = 8394.75$$

The quantity $\sum_{j=1}^{n_1}(y_{1j} - \bar{y}_1)^2$ is calculated in Example 11.2 to be 1762. The estimated variance of $N_1\bar{y}_1$ is then [from Equation (11.8)]

$$\hat{V}(N_1\bar{y}_1) = N_1^2\left(\frac{N_1 - n_1}{N_1}\right)\frac{\sum_{j=1}^{n_1}(y_{1j} - \bar{y}_1)^2}{n_1(n_1 - 1)}$$

$$= (205)^2\left(\frac{205 - 42}{205}\right)\left[\frac{1762}{42(41)}\right] = 34{,}191.19$$

The estimate of the total weekly amount that families with children spend on food, given with a bound on the error of estimation, is

$$N_1\bar{y}_1 \pm 2\sqrt{\hat{V}(N_1\bar{y}_1)}$$

$$8394.75 \pm 2\sqrt{34{,}191.19}$$

$$8394.75 \pm 369.82$$

Frequently, the number of elements in the subpopulation, N_1, is unknown. For example, the exact number of households containing children in a city may be difficult to determine, whereas the total number of households can perhaps be obtained from a city directory. An unbiased estimate of τ can still be obtained even though N_1 is unknown.

Estimator of the subpopulation total τ_1 when N_1 is unknown:

$$\hat{\tau}_1 = \frac{N}{n}\sum_{j=1}^{n_1} y_{1j} \tag{11.9}$$

Estimated variance of $\hat{\tau}_1$:

$$\hat{V}(\hat{\tau}_1) = N^2\left(\frac{N-n}{N}\right)\frac{\sum_{j=1}^{n_1} y_{1j}^2 - \left[\left(\sum_{j=1}^{n_1} y_{1j}\right)^2 \Big/ n\right]}{n(n-1)} \tag{11.10}$$

The variance expression of Equation (11.10) can be written in the form

$$\hat{V}(\hat{\tau}_1) = N^2\left(\frac{N-n}{N}\right)\frac{s_n^2}{n}$$

where s_n^2 is the sample variance calculated from an adjusted sample consisting of replacing all the observations *not* from the subpopulation of interest with zeros. The sample variance is then calculated from all n "observations."

Example 11.4 Suppose that the experimenter in Example 11.3 doubts the accuracy of the preliminary value of N_1. Use the data of Example 11.3 to estimate the total weekly amount spent on food by families with children, without using the value given for N_1.

Solution

The estimator of the total that does not depend on N_1 is $\hat{\tau}_1$, given by Equation (11.9). Thus,

$$\hat{\tau}_1 = \frac{N}{n}\sum_{j=1}^{n_1} y_{1j} = \frac{250}{50}(1720) = 8600$$

Substituting into Equation (11.10) gives the estimated variance of $\hat{\tau}_1$:

$$\begin{aligned}
\hat{V}(\hat{\tau}_1) &= N^2\left(\frac{N-n}{N}\right)\frac{\sum_{j=1}^{n_1} y_{1j}^2 - \frac{1}{n}\left(\sum_{j=1}^{n_1} y_{1j}\right)^2}{n(n-1)} \\
&= (250)^2\left(\frac{250-50}{250}\right)\frac{72{,}200 - (1/50)(1720)^2}{50(49)} \\
&= 265{,}960
\end{aligned}$$

Thus, the estimate of the total weekly amount spent on food, with a bound on the error of estimation, is

$$\hat{\tau}_1 \pm 2\sqrt{\hat{V}(\hat{\tau}_1)}$$

$$8600 \pm 2\sqrt{265{,}960}$$

$$8600 \pm 1031.44$$

The interval is a large bound on the error of estimation and should be reduced by increasing the sample size n.

Note that the variance of $\hat{\tau}_1$, calculated in Example 11.4, is much larger than the variance of $N_1 \bar{y}_1$, calculated in Example 11.3. The variance of τ_1 is larger because the information provided by N_1 is used in $N_1 \bar{y}_1$ but not in $\hat{\tau}_1$. Thus, if N_1 is known or if it can be found with little additional cost, the estimator $N_1 \bar{y}_1$ should be used.

11.4

RANDOM-RESPONSE MODEL

Persons being interviewed often refuse to answer or give correct answers to sensitive questions that may embarrass them or be harmful to them in some way. For example, some persons may not respond truthfully to political questions such as "Are you a Communist?" In this section, we present a method of estimating the proportion of people who have some characteristic of interest without obtaining direct answers from the people interviewed. The method is due to S. L. Warner (1965).

Designate the people in the population who have or do not have the characteristic of interest as groups A and B, respectively. Thus, each person in the population is in either group A or group B. Let p be the proportion of people in group A. The objective is to estimate p without asking each person directly whether or not he belongs to group A. We can estimate p by using a device called a *random-response model*. We start with a stack of cards that are identical except that a fraction, θ, are marked with A and the remaining fraction, $(1 - \theta)$, are marked with B. A simple random sample of n people is selected from the population. Each person in the sample is asked to randomly draw a card from the deck and to state yes if the letter on the card agrees with the group to which he or she belongs or no if the letter on the card is different than the group to which he or she belongs. The card is replaced before the next person draws. The interviewer does not see the card and simply records whether the response is yes or no. Let n_1 be the number of people in the sample who respond with yes. An unbiased estimator $\hat{p}$ of the population proportion p can be found from n_1 and n.

To develop an estimator of p, we can think of the procedure outlined above as consisting of two stages:

1. Select a card.
2. Decide whether or not you belong to the group indicated on the card.

This process can be modeled by a tree diagram:

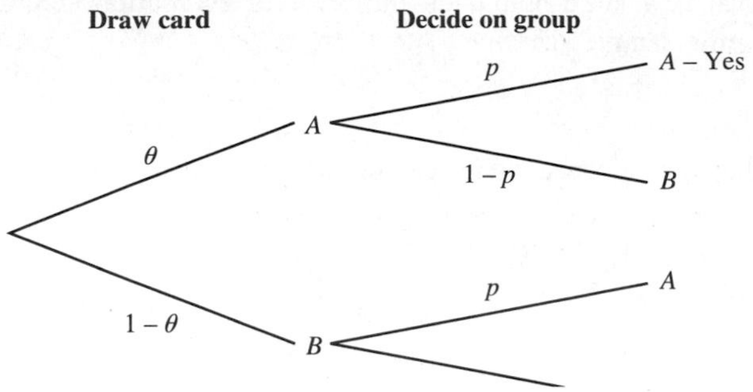

There are then two ways for the interviewer to obtain a yes, as seen by the top and bottom paths. Thus,

$$\begin{aligned} P(\text{yes}) &= \theta p + (1-\theta)(1-p) \\ &= p(2\theta - 1) + (1-\theta) \end{aligned}$$

The sample fraction of yes responses n_1/n, estimates $P(\text{yes})$, and so an estimator $\hat{p}$ of p can be found from the equation

$$\frac{n_1}{n} = \hat{p}(2\theta - 1) + (1-\theta)$$

or

$$\hat{p} = \frac{1}{(2\theta - 1)}\left(\frac{n_1}{n}\right) - \left(\frac{1-\theta}{2\theta - 1}\right) \qquad \theta \neq \frac{1}{2} \tag{11.11}$$

The variance of $\hat{p}$ is easily estimated by methods from Chapter 4 and becomes

$$\begin{aligned} \hat{V}(\hat{p}) &= \frac{1}{(2\theta - 1)^2}\hat{V}\left(\frac{n_1}{n}\right) \\ &= \frac{1}{(2\theta - 1)^2}\frac{1}{n}\left(\frac{n_1}{n}\right)\left(1 - \frac{n_1}{n}\right) \end{aligned} \tag{11.12}$$

for large n and N. Example 11.5 provides an illustration of the use of this technique.

Example 11.5 A study is designed to estimate the proportion of people in a certain district who give false information on income tax returns. Since respondents would not admit to cheating on tax returns, a random-response technique is used. The experimenter constructs a deck of cards in which $\frac{3}{4}$ of the cards are marked F, denoting a falsified return, and $\frac{1}{4}$ are marked C, denoting a correct return. A simple random sample $n = 400$ persons is selected from the large population of taxpayers in the district. In separate interviews, each sampled taxpayer is asked to draw a card from the deck and to respond yes if the letter agrees with the group to which he or she belongs. The experiment results in $n_1 = 120$ yes responses. Estimate p, the proportion of taxpayers in the district who have falsified returns, and place a bound on the error of estimation.

Solution

From Equation (11.11) with $\theta = \frac{3}{4}$,

$$\hat{p} = \frac{1}{2\theta - 1}\left(\frac{n_1}{n}\right) - \left(\frac{1-\theta}{2\theta - 1}\right)$$
$$= 2\left(\frac{120}{400}\right) - \frac{1/4}{1/2} = 0.10$$

and from Equation (11.12),

$$\hat{V}(\hat{p}) = 4\frac{(.3)(.7)}{400} = .0021$$

and

$$2\sqrt{\hat{V}(\hat{p})} = .09$$

Thus, the estimate of p is .10, and we are confident that the estimate lies within .09 of the true p.

Numerous formulations of the random-response model exist. A simple extension of the idea presented above involves two unrelated questions such as

A: Have you ever falsified your tax return?

B: Is the second-to-last digit of your home phone number odd?

The respondent may flip a coin, with the rule being that she answers A if the coin lands heads, and answers B if the coin lands tails. The interviewer does not see the coin, but merely hears the yes or no response. The interviewer never knows which question was answered.

Using a two-stage model, we have

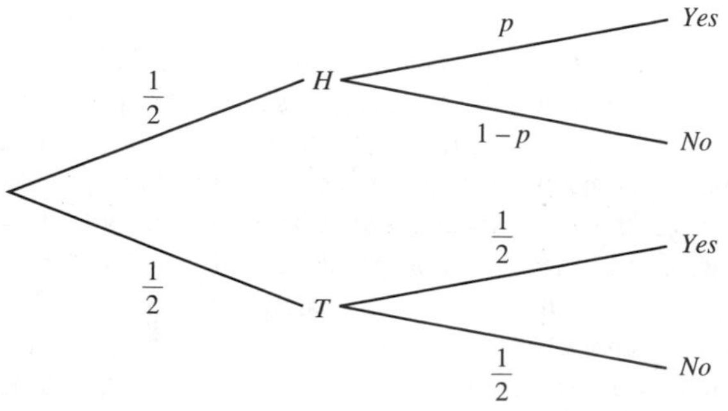

(The last four digits of telephone numbers are essentially random, so the probability of a yes answer to B is $\frac{1}{2}$.) Then

$$P(\text{yes}) = \frac{p}{2} + \frac{1}{4}$$

and

$$\frac{n_1}{n} = \frac{\hat{p}}{2} + \frac{1}{4}$$

or

$$\hat{p} = 2\left(\frac{n_1}{n} - \frac{1}{4}\right) \tag{11.13}$$

It follows that

$$\hat{V}(\hat{p}) = 4\hat{V}\left(\frac{n_1}{n}\right) = \frac{4}{n}\left(\frac{n_1}{n}\right)\left(1 - \frac{n_1}{n}\right) \tag{11.14}$$

Using the scenario of Example 11.5, suppose $n = 400$ and $n_1 = 120$, as given there, for an experiment with two questions as posed above. Then

$$\hat{p} = 2\left(\frac{120}{400} - \frac{1}{4}\right) = .10$$

and

$$\hat{V}(\hat{p}) = \frac{4}{400}(.3)(.7) = .0021$$

as in Example 11.5.

The two methods give approximately the same results, in most cases, with similar precision. The decision as to which method to use is often based on which one might be more successful in eliciting correct responses. The coin flip with two answers is somewhat easier to explain, and the equal likelihood for A and B might make the respondent more comfortable in responding; but either will work well in the hands of a trained interviewer.

Random-response techniques are quite popular and are being used in many areas, including AIDS research (see *Discover*, July 1987, p. 12).

Equations (11.11) through (11.14) are based on the assumption that the population size is large relative to n, so the finite population correction can be ignored. The fraction θ of cards marked A may be arbitrarily chosen by the experimenter but must not equal $\frac{1}{2}$. A value $\theta = 1$ must not be used because respondents will then realize that they are telling whether or not they belong to group A, which is exactly what they do not wish to do. A value of θ between $\frac{1}{2}$ and 1—for example, $\frac{3}{4}$—is usually adequate.

This method generally requires a very large sample size in order to obtain a reasonably small variance of the estimator. A large sample size is needed because each response provides little information on the population proportion p.

The random-response technique presented here is the very simplest of many such techniques. For further information on these techniques, see the papers by Campbell and Joiner (1973) and by Leysieffer and Warner (1976).

Random-response techniques can be used more widely than the simple yes or no type of response situation employed here might indicate. To see how this technique is developed, refer to the paper by Greenberg, Kuebler, Abernathy, and Horvitz (1971).

11.5

SELECTING THE NUMBER OF CALLBACKS

As discussed earlier in the book, nonresponse is an important problem to consider in any survey. If a simple random sample of size n is employed and only $n_1 (n_1 < n)$ responses are obtained, then the two groups (response and nonresponse) can be thought of as constituting a stratified random sample with two strata. Note that this situation is not quite a true stratified random sampling situation since n_1 and $n_2 = n - n_1$ are random variables whose values are determined only after the initial sampling is completed. Nevertheless, thinking in terms of stratified sampling allows us to find an approximately optimal rule for allocating resources to callbacks.

Suppose that out of the n_2 nonrespondents, we decide to make intensive callbacks on r of them, where $r = n_2/k$ for some constant $k > 1$. Also, suppose that it costs c_1 dollars for a standard response and c_2 dollars ($c_2 > c_1$) for a callback response, with c_0 denoting the initial cost of sampling each item. Then the total cost is

$$C = nc_0 + n_1c_1 + rc_2$$

If $\bar{y}_1$ denotes the average of the initial responses and $\bar{y}_2$ the average of the r callback responses, then

$$\bar{y}^* = \frac{1}{n}(n_1\bar{y}_1 + n_2\bar{y}_2) \tag{11.15}$$

is an unbiased estimator of the population mean μ.

A theoretical variance expression for $\bar{y}^*$ can be derived, and then we can find the values of k and n that minimize the expected cost of sampling for a desired fixed value of $V(\bar{y}^*)$—say, V_0. The optimal values of k and n are, for large N, approximately

$$k = \sqrt{\frac{c_2(\sigma^2 - W_2\sigma_2^2)}{\sigma_2^2(c_0 + c_1W_1)}} \tag{11.16}$$

$$n = \frac{N[\sigma^2 + (k-1)W_2\sigma_2^2]}{NV_0 + \sigma^2} \tag{11.17}$$

where W_2 is the nonresponse rate for the population, $W_1 = 1 - W_2$, and σ^2 and σ_2^2 are the variances for the entire population and the nonresponse group, respectively. The variance of $\bar{y}^*$ can be estimated by

$$\hat{V}(\bar{y}^*) = \frac{k-1}{n} W_2 s_2^2 + \left(\frac{N-n}{Nn}\right) s^2$$

where s_2^2 estimates the variance of the nonresponse group and s^2 estimates the overall population variance.

Example 11.6

A mailed questionnaire is to be used to collect data for estimating the average amount per week that a certain group of 1000 college men spends on entertainment. From past experience, the response rate is anticipated to be about 60%. It is thought that $\sigma^2 \approx 120$ and $\sigma_2^2 \approx 80$. (The nonresponse group tends to be those not interested in entertainment and hence spend less and have less variation in spending habits.) Suppose $c_0 = 0$, $c_1 = 1$, and $c_2 = 4$ and a simple random sample is to be used initially. Find n and k so that the variance of the resulting estimator is approximately 5 units.

Solution

Observe that $W_2 = 1 - W_1 = .4$. Then from Equations (11.16) and (11.17),

$$k = \sqrt{\frac{4[120 - .4(80)]}{80(1)(.6)}} = 2.71$$

$$n = \frac{1000[120 + 1.71(.4)(80)]}{1000(5) + 120} = 34.1 \text{ or } 35$$

Since $E(n_2) = nW_2 = 35(.4) = 14$, we can expect that approximately 21 persons will respond initially, and

$$r = \frac{n_2}{k} \approx \frac{14}{2.71} = 5.2 \text{ or } 6$$

callbacks will have to be made.

The importance of making callbacks to correct for nonresponse can be seen easily, if we think about the population of N elements as being divided into a "response" stratum (those who would respond if called upon) of N_1 elements and a "nonresponse" stratum of N_2 elements. Then,

$$\mu = W_1\mu_1 + W_2\mu_2$$

where $W_i = N_i/N$, and the bias of $\bar{y}_1$ as an estimator of μ is

$$E(\bar{y}_1) - \mu = \mu_1 - \mu = \mu_1 - (W_1\mu_1 + W_2\mu_2)$$
$$= W_2(\mu_1 - \mu_2)$$

So $\bar{y}_1$ is biased unless $\mu_1 = \mu_2$, or the proportion of nonresponse W_2 is zero. The sample provides an estimate of μ_1 but no information on μ_2. Thus, callbacks to provide information on μ_2 are essential before a meaningful estimator of μ can be determined.

Even without callbacks, the situation is not entirely hopeless in the case of estimating a proportion p, where

$$p = W_1 p_1 + W_2 p_2$$

Now, p_2 must be between 0 and 1, and we can couple this information with an estimate of p_1 to form an interval estimate of p. Sample data on $\hat{p}_1$ allow us to construct an interval estimate of p_1 as

$$\left(\hat{p}_1 - 2\sqrt{\frac{\hat{p}_1(1-\hat{p}_1)}{n_1 - 1}},\quad \hat{p}_1 + 2\sqrt{\frac{\hat{p}_1(1-\hat{p}_1)}{n_1 - 1}}\right)$$

ignoring finite population corrections, and we will denote this interval by $(\hat{p}_{1,\mathrm{L}}, \hat{p}_{1,\mathrm{U}})$. Using the fact that 0 is a lower limit for p_2, we can construct a lower limit for p as

$$\hat{p}_{\mathrm{L}} = W_1\hat{p}_{1,\mathrm{L}} + W_2(0) = W_1\hat{p}_{1,\mathrm{L}}$$

Similarly, an upper limit for p is

$$\hat{p}_{\mathrm{U}} = W_1\hat{p}_{1,\mathrm{U}} + W_2(1)$$

This interval can be quite large if W_2 is large, but it is sometimes useful if no callbacks can be made. To illustrate, suppose $n_1 = 1000$, $W_2 = .2$, and $\hat{p}_1 = .5$. Then, $\hat{p}_{1,\mathrm{L}} = .465$, $\hat{p}_{1,\mathrm{U}} = .535$,

$$\hat{p}_{\mathrm{L}} = .8(.465) = .372$$

$$\hat{p}_{\mathrm{U}} = .8(.535) + .2 = .628$$

We see that the interval for p is over three times as large as the interval for p_1, even with a large sample. The mere size of the interval produced here may help convince the experimenter that callbacks are indeed necessary.

11.6 SUMMARY

This chapter presented three useful techniques for estimating population parameters when the assumptions underlying the elementary sample survey designs are not valid.

The effect of interviewer bias can be reduced by using interpenetrating subsamples. The estimator of the population mean in this case is given by Equation (11.2), and the estimated variance of this estimator is given by Equation (11.3).

An inadequate frame generates the problem of estimating means and totals over subpopulations. The estimator of the subpopulation mean is given by Equation (11.5), and estimators of the subpopulation total are given by Equations (11.7) and (11.9).

When persons being interviewed will not give correct answers to sensitive questions, a random-response technique can sometimes be used. The method for estimating a population proportion p by using this procedure is explained in Section 11.4.

Sometimes, one can treat the nonrespondents as a separate stratum for purposes of choosing an optimal number of callbacks, as shown in Section 11.5.

EXERCISES

11.1 A researcher is interested in estimating the average yearly medical expense per family in a community of 545 families. The researcher has eight assistants available to do the fieldwork. Skill is required to obtain accurate information on medical expenses because some respondents are reluctant to give detailed information on their health. Since the assistants differ in their interviewing abilities, the researcher decides to use 8 interpenetrating subsamples of 5 families each, with one assistant assigned to each subsample. Hence, a simple random sample of 40 families is selected and divided into 8 random subsamples. The interviews are conducted and yield the results shown in the accompanying table. Estimate the average medical expense per family for the past year and place a bound on the error of estimation.

Subsample	Amount (in dollars) of medical expenses for past year				
1	101	95	310	427	680
2	157	192	108	960	312
3	689	432	187	512	649
4	322	48	93	162	495
5	837	649	152	175	210
6	1015	864	325	470	295
7	837	249	1127	493	218
8	327	419	291	114	287

11.2 An experiment is designed to gauge the emotional reaction to a city's decision on school desegregation. A simple random sample of 50 people is interviewed, and the emotional reactions are given a score from 1 to 10. The scale on which scores are assigned runs from extreme anger to extreme joy. Ten interviewers do the questioning and scoring, with each interviewer working on a random subsample (interpenetrating subsample) of 5 people. Interpenetrating subsamples are used because of the flexible nature of the scoring. The results are given in the accompanying table. Estimate the average score for people in the city and place a bound on the error of estimation.

Subsample	Scores				
1	5	4	6	1	8
2	4	6	5	2	7
3	9	8	9	7	5
4	8	5	4	6	3
5	6	4	5	7	9
6	1	5	6	4	7
7	6	4	3	5	2
8	5	6	7	3	4
9	2	4	4	5	3
10	9	7	8	6	4

11.3 A retail store wants to estimate the average amount of all past-due accounts. The available list of past-due accounts is outdated because some accounts have since been paid. Because drawing up a new list would be expensive, the store uses the outdated list. A simple random sample of 20 accounts is selected from the list, which contains 95 accounts. Of the 20 sampled accounts, 4 have been paid. The 16 past-due accounts contain the following amounts (in dollars): 3.65, 15.98, 40.70, 2.98, 50.00, 60.31, 67.21, 14.98, 10.20, 14.32, 1.87, 32.60, 19.80, 15.98, 12.20, 15.00. Estimate the average amount of past-due accounts for the store and place a bound on the error of estimation.

11.4 For Exercise 11.3, estimate the total amount of past-due accounts for the store and place a bound on the error of estimation.

11.5 An employee of the store in the Exercise 11.3 decides to look through the list of past-due accounts and mark those that have been paid. He finds that only 83 of the 95 accounts are past due. Estimate the total amount of past-due accounts by using this additional information and the data of Exercise 11.3. Place a bound on the error of estimation.

11.6 A study is conducted to estimate the average number of miles from home to place of employment for household heads living in a certain suburban area. A simple random sample of 30 people is selected from the 493 heads of households in the area. While conducting interviews, the experimenter finds some household heads are not appropriate for the study because they are retired or do not go to a place of employment for various reasons. Of the 30 sampled household heads, 24 are appropriate for the study, and the data on miles to place of employment are as follows:

8.5	10.2	25.1	5.0	6.3	7.9	15.8	2.1
9.2	4.2	8.3	4.2	6.7	10.1	15.6	22.1
10.0	6.1	7.9	1.5	8.0	11.0	20.2	9.3

Estimate the average distance between home and place of employment for household heads who commute to a place of employment. Place a bound on the error of estimation.

11.7 For the data of Exercise 11.6, estimate the total travel distance between home and place of employment for all household heads in the suburban area. Place a bound on the error of estimation.

11.8 Suppose you know that 420 out of the 493 household heads (Exercise 11.6) commute to a place of employment. Estimate the total travel distance for all household heads in the suburban area, making use of this additional information. Place a bound on the error of estimation.

11.9 A public health official wants to estimate the proportion of dog owners in a city who have had their dogs vaccinated against rabies. She knows that a dog owner often gives incorrect information about rabies shots out of fear that something might happen to his dog if it has not had the shots. Thus, the official decides to use a random-response technique. She has a stack of cards with .8 of the cards marked A for the group having the shots and .2 marked B for the group not having the shots. A simple random sample of 200 dog owners is selected. Each sampled owner is interviewed and asked to draw a card and to respond with yes if the letter on the card agrees with the group he is in. The official obtained 145 yes responses. Estimate the proportion of dog owners who have had their dogs vaccinated and place a bound on the error of estimation. Assume that the number of dog owners in the city is very large.

11.10 A corporation executive wants to estimate the proportion of corporation employees who have been convicted of a misdemeanor. Since the employees would not want to answer the question directly, the executive uses a random-response technique. A simple random sample of 300 people is selected from a large number of corporation employees. In separate interviews, each employee draws a card from a deck that has .7 of the cards marked "convicted" and .3 marked "not convicted." The employee responds yes if the card agrees with his or her category and no otherwise. The executive obtains 105 yes responses. Estimate the proportion of employees who have been convicted of a misdemeanor and place a bound on the error of estimation.

EXPERIENCES WITH REAL DATA

Select a simple random sample from the appropriate population in at least one of the situations outlined below. Estimate the indicated proportion or average and place a bound on the error by using the appropriate results from Section 11.3 on subpopulations. In each case, assume that the items in the subpopulation cannot be classified as such until after they are observed.

11.1 Estimate the proportion of voters favoring a certain local government proposal from among those who voted in the most recent election.

11.2 Estimate the proportion of students on your campus favoring the quarter system from among those who have been college students on the quarter system and at least one other system.

11.3 Estimate the average amount spent for utilities in the past month for *homeowners* in a certain neighborhood.

11.4 Estimate the average number of words per page among pages that contain no boxed formulas or tables in this book.

CHAPTER 12

SUMMARY

12.1

SUMMARY OF THE DESIGNS AND METHODS

Recall that the objective of statistics is to make inferences about a population from information contained in a sample. This text discusses the design of sample surveys and associated methods of inference for populations containing a finite number of elements. Practical examples have been selected primarily from the fields of business and the social sciences where finite populations of human responses are frequently the target of surveys. Natural resource management examples are also included.

The method of inference employed for most sample surveys is estimation. Thus, we consider appropriate estimators for population parameters and the associated two-standard-deviation bound on the error of estimation. In repeated sampling, the error of estimation will be less than its bound, with probability approximately equal to .95. Equivalently, we construct confidence intervals that, in repeated sampling, enclose the true population parameter approximately 95 times out of 100. The quantity of information pertinent to a given parameter is measured by the bound on the error of estimation.

The material in this text falls naturally into five segments. The first is a review of elementary concepts, the second contains useful sample survey designs, the third considers an estimator that utilizes information obtained on an auxiliary variable, the fourth gives methods of estimating the size of wildlife populations, and the fifth considers methods for making inferences when one or more of the basic assumptions with the standard techniques are not satisfied.

The first segment, presented in Chapters 1–3, reviews the objective of statistics and points to the peculiarities of problems arising in the social sciences, business, and natural resource management that make them different from the traditional type of experiment conducted in the laboratory. These peculiarities primarily involve sampling from finite populations along with a number of difficulties that occur in drawing samples from human populations. The former requires modification of the formulas for the bounds on the error of estimation that are encountered in an introductory course in statistics. The difficulties associated with sampling from human populations suggest specific sample survey designs that reduce the cost of acquiring a specified quantity of information.

In Chapters 4, 5, 7, 8, and 9 we consider specific sample survey designs and their associated methods of estimation. The basic sample survey design, simple random sampling, is presented in Chapter 4. For this design, the sample is selected so that every sample of size n in the population has an equal chance of being chosen. The design does not make a specific attempt to reduce the cost of the desired quantity of information. It is the most basic type of sample survey design, and all other designs are compared with it.

The second type of design, stratified random sampling (Chapter 5), divides the population into homogeneous groups called strata. This procedure usually produces an estimator that possesses a smaller variance than can be acquired by simple random sampling. Thus, the cost of the survey can be reduced by selecting fewer elements to achieve an equivalent bound on the error of estimation.

The third type of experimental design is systematic sampling (Chapter 7), which is usually applied to population elements that are available in a list or line, such as names on file cards in a drawer or people coming out of a factory. A random starting point is selected, and then every kth element thereafter is sampled. Systematic sampling is frequently conducted when collecting a simple random or a stratified random sample is extremely costly or impossible. Once again, the reduction in survey cost is primarily associated with the cost of collecting the sample.

The fourth type of sample survey design is cluster sampling, which is presented in Chapters 8 and 9. Cluster sampling may reduce cost because each sampling unit is a collection of elements usually selected so as to be physically close together. Cluster sampling is most often used when a frame that lists all population elements is not available or when travel costs from element to element are considerable. Cluster sampling reduces the cost of the survey primarily by reducing the cost of collecting the data.

A discussion of ratio, regression, and difference estimators, which utilize information on an auxiliary variable, is covered in the third segment of material, Chapter 6. The ratio estimator illustrates how additional information, frequently acquired at little cost, can be used to reduce the variance of the estimator and, consequently, reduce the overall cost of a survey. It also suggests the possibility of acquiring more sophisticated estimators by using information on more than one auxiliary variable. This chapter on ratio estimation follows naturally the discussion on simple random sampling in Chapter 4. That is, you can take a measurement of y, the response of interest, for each element of the simple random sample and utilize the traditional estimators of Chapter 4. Or as suggested in Chapter 6, you might take a measurement on both y and an auxiliary variable x for each element and utilize the additional information contributed by the auxiliary variable to acquire a better estimator of the parameter. Thus, although it was not particularly stressed, ratio estimators can be employed with any of the designs discussed in the text.

Chapter 10 deals with the specific problems of estimating the size of populations. Two estimators employed use recapture data, which requires that the sampling be done in at least two stages.

The fifth and final segment of material is contained in Chapter 11, which deals with four situations in which some of the basic assumptions of the standard procedures cannot be satisfied. The situations are (1) interviewer biases, which can sometimes be minimized by using interpenetrating subsamples, (2) an inadequate frame, which can sometimes be accounted for by using an estimator for subpopulations of the sampled population, (3) information on sensitive questions, which can be obtained by using a random-response model, and (4) nonresponse, which can be planned for and designed into the survey by treating nonrespondents as a separate stratum.

To summarize, we have presented various elementary sample survey designs along with their associated methods of inference. Treatment of the topics has been directed toward practical applications so that you can see how sample survey design can be employed to make inferences at minimum cost when sampling from finite social, business, or natural resource populations.

12.2 COMPARISONS AMONG THE DESIGNS AND METHODS

With an array of sampling designs and methods of analysis available, we now summarize earlier discussions on how one chooses an appropriate design for a particular problem.

Simple random sampling is the basic building block and point of reference for all other designs discussed in this text. However, few large-scale surveys use only simple random sampling, because other designs often provide greater accuracy or efficiency or both.

Stratified random sampling produces estimators with smaller variance than those from simple random sampling, for the same sample size, when the measurements under study are homogeneous within strata but the stratum means vary among themselves.

The ideal situation for stratified random sampling is to have all measurements within any one stratum equal but have differences occurring as we move from stratum to stratum.

Systematic sampling is used most often simply as a convenience. It is relatively easy to carry out. But this form of sampling may actually be better than simple random sampling, in terms of bounds on the error of estimation, if the correlation between pairs of elements within the same systematic sample is negative. This situation will occur, for example, in periodic data if the systematic sample hits both the high points and the low points of the periodicities. If, in contrast, the systematic sample hits only the high points, the results are very poor. Populations that have a linear trend in the data or that have a periodic structure that is not completely understood may be better sampled by using a stratified design. Economic time series, for example, can be stratified by quarter or month, with a random sample selected from each stratum. The stratified and the systematic sample both force the sampling to be carried out along the whole set of data, but the stratified design offers more random selection and often produces a smaller bound on the error of estimation.

Cluster sampling is generally employed because of cost effectiveness or because no adequate frame for elements is available. However, cluster sampling may be better than either simple or stratified random sampling if the measurements within clusters are heterogeneous and the cluster means are nearly equal. The ideal situation for cluster sampling is, then, to have each cluster contain measurements as different as possible but to have the cluster means equal. This condition is in contrast to that for stratified random sampling in which strata are to be homogeneous but stratum means are to differ.

Another way to contrast the last three designs is as follows. Suppose a population consists of $N = nk$ elements, which can be thought of as k systematic samples each of size n. The nk elements can be thought of as n clusters of size k, and the systematic sample merely selects one such cluster. In this case the clusters should be heterogeneous for optimal systematic sampling. By contrast, the nk elements can also be thought of as n strata of k elements each, and the systematic sample selects one element from each stratum. In this case the strata should be as homogeneous as possible, but the stratum means should differ as much as possible. This design is consistent with the cluster formulation of the problem and once again produces an optimal situation for systematic sampling. So we see that the three sampling designs are different, and yet they are consistent with one another with regard to basic principles.

Some final comments are in order on how to make use of an auxiliary variable x to gain more information on our variable of interest y. Ratio estimation is optimal if the regression of y and x produces a straight line through the origin and if the variation in the y's increases with increasing x. Regression estimation is better than ratio estimation if the regression of y on x does not go through the origin and if the variation in the y's remains relatively constant as x varies. Difference estimation is as good as regression estimation if the regression coefficient is nearly equal to unity.

We now provide some exercises for which you can decide the appropriate method of analysis.

EXERCISES

12.1 A shipment of 6000 automobile batteries is to contain, according to the manufacturer's specifications, batteries weighing approximately 69 pounds each and having positive-plate thicknesses of 120 thousandths of an inch. Thirty batteries were randomly selected from this shipment and tested. The data are recorded in the following table. Do you think either of the manufacturer's specifications is met for this shipment? (Each battery contains 24 positive plates.)

Battery specifications

Battery	Weight (in pounds)	Number of positive plates sampled	Average plate thickness (in thousandths of an inch)	Standard deviation of thicknesses
1	61.5	8	109.6	0.74
2	63.5	16	110.0	1.22
3	63.5	16	107.0	1.83
4	63.8	16	111.6	2.55
5	63.8	17	110.7	1.65
6	64.0	16	108.7	1.40
7	64.0	16	111.4	2.63
8	65.0	13	112.8	2.06
9	64.2	16	107.8	3.35
10	64.5	8	109.9	1.25
11	66.5	16	107.8	3.19
12	63.5	16	110.2	1.22
13	63.8	12	112.0	1.81
14	63.5	12	108.5	1.57
15	64.0	12	110.4	1.68
16	64.0	12	111.8	1.64
17	63.2	12	111.9	1.68
18	66.5	12	112.5	1.00
19	63.0	12	109.2	2.44
20	62.0	12	106.1	2.23
21	63.0	12	112.0	0.95
22	63.5	12	112.8	1.75
23	64.0	12	110.2	2.05
24	63.5	12	108.0	2.37
25	66.5	7	112.4	0.79
26	67.0	12	106.6	2.47
27	66.5	12	110.5	1.62
28	65.5	12	113.3	1.23
29	66.5	12	112.7	1.23
30	66.0	12	110.6	1.68

12.2 The Department of Revenue in a state carefully audits sales tax returns from retail stores. If the department thinks a firm is understating its taxable sales, it can order an audit of the firm's accounts. Just such an audit was ordered for a firm with many retail outlets across the state. Records on taxable sales were kept by each retail store. Hence, the auditors decided to randomly sample sales by store-months. That is, sales records were obtained for randomly selected months at randomly selected stores. The auditors then recorded total taxable sales for comparison with the taxable sales reported by the store. The Department of Revenue wants to estimate the proportional increase in audited taxable sales over reported taxable sales. How can you make this estimate, with a bound on the error, from the data given in the accompanying table for 15 store-months? (Figures are in thousands of dollars.)

Store-month	Audited taxable sales	Reported taxable sales
1	31.5	23.2
2	31.8	22.9
3	21.1	17.6
4	34.7	29.8
5	21.0	16.8
6	40.8	35.1
7	21.3	23.3
8	31.3	26.1
9	19.9	18.8
10	30.9	25.7
11	32.2	29.6
12	32.4	27.1
13	31.7	29.9
14	28.8	31.5
15	30.7	28.4

12.3 The U.S. Geological Survey monitors water flow in U.S. rivers. The data set RIVER in the appendix and on the data disk shows the mean daily flow rates for a specific Florida river over a two-year period, 1977–1979.
(a) Estimate the average daily flow rate by sampling 40 daily measurements over the two-year period. Place a bound on the error of estimation. (Before sampling, you might want to look at the data to see if any trends are apparent.)
(b) Estimate the ratio of the average April flow rate to the average September flow rate and calculate a bound on the error. Is two years enough data to obtain a good estimate here?

12.4 Foresters estimate the net volume of standing trees by measuring the diameter at breast height and the tree height and then observing visible defects and other characteristics of the tree. The actual volume of usable timber can only be found after the tree is felled and processed into boards. For a sample of 20 trees, data on both estimated and actual volume are recorded, along with the species of the tree, in the accompanying table. The total estimated volume for all 180 trees is 60,000 board feet. Use the data in the table to solve the following problems.
(a) Estimate the total actual board feet for the 180 trees.
(b) Estimate the proportion of balsam fir trees in the entire stand.
(c) Estimate the total actual board feet of balsam fir in the stand.
(d) Estimate the total actual board feet of balsam fir if there are 110 balsam fir trees in the stand.
Place bounds on the error of estimation in all four cases.

Species (S = black spruce; F = balsam fir)	Estimated net volume (in board feet)	Actual net volume (in board feet)
F	130	141
S	450	474
S	268	301
F	227	215
F	190	210
F	432	400
S	501	487
F	397	368
F	248	262
S	184	195
S	230	280
F	287	243
F	312	255
F	260	282
S	410	375
S	325	280
F	422	490
S	268	325
F	250	210
F	195	236

12.5 The Environmental Protection Agency and the University of Florida recently cooperated in a large study of the possible effects of drinking water on kidney stone disease.

	Carolinas		*Rockies*	
	New stone	Recurrent stone	New stone	Recurrent stone
Sample size	363	467	259	191
Age	42.2 (10.9)	45.1 (10.2)	42.5 (10.8)	46.4 (9.8)
Calcium (in parts per million)	11.0 (15.1)	11.3 (16.6)	42.4 (31.8)	40.1 (28.4)
Proportion now smoking	.73	.78	.57	.61

Kidney stone patients were sampled in the Carolinas and in the Rocky Mountain States. The patients were divided into "new stones" (the current episode being their first encounter with stone disease) and "recurrent stones." Measurements on three variables of interest, age of patient, amount of calcium in their home drinking water, and smoking activity, are recorded in the accompanying table. (Measurements are averages or proportions; standard deviations are given in parentheses.)

(a) Estimate the average age of all stone patients in the population, and place a bound on the error of estimation.

(b) Estimate the average calcium concentration in drinking water supplies for stone patients in the Carolinas. Place a bound on the error of estimation.

(c) Estimate the average calcium concentration in drinking water supplies for stone patients in the Rockies. Place a bound on the error of estimation. Does the answer here differ considerably from that in part (b)?
(d) Estimate the proportion of smokers among new stones and place a bound on the error.

12.6 In Exercise 12.5, the data were actually collected by first sampling hospitals from the two regions and then sampling stone patients from within hospitals. Explain how you would conduct the analysis asked for in Exercise 12.5 with the data supplied by the hospitals. What additional data would you need?

12.7 Suppose in Exercise 12.6 that the hospitals within regions vary greatly in size. How can you use the information on hospital size advantageously in your sampling design?

12.8 The toxic effects of chemicals on fish are measured in the laboratory by subjecting a certain species of fish to various concentrations of a chemical added to the water. The concentration of chemical that is lethal to 50% of the fish, over the test period, is called the LC 50. Tests in a tank in which water is not renewed during the test process are called static. If new water is constantly coming into the tank, the test is called flow-through. Static tests are cheaper and easier to run, but flow-through tests better approximate reality. Thus, experimenters often estimate a static–to–flow-through conversion factor. From the data given in the accompanying table on 12 static and flow-through tests (the measurements are in milligrams per liter), estimate a factor by which a static-test result should be multiplied to make it comparable to a flow-through–test result. Place a bound on the error of estimation.

Toxicant	LC 50 flow-through	LC 50 static
Malathion	0.5	0.9
DDT	0.8	1.8
Parathion	4.5	2.1
Endrin	5.5	1.3
Azinphosmethyl	1.2	0.2
DDT	3.5	2.3
Parathion	5.0	1.5
Endrin	0.5	3.2
Zectran	83.0	12.0
Chlordane	4.0	10.0
Fenthion	5.8	12.0
Malathion	12.0	90.0

Source: Federal Register, vol. 43, no. 97, May 18, 1978.

12.9 Refer to Exercise 12.8. Can you suggest some improvements in the sampling so as to obtain a better estimate of the conversion factor?

12.10 Refer to Exercise 12.2. Can you suggest a better design for sampling the retail stores? Keep in mind that sales vary from store to store and from month to month.

12.11 Raw sugar is delivered from a grower to a refining mill in bulk form, transported in large tank trucks. The amount paid by the mill for a truckload of sugar depends on the pure-sugar content of the load. This pure-sugar content is determined by laboratory analysis of small test samples, each test sample containing enough raw sugar to fill a test tube. Discuss possible sampling designs to obtain these test samples. (Only a few test samples can be run per truckload of raw sugar.)

12.12 Baled wool from Australia is inspected as it comes into the United States, and an import duty is paid on the basis of pure-wool content. Core samples are taken from bales and analyzed to determine the proportion of the bale that is pure wool. Discuss possible sampling designs for the estimation of pure-wool content in a shipload of bales.

12.13 The Florida Public Service Commission requires companies that sell natural gas to make sure that the meters attached to houses and commercial buildings are operating correctly. However, they will allow a sampling inspection plan rather than a detailed annual examination of every meter. Suppose 20% of the meters owned by a certain company must be checked each year, and the proportion of the company's meters operating correctly must be estimated. (If this proportion is low, the company will be forced to check more meters.) Suggest a sampling plan for this meter inspection policy, keeping the following points in mind: (1) Meters are of varying ages; (2) gas use varies greatly from user to user; and (3) meters are being connected and disconnected continuously.

12.14 You need to estimate the proportion of unsafe tires on automobiles and trucks owned by the University of Florida. Outline how you would collect a sample and estimate this proportion. Include an appropriate variance computation.

12.15 A certain machine in a factory has a sequence of operation times (when it is in service) alternating with down times (when it breaks down and is being repaired). Let operation times be denoted by y_i and down times by d_i. Then the sequence of observations over n cycles looks like this: $y_1, d_1, y_2, d_2, \ldots, y_n, d_n$.
(a) Set up an estimator for the proportion of time that the machine is in operation and show how to calculate an estimate of the variance.
(b) If a week's work contains 40 hours, how would you estimate the total number of hours that the machine is in operation?

12.16 The accompanying table shows the data of sulfate concentrations in a shallow unconfined aquifer that were measured quarterly for a period of seven years. Similarly, chloride concentrations were measured quarterly for nine years at a different site. Estimate the average sulfate concentration over this period, with an approximate 95% confidence interval. Do the same for the average chloride concentration.

Sample no.	Concentration	Sample no.	Concentration
Sulfate (mg/L)			
1	111	12	113
2	107	13	51.1
3	108.7	14	102
4	108.7	15	145
5	109.3	16	87
6	104.7	17	112
7	104.7	18	111
8	108	19	104
9	108	20	151
10	109.3	21	103
11	114.5	22	113

(continued)

(*continued*)

Sample no.	Concentration	Sample no.	Concentration
23	113	25	101
24	125	26	108
Chloride (mg/L)			
1	38	18	—
2	40	19	41
3	35	20	—
4	37	21	35
5	32	22	49
6	37	23	64
7	37	24	73
8	—	25	67
9	32	26	67
10	45	27	—
11	38	28	59
12	33.8	29	73
13	14	30	—
14	—	31	92.5
15	39	32	45.5
16	46	33	40.4
17	48	34	33.9
		35	28.1

Source: Harris, Loftis, and Montgomery, *Ground Water* 25, no. 2 (1987): 185–193.

12.17 The following table gives average height and weight measurements for children.

Average height and weight of children

	Boys		***Girls***	
Age (yrs)	**Height (cm)**	**Weight (kg)**	**Height (cm)**	**Weight (kg)**
1	73.6	9.5	73.6	9.1
2	83.8	11.8	83.8	11.3
3	91.4	14.0	91.4	13.6
4	99.0	15.4	99.0	15.0
5	106.6	17.7	104.1	17.2
6	114.2	20.9	111.7	20.4
7	119.3	23.1	119.3	22.2
8	127.0	25.9	127.0	25.4
9	132.0	28.6	132.0	28.1
10	137.1	31.3	137.1	31.3
11	142.2	34.9	142.2	34.9
12	147.3	37.7	147.3	39.0
13	152.4	41.7	152.4	45.5
14	157.5	48.5	157.5	48.5

Source: The World Almanac & Book of Facts, 1988 edition. Copyright © 1987 Newspaper Enterprise Association, Inc., New York.

(a) Estimate the ratio of height to weight for boys under the age of 15 for an approximate 95% confidence interval.
(b) Estimate the ratio of height to weight for girls under the age of 15.
(c) Do the ratios appear to differ significantly?

12.18 The following summaries come from selected Gallup polls for 1994. The target sample size is generally around 1000 respondents, but some specific sample sizes are provided for certain questions. The results are reported in terms of sample percentages.

Choose some questions of interest to you and produce interval estimates of true population proportions or differences between proportions. For example, you may want to look at questions such as the following:

> Does preference for beer change with age?
>
> Has the proportion of those who drink increased since 1950?
>
> Do males feel differently from females on the question of restricting smoking to designated areas?

Have fun!

SMOKING IN PUBLIC PLACES

QUESTION: What is your opinion regarding smoking in public places? First, in [place]—should they set aside certain areas, should they totally ban smoking, or should there be no restrictions on smoking? Next, in [place]. . .(RANDOM ORDER)

	Hotels & Motels?			Workplaces?			Restaurants?			
	Designated areas	*Ban altogether*	*No restrictions*	*Designated areas*	*Ban altogether*	*No restrictions*	*Designated areas*	*Ban altogether*	*No restrictions*	*No. of interviews*
National	68%	20%	10%	63%	32%	4%	57%	38%	4%	1007
Sex										
Male	63	21	15	62	31	6	55	38	6	509
Female	72	19	7	65	32	2	59	38	2	498
Age										
18–29 years	70	15	14	65	30	5	57	39	3	230
30–49 years	69	21	9	62	34	3	56	40	3	424
50–64 years	68	23	8	65	33	2	57	41	2	173
65 and older	59	23	12	62	28	6	57	32	7	170
Region										
East	70	19	9	65	31	4	54	43	2	246
Midwest	64	24	9	60	35	3	55	40	3	256
South	66	18	14	68	27	5	63	31	5	306
West	70	20	10	59	35	4	54	41	4	199

(continued)

(*continued*)

	Hotels & Motels?			Workplaces?			Restaurants?			
	Desig-nated areas	*Ban alto-gether*	*No restric-tions*	*Desig-nated areas*	*Ban alto-gether*	*No restric-tions*	*Desig-nated areas*	*Ban alto-gether*	*No restric-tions*	*No. of inter-views*
Community										
Urban	66	22	11	62	35	3	57	40	2	431
Suburban	71	17	10	66	29	4	59	37	3	321
Rural	66	21	10	62	30	5	55	35	8	249
Race										
White	68	20	10	64	31	4	58	37	4	844
Nonwhite	66	20	12	59	37	4	52	43	3	156
Education										
College postgrad.	67	30	3	44	54	2	39	57	4	140
College graduate	65	29	5	46	51	3	44	52	3	273
College incomplete	69	21	9	67	29	3	54	43	3	278
No college	68	16	13	69	24	5	64	30	4	449
Politics										
Republicans	66	24	9	61	35	3	55	42	2	283
Democrats	71	18	9	67	29	3	63	32	4	341
Independents	66	19	13	62	32	5	53	41	4	383
Clinton approval										
Approve	69	19	11	63	33	3	57	40	2	508
Disapprove	67	21	11	65	29	5	59	35	5	417
Income										
$50,000 and over	68	25	7	57	40	3	50	46	4	259
$30,000–49,999	68	21	10	63	33	3	57	40	3	253
$20,000–29,999	67	18	14	66	30	4	56	37	6	187
Under $20,000	68	16	13	67	26	13	64	31	3	247
Smoking status										
Never smoked	62	29	6	53	43	3	44	52	2	479
Ex-smoker	67	20	12	64	30	6	57	36	6	267
Smoker	77	18	4	81	13	5	80	16	3	261

Note: "No opinion" omitted.

Source: *The Gallup Poll Monthly*, March 1994.

Methodology

The results are based on telephone interviews with a randomly selected national sample of 1007 adults, 18 years and older, conducted March 11–13, 1994. For results based on samples of this size, one can say with 95% confidence that the error attributable to sampling and other random effects could be plus or minus 3 percentage points.

The poll includes 261 current smokers and 746 nonsmokers. Results based on these subgroups have margins of sampling error of plus or minus 7 and 4 percentage points, respectively.

In addition to sampling error, question wording and practical difficulties in conducting surveys can introduce error or bias into the findings of public opinion polls.

Have you, yourself, smoked any cigarettes in the past week?

Cigarette Smoking—Recent Trend

	Yes	No
1994 Mar	27%	73%
1991	28	72
1990	27	73
1989	27	73
1988	32	68
1987	30	70

In general, how harmful do you feel second-hand smoke is to adults—very harmful, somewhat harmful, not too harmful or not at all harmful?

	1994			
	Total	Smokers	Ex-smokers	Never smoked
Very	36%	14%	33%	51%
Somewhat	42	49	44	37
Not too	12	20	13	7
Not at all	6	12	6	2
Depends (vol.)	1	*	1	1
No opinion	3	5	3	2
	100%	100%	100%	100%

* Less than 0.5%

Do you think there should or should not be a complete ban on cigarette advertising?

Ban Advertising—Trend

	Total	Smokers	Ex-smokers	Never smoked
1994				
Yes	45%	40%	40%	51%
No	51	56	56	45
1990				
Yes	49			
No	48			

Note: "No opinion" omitted.

PREFERRED TYPE OF SPIRITS

(based on those who use alcoholic beverages)

Do you most often drink liquor, wine, or beer?

	Beer	Wine	Liquor	All equally (vol.)	Wine coolers (vol.)	No opinion	No. of interviews
National	47%	29%	18%	3%	1%	2%	663
Sex							
Male	64	17	15	4	*	0	358
Female	30	42	22	2	1	3	305
Age							
18–29 years	69	17	10	2	*	2	156
30–49 years	50	29	15	4	2	*	285
50–64 years	34	40	21	1	0	4	119
65 and older	18	39	39	4	0	0	97
Region							
East	44	34	14	5	1	2	173
Midwest	56	23	16	2	2	1	184
South	47	30	20	2	0	1	168
West	41	31	24	2	0	2	138
Community							
Urban	48	33	13	2	1	3	264
Suburban	40	31	25	3	1	*	231
Rural	55	22	17	5	0	1	166
Race							
White	47	30	18	3	1	1	593
Nonwhite	50	23	16	7	0	4	63
Education							
College postgraduate	38	45	12	5	0	0	117
College graduate	36	46	14	3	0	1	239
Some college	49	27	19	4	*	1	196
No college	54	20	20	3	1	2	226
Ideology							
Liberal	56	27	15	2	0	0	126
Moderate	47	29	18	4	1	1	264
Conservative	44	32	19	3	1	1	256
Income							
$75,000 and over	36	46	13	2	0	3	104
$50,000 and over	43	37	17	2	0	1	227
$30,000–49,999	54	27	14	4	1	0	175
$20,000–29,999	50	23	22	2	1	2	93
Under $20,000	46	23	23	4	1	3	137
Family drink problems							
Have experienced	47	27	22	2	0	2	143
Have not	47	30	17	4	1	1	519
Drinks in past week							
7 or less	48	33	15	2	1	1	300
8 or more	74	10	12	4	0	0	84

* Less than 1%.

Source: The Gallup Poll Monthly, June 1994.

I have a few questions about alcoholic beverages.... Do you have occasion to use alcoholic beverages such as liquor, wine or beer, or are you a total abstainer?

Percent Who Drink, by gender–trend

	Total	Men	Women
1994 Jun 3–6	65%	70%	61%
1992	64	72	57
1990	64	64	51
1989	56	64	48
1988	63	72	55
1987	65	72	57
1985	67	72	62
1984	64	73	57
1983	65	71	58
1982	65	69	61
1981	70	75	66
1979	69	74	64
1978	71	75	64
1977	71	77	65
1976	71	NA	NA
1974	68	77	61
1969	64	NA	NA
1966	65	70	61
1964	63	NA	NA
1960	62	NA	NA
1958	55	66	45
1957	58	67	50
1956	60	NA	NA
1952	60	68	53
1951	59	70	46
1950	60	NA	NA
1949	58	66	49
1947	63	72	54
1946	67	NA	NA
1945	67	75	60
1939	58	70	45

[Sample A, 508 respondents, ±5%] *Finally, Please tell me whether you agree strongly, agree somewhat, disagree somewhat, or disagree strongly with this statement: The use of alcohol by pregnant women can cause birth defects.*

Alcohol and Birth Defects–Trend

	Jan 1982	Apr 1987	Jun 1994
Agree strongly	57%	69%	81%
Agree somewhat	30	21	10
Disagree somewhat	3	4	3
Disagree strongly	2	1	3
No opinion	8	5	3
	100%	100%	100%

THE CRIME BILL: ASSAULT WEAPONS AND THE NRA

In your view, should Congress pass the crime bill in its original form as Clinton supports it, or should Congress change the bill to remove the ban on assault weapons?

Do you think the National Rifle Association has had mostly a positive or mostly a negative influence on the crime bill debate in the Congress?

	Assault weapons?				Impact of NRA?			
	Keep ban	Remove ban	Defeat entire bill (vol.)	No opinion	Positive	Negative	No opinion	No. of interviews
National	57%	30%	3%	10%	32%	56%	12%	1010
Sex								
Male	54	35	3	8	38	53	9	515
Female	60	25	3	12	27	58	15	496
Age								
18–29 years	56	37	1	6	36	55	9	180
30–49 years	58	30	2	10	32	58	10	434
50–64 years	63	26	4	7	33	59	8	219
65 and older	51	29	4	16	27	47	26	164
Region								
East	58	29	1	12	28	60	12	245
Midwest	55	32	3	10	36	52	12	260
South	59	26	3	12	34	51	15	308
West	55	33	5	7	31	62	7	198
Community								
Urban	62	25	3	10	27	61	12	392
Suburban	56	30	3	11	30	58	12	389
Rural	50	38	2	10	43	43	14	226
Race								
White	56	30	3	11	33	55	12	888
Nonwhite	68	27	0	5	28	60	12	116
Education								
College postgrad.	61	22	5	12	22	75	3	156
College graduate	62	22	6	10	24	69	7	318
Some college	59	29	3	9	32	58	10	296
No college	54	34	1	11	36	48	16	394
Politics								
Republicans	38	43	7	12	37	47	16	300
Democrats	71	19	2	8	28	63	9	279
Independents	60	29	1	10	32	56	12	432
Ideology								
Liberal	62	28	*	10	27	66	7	148
Moderate	67	22	2	9	30	60	10	441
Conservative	44	40	5	11	38	48	14	388

(continued)

(*continued*)

	Assault weapons?				Impact of NRA?			
	Keep ban	Remove ban	Defeat entire bill (vol.)	No opinion	Positive	Negative	No opinion	No. of interviews
Clinton approval								
Approve	72	20	*	8	27	64	9	387
Disapprove	44	40	5	11	38	50	12	546
Income								
$75,000 and over	61	23	3	13	23	71	6	119
$50,000 and over	58	28	3	11	28	65	7	266
$30,000–49,999	63	26	3	8	33	58	9	294
$20,000–29,999	58	33	2	7	29	55	16	162
Under $20,000	51	34	3	12	37	47	16	216
Crime bill								
Favor	77	17	1	5	24	58	8	476
Oppose	31	55	7	7	48	44	8	304

* Less than 1%.

Source: The Gallup Poll Monthly, August 1994.

APPENDIX 1

REFERENCES AND BIBLIOGRAPHY

Bailey, A. D. 1981. *Statistical Auditing*. New York: Harcourt Brace Jovanovich.

Bailey, N. T. J. 1951. "On Estimating the Size of Mobile Populations from Recaptive Data." *Biometrika*, 38:292–306.

Bergsten, J. W. 1979. "Some Methodological Results from Four Statewide Telephone Surveys Using Random Digit Dialing." *American Statistical Association Proceedings of the Section on Survey Research Methods*, pp. 239–243.

Bradburn, N. N., Rips, L. J., and Shevell, S. K. 1987. "Answering Autobiographical Questions: The Impact of Memory and Inference on Surveys." *Science*, April 10, pp. 157–161.

Bryson, M. C. 1976. "The Literary Digest Poll: Making of a Statistical Myth." *American Statistician*, 30(4):184–185.

Bureau of Labor Statistics, Handbook of Methods, vols. I and II. 1982. Washington, D.C.: U.S. Department of Labor.

Campbell, C., and Joiner, B. 1973. "How to Get the Answer Without Being Sure You Asked the Question." *American Statistician*, 27:229–231.

Careers in Statistics. 1980. Washington, D.C.: American Statistical Association.

Chapman, D. G. 1952. "Inverse, Multiple and Sequential Sample Censuses." *Biometrics*, 8:286–306.

Cochran, W. G. 1950. "Estimation of Bacterial Densities by Means of the 'Most Probable Number.'" *Biometrics*, 6:105.

———. 1977. *Sampling Techniques*, 3rd ed. New York: Wiley.

Crossen, C. 1994. *Tainted Truth: The Manipulation of Fact in America*. New York: Simon & Schuster.

Deming, W. E. 1960. *Sample Design in Business Research*. New York: Wiley.

Frankel, L. R. 1976. "Statisticians and People—The Statistician's Responsibility." *Journal of the American Statistical Association*, 7:9–16.

Gallup, George. 1972. *The Sophisticated Poll Watcher's Guide*. Princeton, N.J.: Princeton Opinion Press.

Greenberg, B. G., Kuebler, R. R., Abernathy, J. R., and Horvitz, D. G. 1971. "Application of Randomized Response Technique in Obtaining Quantitative Data." *Journal of the American Statistical Association*, 66:245–250.

Groves, Robert M. 1989. *Survey Errors and Survey Costs*. New York: Wiley.

Hansen, M. H., Hurwitz, W. N., and Madow, W. G. 1953. *Sample Survey Methods and Theory*, vol. 1. New York: Wiley.

Harper, W. B., Westfall, R., and Stasch, S. F. 1977. *Marketing Research*. Homewood, Ill.: Irwin.

Jessen, Raymond T. 1978. *Statistical Survey Techniques*. New York: Wiley.

Jones, H. L. 1956. "Investigation of the Properties of a Sample Mean by Employing Random Subsample Means." *Journal of the American Statistical Association*, 51:54–83.

Kinnear, T. C., and Taylor, J. R. 1983. *Marketing Research, An Applied Approach*. New York: McGraw-Hill.

Kish, L. 1965. *Survey Sampling*. New York: Wiley.

Levy, P. S., and Lemeshow, S. 1980. *Sampling for Health Professionals*. Belmont, Calif: Lifetime Learning.

Leysieffer, F., and Warner, S. 1976. "Respondent Jeopardy and Optimal Designs in Randomized Response Models." *Journal of the American Statistical Association*, 71:649–656.

Mendenhall, W. and Beaver, R. 1994. *Introduction to Probability and Statistics*, 9th ed. Boston: Duxbury Press.

Mosby, H. S., ed. 1969. *Wildlife Investigational Techniques*, 3rd ed. Washington, D.C.: Wildlife Society.

Raj, D. 1968. *Sampling Theory*. New York: McGraw-Hill.

Ray, A. A., ed. 1982. *SAS Users Guide: Statistics*. Cary, N.C.: SAS Institute.

Roberts, D. 1978. *Statistical Auditing*. New York: American Institute of Certified Accountants.

Ryan, T. A., Joiner, B. L., and Ryan, B. F. 1976. *Minitab Student Handbook*. Boston: Duxbury Press.

Schuman, H., and Presser, S. 1981. *Questions and Answers in Attitude Surveys*. New York: Academic Press.

Stephan, F. F., and McCarthy, P. M. 1958. *Sampling Opinions, An Analysis of Survey Procedure*, New York: Wiley.

Sudman, S. 1976. *Applied Sampling*. New York: Academic Press.

Swindel, B. F. 1983. "Choice of Size and Number of Quadrats to Estimate Density from Frequency in Poisson and Binomially Dispersed Populations." *Biometrics*, 39:455.

Tanur, J., ed. 1989. *Statistics: A Guide to the Unknown*. Pacific Grove, Calif.: Wadsworth Advanced Books and Software.

Thompson, S. K. 1992. *Sampling*. New York: Wiley.

Warner, S. L. 1965. "Randomized Response: A Survey Technique for Eliminating Evasive Answer Bias." *Journal of the American Statistical Association*, 60:63–69.

Weeks, M. F., Jones, B. L., Folsom, R. E., and Benrud, C. H. 1980. "Optimal Times to Contact Sample Households." *Public Opinion Quarterly*, 44:101–114.

Williams, B. 1978. *A Sampler on Sampling*. New York: Wiley.

TABLES

TABLE 1 **Normal curve areas**

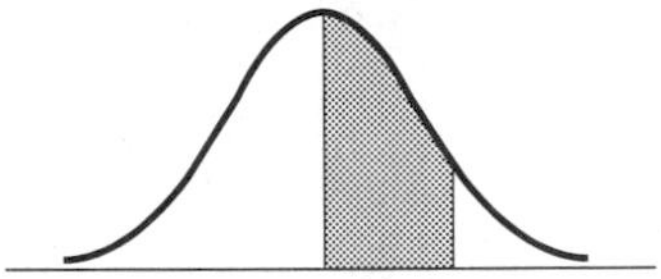

z	.00	.01	.02	.03	.04	.05	.06	.07	.08	.09
0.0	.0000	.0040	.0080	.0120	.0160	.0199	.0239	.0279	.0319	.0359
0.1	.0398	.0438	.0478	.0517	.0557	.0596	.0636	.0675	.0714	.0753
0.2	.0793	.0832	.0871	.0910	.0948	.0987	.1026	.1064	.1103	.1141
0.3	.1179	.1217	.1255	.1293	.1331	.1368	.1406	.1443	.1480	.1517
0.4	.1554	.1591	.1628	.1664	.1700	.1736	.1772	.1808	.1844	.1879
0.5	.1915	.1950	.1985	.2019	.2054	.2088	.2123	.2157	.2190	.2224
0.6	.2257	.2291	.2324	.2357	.2389	.2422	.2454	.2486	.2517	.2549
.07	.2580	.2611	.2642	.2673	.2704	.2734	.2764	.2794	.2823	.2852
0.8	.2881	.2910	.2939	.2967	.2995	.3023	.3051	.3078	.3106	.3133
0.9	.3159	.3186	.3212	.3238	.3264	.3289	.3315	.3340	.3365	.3389
1.0	.3413	.3438	.3461	.3485	.3508	.3531	.3554	.3577	.3599	.3621
1.1	.3643	.3665	.3686	.3708	.3729	.3749	.3770	.3790	.3810	.3830
1.2	.3849	.3869	.3888	.3907	.3925	.3944	.3962	.3980	.3997	.4015
1.3	.4032	.4049	.4066	.4082	.4099	.4115	.4131	.4147	.4162	.4177
1.4	.4192	.4207	.4222	.4236	.4251	.4265	.4279	.4292	.4306	.4319
1.5	.4332	.4345	.4357	.4370	.4382	.4394	.4406	.4418	.4429	.4441
1.6	.4452	.4463	.4474	.4484	.4495	.4505	.4515	.4525	.4535	.4545
1.7	.4554	.4564	.4573	.4582	.4591	.4599	.4608	.4616	.4625	.4633
1.8	.4641	.4649	.4656	.4664	.4671	.4678	.4686	.4693	.4699	.4706
1.9	.4713	.4719	.4726	.4732	.4738	.4744	.4750	.4756	.4761	.4767
2.0	.4772	.4778	.4783	.4788	.4793	.4798	.4803	.4808	.4812	.4817
2.1	.4821	.4826	.4830	.4834	.4838	.4842	.4846	.4850	.4854	.4857
2.2	.4861	.4864	.4868	.4871	.4875	.4878	.4881	.4884	.4887	.4890
2.3	.4893	.4896	.4898	.4901	.4904	.4906	.4909	.4911	.4913	.4916
2.4	.4918	.4920	.4922	.4925	.4927	.4929	.4931	.4932	.4934	.4936
2.5	.4938	.4940	.4941	.4943	.4945	.4946	.4948	.4949	.4951	.4952
2.6	.4953	.4955	.4956	.4957	.4959	.4960	.4961	.4962	.4963	.4964
2.7	.4965	.4966	.4967	.4968	.4969	.4970	.4971	.4972	.4973	.4974
2.8	.4974	.4975	.4976	.4977	.4977	.4978	.4979	.4979	.4980	.4981
2.9	.4981	.4982	.4982	.4982	.4984	.4984	.4985	.4985	.4986	.4986
3.0	.4987	.4987	.4987	.4988	.4988	.4989	.4989	.4989	.4990	.4990

Abridged from Table 1 of *Statistical Tables and Formulas* by A. Hald (New York: John Wiley & Sons, Inc., 1952). Reproduced by permission of A. Hald and the publishers, John Wiley & Sons, Inc.

TABLE 2 **Random numbers**

Line/ Col.	(1)	(2)	(3)	(4)	(5)	(6)	(7)	(8)	(9)	(10)	(11)	(12)	(13)	(14)
1	10480	15011	01536	02011	81647	91646	69179	14194	62590	36207	20969	99570	91291	90700
2	22368	46573	25595	85393	30995	89198	27982	53402	93965	34095	52666	19174	39615	99505
3	24130	48360	22527	97265	76393	64809	15179	24830	49340	32081	30680	19655	63348	58629
4	42167	93093	06243	61680	07856	16376	39440	53537	71341	57004	00849	74917	97758	16379
5	37570	39975	81837	16656	06121	91782	60468	81305	49684	60672	14110	06927	01263	54613
6	77921	06907	11008	42751	27756	53498	18602	70659	90655	15053	21916	81825	44394	42880
7	99562	72905	56420	69994	98872	31016	71194	18738	44013	48840	63213	21069	10634	12952
8	96301	91977	05463	07972	18876	20922	94595	56869	69014	60045	18425	84903	42508	32307
9	89579	14342	63661	10281	17453	18103	57740	84378	25331	12565	58678	44947	05585	56941
10	85475	36857	53342	53988	53060	59533	38867	62300	08158	17983	16439	11458	18593	64952
11	28918	69578	88231	33276	70997	79936	56865	05859	90106	31595	01547	85590	91610	78188
12	63553	40961	48235	03427	49626	69445	18663	72695	52180	20847	12234	90511	33703	90322
13	09429	93969	52636	92737	88974	33488	36320	17617	30015	08272	84115	27156	30613	74952
14	10365	61129	87529	85689	48237	52267	67689	93394	01511	26358	85104	20285	29975	89868
15	07119	97336	71048	08178	77233	13916	47564	81056	97735	85977	29372	74461	28551	90707
16	51085	12765	51821	51259	77452	16308	60756	92144	49442	53900	70960	63990	75601	40719
17	02368	21382	52404	60268	89368	19885	55322	44819	01188	65255	64835	44919	05944	55157
18	01011	54092	33362	94904	31273	04146	18594	29852	71585	85030	51132	01915	92747	64951
19	52162	53916	46369	58586	23216	14513	83149	98736	23495	64350	94738	17752	35156	35749
20	07056	97628	33787	09998	42698	06691	76988	13602	51851	46104	88916	19509	25625	58104
21	48663	91245	85828	14346	09172	30168	90229	04734	59193	22178	30421	61666	99904	32812
22	54164	58492	22421	74103	47070	25306	76468	26384	58151	06646	21524	15227	96909	44592
23	32639	32363	05597	24200	13363	38005	94342	28728	35806	06912	17012	64161	18296	22851
24	29334	27001	87637	87308	58731	00256	45834	15398	46557	41135	10367	07684	36188	18510
25	02488	33062	28834	07351	19731	92420	60952	61280	50001	67658	32586	86679	50720	94953
26	81525	72295	04839	96423	24878	82651	66566	14778	76797	14780	13300	87074	79666	95725
27	29676	20591	68086	26432	46901	20849	89768	81536	86645	12659	92259	57102	80428	25280
28	00742	57392	39064	66432	84673	40027	32832	61362	98947	96067	64760	64584	96096	98253
29	05366	04213	25669	26422	44407	44048	37937	63904	45766	66134	75470	66520	34693	90449
30	91921	26418	64117	94305	26766	25940	39972	22209	71500	64568	91402	42416	07844	69618
31	00582	04711	87917	77341	42206	35126	74087	99547	81817	42607	43808	76655	62028	76630
32	00725	69884	62797	56170	86324	88072	76222	36086	84637	93161	76038	65855	77919	88006
33	69011	65795	95876	55293	18988	27354	26575	08625	40801	59920	29841	80150	12777	48501
34	25976	57948	29888	88604	67917	48708	18912	82271	65424	69774	33611	54262	85963	03547
35	09763	83473	73577	12908	30883	18317	28290	35797	05998	41688	34952	37888	38917	88050
36	91567	42595	27958	30134	04024	86385	29880	99730	55536	84855	29080	09250	79656	73211
37	17955	56349	90999	49127	20044	59931	06115	20542	18059	02008	73708	82517	36103	42791
38	46503	18584	18845	49618	02304	51038	20655	58727	28168	15475	56942	53389	20562	87338
39	92157	89634	94824	78171	84610	82834	09922	25417	44137	48413	25555	21246	35509	20468
40	14577	62765	35605	81263	39667	47358	56873	56307	61607	49518	89656	20103	77490	18062
41	98427	07523	33362	64270	01638	92477	66969	98420	04880	45585	46565	04102	46880	45709
42	34914	63976	88720	82765	34476	17032	87589	40836	32427	70002	70663	88863	77775	69348
43	70060	28277	39475	46473	23219	53416	94970	25832	69975	94884	19661	72828	00102	66794
44	53976	54914	06990	67245	68350	82948	11398	42878	80287	88267	47363	46634	06541	97809
45	76072	29515	40980	07391	58745	25774	22987	80059	39911	96189	41151	14222	60697	59583

(*continues*)

TABLE 2 (continued)

Line/ Col.	(1)	(2)	(3)	(4)	(5)	(6)	(7)	(8)	(9)	(10)	(11)	(12)	(13)	(14)
46	90725	52210	83974	29992	65831	38857	50490	83765	55657	14361	31720	57375	56228	41546
47	64364	67412	33339	31926	14883	24413	59744	92351	97473	89286	35931	04110	23726	51900
48	08962	00358	31662	25388	61642	34072	81249	35648	56891	69352	48373	45578	78547	81788
49	95012	68379	93526	70765	10592	04542	76463	54328	02349	17247	28865	14777	62730	92277
50	15664	10493	20492	38391	91132	21999	59516	81652	27195	48223	46751	22923	32261	85653
51	16408	81899	04153	53381	79401	21438	83035	92350	36693	31238	59649	91754	72772	02338
52	18629	81953	05520	91962	04739	13092	97662	24822	94730	06496	35090	04822	86774	98289
53	73115	35101	47498	87637	99016	71060	88824	71013	18735	20286	23153	72924	35165	43040
54	57491	16703	23167	49323	45021	33132	12544	41035	80780	45393	44812	12515	98931	91202
55	30405	83946	23792	14422	15059	45799	22716	19792	09983	74353	68668	30429	70735	25499
56	16631	35006	85900	98275	32388	52390	16815	69298	82732	38480	73817	32523	41961	44437
57	96773	20206	42559	78985	05300	22164	24369	54224	35083	19687	11052	91491	60383	19746
58	38935	64202	14349	82674	66523	44133	00697	35552	35970	19124	63318	29686	03387	59846
59	31624	76384	17403	53363	44167	64486	64758	75366	76554	31601	12614	33072	60332	92325
60	78919	19474	23632	27889	47914	02584	37680	20801	72152	39339	34806	08930	85001	87820
61	03931	33309	57047	74211	63445	17361	62825	39908	05607	91284	68833	25570	38818	46920
62	74426	33278	43972	10119	89917	15665	52872	73823	73144	88662	88970	74492	51805	99378
63	09066	00903	20795	95452	92648	45454	09552	88815	16553	51125	79375	97596	16296	66092
64	42238	12426	87025	14267	20979	04508	64535	31355	86064	29472	47689	05974	52468	16834
65	16153	08002	26504	41744	81959	65642	74240	56302	00033	67107	77510	70625	28725	34191
66	21457	40742	29820	96783	29400	21840	15035	34537	33310	06116	95240	15957	16572	06004
67	21581	57802	02050	89728	17937	37621	47075	42080	97403	48626	68995	43805	33386	21597
68	55612	78095	83197	33732	05810	24813	86902	60397	16489	03264	88525	42786	05269	92532
69	44657	66999	99324	51281	84463	60563	79312	93454	68876	25471	93911	25650	12682	73572
70	91340	84979	46949	81973	37949	61023	43997	15263	80644	43942	89203	71795	99533	50501
71	91227	21199	31935	27022	84067	05462	35216	14486	29891	68607	41867	14951	91696	85065
72	50001	38140	66321	19924	72163	09538	12151	06878	91903	18749	34405	56087	82790	70925
73	65390	05224	72958	28609	81406	39147	25549	48542	42627	45233	57202	94617	23772	07896
74	27504	96131	83944	41575	10573	08619	64482	73923	36152	05184	94142	25299	84387	34925
75	37169	94851	39117	89632	00959	16487	65536	49071	39782	17095	02330	74301	00275	48280
76	11508	70225	51111	38351	19444	66499	71945	05422	13442	78675	84081	66938	93654	59894
77	37449	30362	06694	54690	04052	53115	62757	95348	78662	11163	81651	50245	34971	52924
78	46515	70331	85922	38329	57015	15765	97161	17869	45349	61796	66345	81073	49106	79860
79	30986	81223	42416	58353	21532	30502	32305	86482	05174	07901	54339	58861	74818	46942
80	63798	64995	46583	09785	44160	78128	83991	42865	92520	83531	80377	35909	81250	54238
81	82486	84846	99254	67632	43218	50076	21361	64816	51202	88124	41870	52689	51275	83556
82	21885	32906	92431	09060	64297	51674	64126	62570	26123	05155	59194	52799	28225	85762
83	60336	98782	07408	53458	13564	59089	26445	29789	85205	41001	12535	12133	14645	23541
84	43937	46891	24010	25560	86355	33941	25786	54990	71899	15475	95434	98227	21824	19585
85	97656	63175	89303	16275	07100	92063	21942	18611	47348	20203	18534	03862	78095	50136
86	03299	01221	05418	38982	55758	92237	26759	86367	21216	98442	08303	56613	91511	75928
87	79626	06486	03574	17668	07785	76020	79924	25651	83325	88428	85076	72811	22717	50585
88	85636	68335	47539	03129	65651	11977	02510	26113	99447	68645	34327	15152	55230	93448
89	18039	14367	61337	06177	12143	46609	32989	74014	64708	00533	35398	58408	13261	47908
90	06362	15656	60627	36478	65648	16764	53412	09013	07832	41574	17639	82163	60859	75567

DERIVATION OF SOME MAIN RESULTS

In this section, we present the mathematical derivation of some of the main formulas used throughout the text. We assume the reader has some knowledge of probability theory, so that expectations, variances, and covariances can be manipulated with little explanation.

Let y_i denote a random variable with probability distribution $p(y)$. Then we have the following definitions from elementary probability theory:

$$E(y) = \sum_y yp(y) = \mu$$

$$E[g(y)] = \sum_y g(y)p(y)$$

$$V(y) = E(y-\mu)^2 = \sum_y (y-\mu)^2 p(y) = \sigma^2$$

where E denotes expected value, V denotes variance, and $g(y)$ is a function of y.

Suppose $y_1, y_2, \ldots, y_n$ denotes a sample of size n and $a_1, a_2, \ldots, a_n$ are constants. If

$$U = \sum_{i=1}^{n} a_i y_i$$

then

$$E(U) = \sum_{i=1}^{n} a_i E(y_i) \tag{A.1}$$

and

$$V(U) = \sum_{i=1}^{n} a_i^2 V(y_i) + 2 \sum\sum_{i<j} a_i a_j \operatorname{cov}(y_i, y_j) \tag{A.2}$$

where cov denotes covariances. If the y_i's are uncorrelated, then

$$V(U) = \sum_{i=1}^{n} a_i^2 V(y_i) \tag{A.3}$$

SIMPLE RANDOM SAMPLING

Suppose $y_1, y_2, \ldots, y_n$ denotes a simple random sample from a population of values $\{u_1, u_2, \ldots, u_N\}$. Considering y_i by itself (a simple random sample of size 1), we have

$$E(y_i) = \sum_{i=1}^{N} u_i \left(\frac{1}{N}\right) = \mu$$

and

$$V(y_i) = \sum_{i=1}^{N} (u_i - \mu)^2 \left(\frac{1}{N}\right) = \sigma^2$$

By Equation (A.1),

$$E(\bar{y}) = \frac{1}{n} \sum_{i=1}^{n} E(y_i) = \frac{1}{n} \sum_{i=1}^{n} \mu = \mu$$

Also,

$$\begin{aligned}
\operatorname{cov}(y_i, y_j) &= E[(y_i - \mu)(y_j - \mu)] = E(y_i y_j) - \mu^2 \\
&= \sum_{i \neq j}^{N} u_i u_j \left[\frac{1}{N(N-1)}\right] - \frac{1}{N^2}\left(\sum_{i=1}^{N} u_i\right)^2 \\
&= \frac{1}{N}\left[\sum_{i \neq j}^{N} \frac{u_i u_j}{N-1} - \frac{1}{N}\left(\sum_{i=1}^{N} u_i\right)^2\right] \\
&= \frac{1}{N}\left[\frac{\left(\sum_{i=1}^{N} u_i\right)^2 - \sum_{i=1}^{N} u_i^2}{N-1} - \frac{1}{N}\left(\sum_{i=1}^{N} u_i\right)^2\right] \\
&= -\frac{1}{N}\left[\frac{1}{N-1}\sum_{i=1}^{N} u_i^2 - \frac{1}{N(N-1)}\left(\sum_{i=1}^{N} u_i\right)^2\right] \\
&= -\frac{1}{N(N-1)}\sum_{i=1}^{N}(u_i - \mu)^2 = -\frac{1}{N-1}\sigma^2
\end{aligned}$$

Using this fact and Equation (A.2), we can find the variance of $\bar{y}$. We have

$$\begin{aligned}
V(\bar{y}) = V\left(\frac{1}{n}\sum_{i=1}^{n} y_i\right) &= \frac{1}{n^2}\left[\sum_{i=1}^{n} \sigma^2 + 2\sum_{i<j}\sum \operatorname{cov}(y_i, y_j)\right] \\
&= \frac{1}{n^2}\left[\sum_{i=1}^{n} \sigma^2 + 2\sum_{i<j}\sum \frac{-\sigma^2}{N-1}\right] \\
&= \frac{1}{n^2}\left\{n\sigma^2 - \frac{2\sigma^2}{N-1}\left(\frac{n(n-1)}{2}\right)\right\}
\end{aligned}$$

since there are $n(n-1)/2$ pairs (i, j) selected from the integers $1, 2, \ldots, n$ so that $i < j$. Therefore

$$V(\bar{y}) = \frac{\sigma^2}{n}\left(\frac{N-n}{N-1}\right)$$

We now show that $[(N-n)/N](s^2/n)$ is an unbiased estimator of $V(\bar{y})$. We have

$$\begin{aligned}
E(s^2) &= E\left[\left(\frac{1}{n-1}\right)\sum_{i=1}^{n}(y_i-\bar{y})^2\right] \\
&= \left(\frac{1}{n-1}\right)E\left\{\sum_{i=1}^{n}[(y_i-\mu)-(\bar{y}-\mu)]^2\right\} \\
&= \left(\frac{1}{n-1}\right)E\left[\sum_{i=1}^{n}(y_i-\mu)^2-n(\bar{y}-\mu)^2\right] \\
&= \frac{1}{n-1}\left[\sum_{i=1}^{n}E(y_i-\mu)^2-nE(\bar{y}-\mu)^2\right] \\
&= \frac{1}{n-1}[n\sigma^2-nV(\bar{y})] = \frac{1}{n-1}\left[n\sigma^2-n\left(\frac{N-n}{N-1}\right)\left(\frac{\sigma^2}{n}\right)\right] \\
&= \frac{\sigma^2}{n-1}\left(n-\frac{N-n}{N-1}\right) = \frac{N}{N-1}\sigma^2
\end{aligned}$$

Therefore,

$$\begin{aligned}
E\left[\left(\frac{N-n}{N}\right)\left(\frac{s^2}{n}\right)\right] &= \left(\frac{N-n}{N}\right)\left(\frac{1}{n}\right)\left(\frac{N}{N-1}\right)\sigma^2 \\
&= \left(\frac{N-n}{N-1}\right)\left(\frac{\sigma^2}{n}\right) = V(\bar{y})
\end{aligned}$$

which was to be shown.

This derivation results in Equations (4.2), (4.3), and (4.4). Now $\hat{\tau} = N\bar{y}$ is an unbiased estimator of τ by Equation (A.1), and Equation (4.6) follows from Equation (A.3) and results already shown.

Since $\hat{p}$ is actually a $\bar{y}$ for $\{0, 1\}$ data, $\hat{p}$ is an unbiased estimator of p, and Equation (4.16) follows directly, after observing that

$$\frac{s^2}{n} = \frac{\hat{p}(1-\hat{p})}{n-1}$$

for the $\{0, 1\}$ data.

STRATIFIED RANDOM SAMPLING

In stratified random sampling,

$$\bar{y}_{st} = \sum_{i=1}^{L}\left(\frac{N_i}{N}\right)\bar{y}_i$$

is of the same form as U, and the $\bar{y}_i$'s are independently selected through simple random sampling. Thus,

$$E(\bar{y}_{\text{st}}) = \sum_{i=1}^{L} \left(\frac{N_i}{N}\right) E(\bar{y}_i) = \sum_{i=1}^{L} \left(\frac{N_i}{N}\right) \mu_i = \mu$$

by Equation (A.1), and

$$V(\bar{y}_{\text{st}}) = \sum_{i=1}^{L} \left(\frac{N_i}{N}\right)^2 V(\bar{y}_i)$$

by Equation (A.3), and Equation (5.2) follows.

In the sample size and allocation formulas of Chapter 5, we set $N/(N-1)$ equal to 1 for convenience.

RATIO ESTIMATION

The ratio estimator r is approximately an unbiased estimator of $R = (\mu_y/\mu_x)$ if n is reasonably large. That is, $E(\bar{y}/\bar{x})$ is approximately R. Hence,

$$V(r) = V\left(\frac{\bar{y}}{\bar{x}}\right) \doteq E\left(\frac{\bar{y}}{\bar{x}} - R\right)^2 = E\left(\frac{\bar{y} - R\bar{x}}{\bar{x}}\right)^2$$

$$\doteq E\left(\frac{\bar{y} - R\bar{x}}{\mu_x}\right)^2 = \frac{1}{\mu_x^2} E(\bar{y} - R\bar{x})^2 = \frac{1}{\mu_x^2} V(\bar{y} - R\bar{x})$$

since $E(\bar{y} - R\bar{x}) = 0$. Since $\bar{y} - R\bar{x}$ is the sample mean of quantities $y_i - Rx_i$, with $E(y_i - Rx_i) = 0$, then $V(\bar{y} - R\bar{x})$ can be estimated by

$$\left(\frac{N-n}{Nn}\right)\left(\frac{1}{n-1}\right) \sum_{i=1}^{n} (y_i - Rx_i)^2$$

If R is replaced by r in the latter expression, Equation (6.2) follows. Variance expressions (6.5) and (6.7) follow by using Equation (A.3).

SINGLE-STAGE CLUSTER SAMPLING

The estimator of Equation (8.1) from cluster sampling is a ratio estimator, and its variance (8.2) follows from results derived above. The variance expression (8.5) then comes about by applying Equation (A.3). The estimator of τ given in Equation (8.7) is simply based on a sample mean of cluster totals, and Equation (8.8) follows from basic principles used above.

TWO-STAGE CLUSTER SAMPLING

Since this situation requires careful manipulation of between-cluster variances and within-cluster variances, we illustrate the derivations only for the case in which all clusters are of the same size. That is, we assume

$$m_1 = m_2 = \cdots = m_N = m$$

and

$$M_1 = M_2 = \cdots = \bar{M} = \frac{M}{N}$$

In this case,

$$\hat{\mu} = \left(\frac{N}{M}\right)\left(\frac{1}{n}\right)\sum_{i=1}^{n} M_i \bar{y}_i = \sum_{i=1}^{n} \bar{y}_i / n$$

We find the mean and variance of $\hat{\mu}$ by first fixing the n clusters in the sample and then averaging over all possible samples of n clusters. Expectation and variance operations when the n clusters are fixed will be denoted by E_2 and V_2, respectively. Similarly, expectations and variances over all possible samples of n clusters will be denoted by E_1 and V_1. (When the n clusters are fixed, the cluster sample looks like a stratified random sample.)

Now

$$E(\hat{\mu}) = E_1[E_2(\hat{\mu})] = E_1\left[\frac{1}{n}\sum_{i=1}^{n} E_2(\bar{y}_i)\right]$$
$$= E_1\left(\frac{1}{n}\sum_{i=1}^{n} \mu_i\right)$$

where μ_i is the mean of cluster i. Since the expected value of a sample mean is the corresponding population mean in simple random sampling,

$$E_1\left(\frac{1}{n}\sum_{i=1}^{n} \mu_i\right) = \frac{1}{N}\sum_{i=1}^{N} \mu_i = \frac{1}{N\bar{M}}\sum_{i=1}^{N} \bar{M}\mu_i$$
$$= \frac{1}{M}\sum_{i=1}^{N} \tau_i = \frac{\tau}{M} = \mu$$

where τ_i is the total for cluster i. Thus, $\hat{\mu}$ is an unbiased estimator of μ.

From a basic result in probability theory,

$$V(\hat{\mu}) = V_1[E_2(\hat{\mu})] + E_1[V_2(\hat{\mu})]$$

Now

$$V_1[E_2(\hat{\mu})] = V_1\left(\frac{1}{n}\sum_{i=1}^{n} \mu_i\right) = \left(\frac{N-n}{N}\right)\left(\frac{1}{n}\right)\left(\frac{1}{N-1}\right)\sum_{i=1}^{N} (\mu_i - \bar{\mu})^2 \qquad \text{(A.4)}$$

where $\bar{\mu} = (1/N)\sum_{i=1}^{N} \mu_i$. This expression follows from basic results explored earlier. Also,

$$E_1[V_2(\hat{\mu})] = E_1\left[V_2\left(\frac{1}{n}\sum_{i=1}^{n} \bar{y}_i\right)\right] = E_1\left[\frac{1}{n^2}\sum_{i=1}^{n} V_2(\bar{y}_i)\right]$$

$$= \left(\frac{1}{n}\right)\left(\frac{1}{N}\right)\sum_{i=1}^{N} V(\bar{y}_i) \tag{A.5}$$

where $V(\bar{y}_i)$ is the usual variance of a sample mean for a simple random sample of m elements from $\bar{M}$ elements.

We must now estimate the two parts of $V(\mu)$. For the first part, we might start with

$$\left(\frac{1}{\bar{M}}\right)^2 s_b^2 = \frac{1}{n-1}\sum_{i=1}^{n} (\bar{y}_i - \hat{\mu})^2$$

We have, as its expected value,

$$E_1\left\{E_2\left[\left(\frac{1}{n-1}\right)\sum_{i=1}^{n} (\bar{y}_i - \hat{\mu})^2\right]\right\}$$

$$= E_1\left\{\left(\frac{1}{n-1}\right)E_2\left[\sum_{i=1}^{n} \bar{y}_i^2 - \frac{1}{n}\left(\sum_{i=1}^{n} \bar{y}_i\right)^2\right]\right\}$$

$$= E_1\left\{\left(\frac{1}{n-1}\right)E_2\left[\sum_{i=1}^{n} \bar{y}_i^2 - \frac{1}{n}\left(\sum_{i=1}^{n} \bar{y}_i^2 + 2\sum\sum_{i<j} \bar{y}_i\bar{y}_j\right)\right]\right\}$$

$$= E_1\left\{\left(\frac{1}{n-1}\right)E_2\left[\left(1 - \frac{1}{n}\right)\sum_{i=1}^{n} \bar{y}_i^2 - \frac{2}{n}\sum\sum_{i<j} \bar{y}_i\bar{y}_j\right]\right\}$$

$$= E_1\left\{\left(\frac{1}{n-1}\right)\left[\left(1 - \frac{1}{n}\right)\sum_{i=1}^{n} (V_2(\bar{y}_i) + \mu_i^2) - \frac{2}{n}\sum\sum_{i<j} \mu_i\mu_j\right]\right\}$$

$$= E_1\left\{\left(\frac{1}{n-1}\right)\left[\left(1 - \frac{1}{n}\right)\sum_{i=1}^{n} V_2(\bar{y}_i) + \sum_{i=1}^{n} \mu_i^2 - \frac{1}{n}\left(\sum_{i=1}^{n} \mu_i\right)^2\right]\right\}$$

$$= E_1\left\{\frac{1}{n}\sum_{i=1}^{n} V_2(\bar{y}_i) + \left(\frac{1}{n-1}\right)\left[\sum_{i=1}^{n} \mu_i^2 - \frac{1}{n}\left(\sum_{i=1}^{n} \mu_i\right)^2\right]\right\}$$

$$= \frac{1}{N}\sum_{i=1}^{n} V(\bar{y}_i) + \frac{1}{N-1}\sum_{i=1}^{n} (\mu_i - \bar{\mu})^2$$

Thus,

$$\left(\frac{N-n}{N}\right)\left(\frac{1}{n\bar{M}^2}\right)s_b^2$$

estimates Equation (A.4) plus a term

$$\left(\frac{N-n}{N}\right)\left(\frac{1}{nN}\right)\sum_{i=1}^{N} V(\bar{y}_i) \tag{A.6}$$

and we need to find an estimate of (A.5)–(A.6) in order to estimate $V(\hat{\mu})$. But

$$\begin{aligned}(\text{A.5}) - (\text{A.6}) &= \frac{1}{nN}\sum_{i=1}^{N} V(\bar{y}_i)\left(1-\frac{N-n}{N}\right)\\ &= \left(\frac{1}{N}\right)\left(\frac{1}{N}\right)\sum_{i=1}^{N} V(\bar{y}_i)\end{aligned}$$

which can be estimated unbiasedly by

$$\left(\frac{1}{N}\right)\left(\frac{1}{n}\right)\sum_{i=1}^{n}\left(\frac{\bar{M}-m}{\bar{M}}\right)\left(\frac{s_i^2}{m}\right)$$

where

$$s_i^2 = \frac{1}{m-1}\sum_{i=1}^{m}(y_{ij}-\bar{y}_i)^2$$

The estimator of $V(\hat{\mu})$ is, then,

$$\hat{V}(\hat{\mu}) = \left(\frac{N-n}{N}\right)\left(\frac{1}{n\bar{M}^2}\right)s_b^2 + \frac{1}{nN}\sum_{i=1}^{n}\left(\frac{\bar{M}-m}{\bar{M}}\right)\left(\frac{1}{m-1}\right)s_i^2$$

This equation is equivalent to Equation (9.2) in the case of equal cluster sizes. The case for unequal cluster sizes is derived analogously.

APPENDIX 2

MACROS FOR MINITAB

The macros displayed below are demonstrated in Chapters 4 and 5. They also appear on the data disk within those chapters. Two notes of caution are in order. The macros use columns 50–100 of the data display, so all data must be in the first 49 columns. The macros for estimating means and proportions in stratified sampling make use of the matrix command in MINITAB. Therefore, they will not produce all results needed if this command is not in the version of MINITAB being used.

ESTIMATING A MEAN IN SIMPLE RANDOM SAMPLING: SRSMEAN.MTB

```
#
NOECHO
Let k51 = 1
let k75 = 55
Note Input the number of columns of data.
```

```
Set 'terminal' c52;
Nobs = 1.
Copy c52 k50
Exec 'varcolm' k50
let k76 = k50 + 54
copy c55-ck76 M1
Transpose M1 M1
copy M1 c90-c95
Name c90 'samsize' c91 'popsize' c92 'mean' c93 'variance' c94 'std dev' c95 'bound'
Note
Note The following table lists the sample size, population
Note size, sample mean, estimated variance of the sample mean,
Note estimated standard deviation of the sample mean,
Note and the bound on the error of estimation for each column:
print c90-c95
end
ECHO
```

CALCULATIONS NESTED WITHIN SRSMEAN: VARCOLM.MTB

```
#
Note Input the sample size followed by the population size.
Set 'terminal' c50;
Nobs = 2.
Copy c50 k11 k22
let k1 = Sum(ck51)
let k2 = Count(ck51)
Let K3 = K1/K2
Let c51 = ck51**2
Let k10 = sum(c51)
Let k12 = (k10 - k2*(k3**2)) / (k2 - 1)
let k4 = (k12/k2) * ((k22-k2)/k22)
let k5 = sqrt(k4)
let k6 = 2 * sqrt(k4)
let k51 = k51 + 1
copy k2 k22 k3 k4 k5 k6 ck75
let k75 = k75 + 1
Note
 Note The sample size is
print k2
 Note
```

```
Note The population size is
print k22
Note
 Note The sample mean is
print k3
Note
 Note The estimated variance of the sample mean is
print k4
Note
 Note The estimated standard deviation of the sample mean is
print k5
Note
 Note The bound on the error of estimation is
print k6
Note
 end
```

ESTIMATING A MEAN IN STRATIFIED RANDOM SAMPLING: STMEAN.MTB

```
#
Exec 'mvarcolm'
NOECHO
let k7 = sum(c91)
let c97 = c91 * c92
let k8 = sum(c97)
let k9 = k8/k7
Note The estimator of the mean, where stratified sampling is used, is
print k9
Note
let c98 = (c91**2) * c93
let k13 = sum(c98) / (k7**2)
Note The estimated variance of the mean, using stratified sampling, is
print k13
Note
let k14 = sqrt(k13)
Note The estimated standard deviation of the mean, using stratified
Note sampling, is
print k14
Note
let k15 = 2 * k14
Note The bound on the error of estimation, using stratified sampling, is
print k15
```

```
Note
end
ECHO
```

NESTED COMMANDS: MVARCOLM.MTB

```
#
NOECHO
let k75 = 55
Note Input the column number of the first column of data
Note followed by the number of the last column of data.
Set 'terminal' c52;
Nobs = 2.
Copy c52 k51 k52
Let k53 = k52 - k51 + 1
Exec 'varcolm' k53          #Executes Macro varcolm k53 times
let k76 = k53 + 54          #Converts the output into a more
copy c55-ck76 M1            #easily accessible tabular form
Transpose M1 M1
copy M1 c90-c95
Name c90 'samsize' c91 'popsize' c92 'mean' c93 'variance' c94 'std dev' c95 'bound'
Note
Note The following table lists the sample size, population
Note size, sample mean, estimated variance of the sample mean,
Note estimated standard deviation of the sample mean,
Note and the bound on the error of estimation for each column:
Note
print c90-c95
end
ECHO
```

NESTED COMMANDS: VARCOLM.MTB

```
#
Note Input the sample size followed by the population size.
Set 'terminal' c50;
Nobs = 2.
Copy c50 k11 k22
let k1 = Sum(ck51)
```

```
let k2 = Count(ck51)
Let K3 = K1 / K2
Let c51 = ck51**2
let k10 = sum(c51)
Let k12 = (k10 - k2*(k3**2)) / (k2 - 1)
let k4 = (k12/k2) * ((k22-k2)/k22)
let k5 = sqrt(k4)
let k6 = 2 * sqrt(k4)
let k51 = k51 + 1
copy k2 k22 k3 k4 k5 k6 ck75
let k75 = k75 + 1
Note
 NOTE The sample size is
print k2
 Note
Note The population size is
print k22
Note
 Note The sample mean is
print k3
Note
 Note The estimated variance of the sample mean is
print k4
Note
 Note The estimated standard deviation of the sample mean is
print k5
Note
 Note The bound on the error of estimation is
print k6
Note
 end
```

ESTIMATING A PROPORTION IN STRATIFIED RANDOM SAMPLING: STPROP.MTB

```
#
Exec 'mvarcolp'
NOECHO
copy M1 c90-c95
let k7 = sum(c91)
let c97 = c91 * c92
```

```
let k8 = sum(c97)
let k9 = k8/k7
Note The estimator of the proportion, where stratified sampling is used, is
print k9
Note
let c98 = (c91**2) * c93
let k13 = sum(c98) / (k7**2)
Note The estimated variance of the proportion, using stratified sampling, is
print k13
Note
let k14 = sqrt(k13)
Note The estimated standard deviation of the proportion, using stratified
Note sampling, is
print k14
Note
let k15 = 2 * k14
Note The bound on the error of estimation, using stratified sampling, is
print k15
Note
end
ECHO
```

NESTED COMMANDS: MVARCOLP.MTB

```
#
NOECHO
let k75 = 55
Note Input the number of strata.
Set 'terminal' c52;
Nobs = 1.
Copy c52 k53
Exec 'varcolpr' k53
let k76 = k53 + 54
copy c55-ck76 M1
Transpose M1 M1
copy M1 c90-c95
Name c90 'samsize' c91 'popsize' c92 'samprop' c93 'variance' c94 'std dev' c95
'bound'
Note
Note The following table lists the sample size, population
Note size, sample proportion, estimated variance of the
Note sample proportion, estimated standard deviation of
```

```
Note the sample proportion, and the bound on the error of
Note estimation for each column:
Note
print c90-c95
end
ECHO
```

NESTED COMMANDS: VARCOLPR.MTB

```
#
Note Input the sample size followed by the population size.
Set 'terminal' c50;
Nobs = 2.
Copy c50 k11 k22
Note Input the sample proportion.
Set 'terminal' c55;
Nobs = 1.
Copy c55 k1
let k2 = ((k22 - k11)/k22) * (k1*(1 - k1)/(k11 - 1))
let k3 = sqrt(k2)
let k4 = 2 * k3
let k51 = k51 + 1
copy k11 k22 k1 k2 k3 k4 ck75
let k75 = k75 + 1
 NOTE The sample size is
print k11
 Note
Note The population size is
print k22
Note
 Note The sample proportion is
print k1
Note
 Note The estimated variance of the sample proportion is
print k2
Note
 Note The estimated standard deviation of the sample proportion is
print k3
Note
 Note The bound on the error of estimation is
print k4
```

```
Note
  end
```

ALLOCATING THE SAMPLE FOR ESTIMATING A MEAN: STALLOCM.MTB

```
#
NOECHO
Note Input the number of strata.
Set 'terminal' c30;
Nobs = 1.
Copy c30 k30
Note Input the population size for each stratum.
Set 'terminal' c31;
Nobs = k30.
Note Input the population VARIANCE for each stratum.
Set 'terminal' c32;
Nobs = k30.
Note Input the cost of obtaining an observation from each stratum.
Set 'terminal' c33;
Nobs = k30.
Note Input the actual number desired for D.
Set 'terminal' c39;
Nobs = 1.
Copy c39 k39
let c34 = ((c31) * (sqrt(c32))) / (sqrt(c33))
let k34 = sum(c34)
let c36 = c34 / k34
let c40 = ((c31**2) * c32) / c36
let k40 = sum (c40)
let k41 = sum (c31)
let c42 = c31 * c32
let k42 = sum (c42)
let k43 = k40 / (((k41**2) * k39) + k42)
let c37 = k43 * c36
Note The weights for each stratum are given below:
print c36
Note
Note The approximate sample size required to estimate mu or tau
Note with a bound B on the error of estimation is
print k43
Note
Note The approximate individual sample sizes for each
```

```
Note stratum are given below:
print c37
Note
end
ECHO
```

ALLOCATING THE SAMPLE FOR ESTIMATING A PROPORTION: STALLOCP.MTB

```
#
NOECHO
Note Input the number of strata.
Set 'terminal' c30;
Nobs = 1.
Copy c30 k30
Note Input the population size for each stratum.
Set 'terminal' c31;
Nobs = k30.
Note Input the population proportion for each stratum.
Set 'terminal' c32;
Nobs = k30.
Note Input the cost of obtaining an observation from each stratum.
Set 'terminal' c33;
Nobs = k30.
Note Input the actual number desired for D.
Set 'terminal' c38;
Nobs = 1.
Copy c38 k38
let c34 = ((c31) * (sqrt(c32*(1-c32)))) / (sqrt(c33))
let k34 = sum(c34)
let c36 = c34 / k34
let c39 = c32 * (1 – c32)
let c40 = ((c31**2) * c39) / c36
let k40 = sum (c40)
let k41 = sum (c31)
let c42 = c31 * c39
let k42 = sum (c42)
let k43 = k40 / (((k41**2) * k38) + k42)
let c37 = k43 * c36
Note The weights for each stratum are given below:
print c36
Note
```

```
Note The approximate sample size required to estimate p or tau
Note with a bound B on the error of estimation is
print k43
Note
Note The approximate individual sample sizes for each
Note stratum are given below:
print c37
end
ECHO
```

CALCULATING THE SAMPLE SIZE FOR ESTIMATING A MEAN: STSAMPSM.MTB

```
#
NOECHO
Note Input the number of strata.
Set 'terminal' c60;
Nobs = 1.
Copy c60 k60
Note Input the weight for each stratum.
Set 'terminal' c61;
Nobs = k60.
Note Input the population VARIANCE for each stratum.
Set 'terminal' c62;
Nobs = k60.
Note Input the population size for each stratum.
Set 'terminal' c63;
Nobs = k60.
Note Input the actual number desired for D.
Set 'terminal' c64;
Nobs = 1.
Copy c64 k64
let c65 = ((c63**2) * c62) / c61
let k65 = sum (c65)
let k63 = sum (c63)
let c66 = c63 * c62
let k66 = sum (c66)
let k67 = k65 / (((k63**2) * k64) + k66)
Note
Note The approximate sample size required to estimate mu or tau
Note with a bound B on the error of estimation is
```

```
print k67
Note
let c70 = k67 * c61
Note The approximate individual samples sizes for each
Note stratum are given below:
print c70
end
ECHO
```

CALCULATING THE SAMPLE SIZE FOR ESTIMATING A PROPORTION: STSAMPSP.MTB

```
#
NOECHO
Note Input the number of strata.
Set 'terminal' c60;
Nobs = 1.
Copy c60 k60
Note Input the weight for each stratum.
Set 'terminal' c61;
Nobs = k60.
Note Input the population proportion for each stratum.
Set 'terminal' c59;
Nobs = k60.
Note Input the population size for each stratum.
Set 'terminal' c63;
Nobs = k60.
Note Input the actual number desired for D.
Set 'terminal' c64;
Nobs = 1.
Copy c64 k64
let c62 = c59 * (1 - c59)
let c65 = ((c63**2) * c62) / c61
let k65 = sum (c65)
let k63 = sum (c63)
let c66 = c63 * c62
let k66 = sum (c66)
let k67 = k65 / (((k63**2) * k64) + k66)
Note
Note The approximate sample size required to estimate p or tau
Note with a bound B on the error of estimation is
```

```
print k67
Note
let c70 = k67 * c61
Note The approximate individual sample sizes for each
Note stratum are given below:
print c70
end
ECHO
```

APPENDIX 3

MACROS FOR SAS

1

INTRODUCTION

GENERAL INSTRUCTIONS

The SAS sampling macros are included on the data disk in the directory MACROS.SAS. They can be used to select samples or calculate estimates and standard errors from sample data sets for the basic sampling designs described in the textbook. This section includes a general description of the macros and some examples. The data sets used in the examples are in the directory DATA.SAS.

Calling a Macro

A macro, a subprogram in the SAS macro language, performs a specific task. Each macro is called from a main SAS program written by the user and is invoked by one statement of the form:

```
%NAME( Keyword1=char1, Keyword2=char2,..., );
```

where NAME is the name of the macro. The macro name is followed by a sequence of keyword parameter assignments enclosed in parentheses. The keyword assignments are separated by commas and may appear in any order in the parameter list. Each keyword is followed by an equals sign and a character string. The character string following the equals sign may represent the name of a SAS data set or variable, a constant, or a particular option. A list of keywords available for each macro appears in the file GLOSSARY.SAS on the data disk.

Organization of Files and the %include Statement

Each macro is in a file with the same name as the macro. For example, the macro SRS, which selects a simple random sample, is in the file SRS; the macro EST_SRS, which computes estimates from such a sample, is in the file EST_SRS. Each macro must be included in the SAS program with a separate %include statement. For the examples given here, the macros are assumed to be in the same directory as the program. In general, the full path name for the macro must be included within single quotes in the %include statement. The following program uses the EST_SRS macro to estimate the population mean from the simple random sample in Example 4.3 of the text:

```
%include 'est_srs';
options ls=72 nodate formdlim='-';

data accounts
input owed @@;
cards;
33.50 32.00 52.00 43.00 40.00 41.00 45.00 42.50 39.00
;
%est_srs(sample=accounts, npop=484, response=owed,
param=mean);
run;
```

The statement %EST_SRS( ) invokes the macro. The assignment SAMPLE= ACCOUNTS informs the macro that the sample data are stored in the data set ACCOUNTS. The option NPOP=484 indicates the population size, and PARAM=MEAN requests the estimate of the mean of the population of values of the response variable OWED. The output from this program is as follows:

```
-----------------------------------------------------------------

            Estimate of the Population Mean
             Simple Random Sampling Design
                Response Variable=owed

                 Standard                    s^2      Sample
      Estimate     Error       Bound       (owed)      Size

      40.8889    1.97232     3.94463     35.6736        9
-----------------------------------------------------------------
```

The output includes the estimate, the standard error, and the two-standard-error bound. The entry s^2 is the sample variance, $s^2 = \sum_{i=1}^{n}(y_i - \bar{y})^2/(n-1)$, for the values of the response variable "owed." The sample size shown indicates the number of non-missing observations in the sample.

SAMPLE SELECTION MACROS

There are ten sample selection macros. Each macro selects a sample from a finite population stored in a SAS data set. The sampling design is indicated by the name of the macro. Each macro has a specific set of keyword parameters, but all the selection macros have five keyword parameters in common, as described next.

Frame

FRAME is the keyword that indicates the SAS data set containing the frame for sample selection. Appendix Data Set 2 shows information concerning gas mileages for 1993 cars. The data are stored in a file called CARS93.DAT on the data disk. For Exercise 4.2(a) in Experiences with Real Data, the following program selects a simple random sample of $n = 5$ cars from this data set:

```
%include 'srs';
options ls=72 nodate formdlim='-';

data cars; infile 'cars93.dat';
input make $ model $ type $ mpg_city 45-46 mpg_hwy 48-49;

%srs(frame=cars, npop=57, n=5);
run;
```

The assignment FRAME = CARS indicates that the sample is to be selected from the data set CARS, which is obtained by reading the data file CARS93.DAT. The output is the following:

```
------------------------------------------------------------------

                    Simple Random Sample
                  Output Data Set=sample

   OBS   MAKE        MODEL      TYPE    MPG_CITY   MPG_HWY

    1    Audi        100        Mid        19         26
    2    Cadillac    Seville    Mid        16         25
    3    Geo         Storm      Sport      30         36
    4    Hyundai     Scoupe     Sport      26         34
    5    Lincoln     Continen   Mid        17         26
------------------------------------------------------------------
```

Noprint

Each sample selection macro has an option to suppress printing of the sample. This option is specified by placing the word NOPRINT in the first position, before any keyword parameters. The NOPRINT option does *not* have a trailing equal sign. If NOPRINT is omitted, then the full sample is printed. For example, the following statement selects a simple random sample of size 5 from a data set CARS93.DAT, with 57 observations, and prints the sample data:

```
%srs(frame=cars, npop=57, n=5);.
```

The next statement selects the sample, but the printing of the sample data is suppressed;

```
%srs(noprint, frame=cars, npop=57, n=5);
```

Seed

The macros use the SAS uniform random number generator RANUNI. The seed value is set to zero by default, and the clock time is used to generate the starting seed. If the user wishes to generate the same sample on a second run, then the seed may be set to some constant value using the SEED option. In the preceding example, the following statement sets the seed to a particular value, the sample returned for a particular seed may vary with the installation:

```
%srs(noprint, frame=cars, npop=57, n=5, seed=137897);
```

Setup

In a simple random sample, the population size and sample size are simple constants. Their values may be specified as constants in the parameter list of the macro. If these parameters have more than one value (such as the strata sizes for a stratified random sample), then these numbers must be stored as observations of a SAS variables in a second data set. In this case, the keyword SETUP indicates the name of the data set; to specify the names of these variables, assign keyword parameters equal to the variable names. The following program selects a stratified random sample of cars in the CARS93.DAT data file, in which two strata have been formed: US and foreign manufacturers. The frame has $N_1 = 23$ US-manufactured cars and $N_2 = 34$ foreign-manufactured cars:

```
%include 'strs';
options ls=72 nodate formdlim='-';

data cars; infile 'cars93.dat';
input make $ model $ mpg_city 45-46 mpg_hwy 48-49 us 73;
```

```
data info; input us capn ni;
cards;
0  23  4
1  34  6

;
%strs(frame=cars, strata=us, setup=info, npop=capn,
n=ni);
run;
```

Note that SETUP = INFO indicates that the population parameters are in the data set INFO. The keywords NPOP = CAPN and N = NI indicate the names of the variables for the strata and sample sizes, respectively. The data set INFO must contain the strata variable so that the parameters can be paired with their corresponding strata. The output is the following:

```
------------------------------------------------------------

                 Stratified Random Sample
                 Output Data Set=sample

     OBS   MAKE       MODEL      MPG_CITY   MPG_HWY   US

       1   Audi       90         20         26         0
       2   Geo        Storm      30         36         0
       3   Honda      Prelude    24         31         0
       4   Mazda      RX-7       17         25         0
       5   Cadillac   Seville    16         25         1
       6   Chevrole   Lumina     21         29         1
       7   Dodge      Shadow     23         29         1
       8   Dodge      Dynasty    21         27         1
       9   Ford       Tempo      22         27         1
      10   Lincoln    Continen   17         26         1
------------------------------------------------------------
```

In two-stage designs, the SETUP option is reserved for second-stage parameters. First-stage parameters must be entered as constants in the parameter list. The SETUP option is useful if the parameters are stored as values of variables in a data set that is the result of previous computations.

Sample

Each of the sample selection macros creates a data set containing the observations selected from the frame. The default name of this data set is SAMPLE. The name can be changed using the optional keyword SAMPLE. In the following statement, the sample data are output to a data set called PILOT:

```
%strs(frame=cars, strata=us, setup=info, npop=capn,
n=ni, sample=pilot)
```

ESTIMATION MACROS

The directory MACROS.SAS contains macros for computation of all commonly used estimators in the text. The sample data must be stored in a SAS data set containing the variable of interest, called the response variable. In each estimation macro, you must indicate the response variable and the parameter to be estimated. The next program uses macro EST_SRS to estimate both the mean number of cavities per child and total cavities, in a population of 5000 children (see Exercise 4.11):

```
%include 'est_srs';
options ls=72 nodate formdlim='-';

data teeth; input cavities @@;
cards;
0 4 2 3 2 0 3 4 1 1
;
%est_srs(sample=teeth, npop=5000, response=cavities,
param=mean total);
run;
```

Examples of more complicated estimators are given in Section 3, "Cluster Designs."

THE ALLOC AND STATS MACROS

The macro ALLOC can be used to determine sample sizes and allocations under various allocation schemes for stratified random sample designs. The input data set consists of a set of summary statistics or parameters describing the strata. The macro STATS computes basic statistics from a data set and can be used to summarize data sets from a stratified random sample. For example, the output from STATS, based on a pilot sample, can be used as input for ALLOC. Several examples of the macro ALLOC appear in the Section 3, "Cluster Designs."

ADDITIONAL INFORMATION

Omission of Keyword Parameters

Some of the required keyword parameter assignments can be omitted from the argument list of a macro if the names of SAS data sets and variables in the main program are the same as the keyword parameters. These parameters include FRAME, STRATA, CLUSTER, SIZE, X, and second-stage parameters in two-stage sampling designs. The preceding program could have been written more simply as

```
%include 'strs';
options ls=72 nodate formdlim='-';

data frame; infile 'cars93.dat';
input make $ model $ mpg_city 45-46 mpg_hwy 48-49 strata
73;

data info; input strata $ npop n;
cards;
 0  23  4
 1  34  6

;
%strs(setup=info);
run;
```

The parameter SETUP=, when required, should not be omitted from the argument list. If in doubt, always include all required keyword parameters in the argument list of the macro.

The Autocall Facility

The autocall facility is available in SAS 6.0 and later versions. With this facility, macros can be stored in a separate directory and invoked as they are needed by specifying this directory. This relieves you from specifying the name of each macro in the include statement. With the autocall facility, all macros must be stored in the directory identified as the autocall library in the main SAS program with the SASAUTOS= system option. The main SAS program begins with two statements of the following general form:

```
filename sasmacs 'directory.nam';
options sasautos=sasmacs mrecall mautosource ls=72
nodate;
```

where SASMACS is a SAS name chosen by the user and DIRECTORY.NAM represents the full path name of the directory containing the SAS macros. The use of this facility varies with the installation, so check with a system administrator for details.

Notes

Omission of the commas separating the keyword parameters can cause unpredictable results.

All variables and data set names used in the macro source code have trailing underscores (for example, ybar_, flag_, temp1_). You will not encounter a conflict in naming variables and data sets in the main program, if trailing underscores are avoided.

2 EXAMPLES

In this section, examples illustrate the most commonly used macros. The data are from examples or exercises in the text. SAS options and run statements will be omitted in subsequent examples.

SIMPLE RANDOM SAMPLES

Two examples using simple random samples were included in the introduction. This example indicates how a proportion can be estimated from raw data using the PARAM = MEAN option of the EST_SRS macro. Exercise 4.31 calls for an estimate of the proportion of a firm's accounts that fail to comply with stated procedures. The data are in data file EXER4-30.DAT.

```
/*-----------------------------------------------------------

 EXERCISE 4.31 Estimating a proportion from raw data
-----------------------------------------------------------*/
 %include 'est_srs';

 data firm; infile 'exer4_30.dat';
 input account amount complnc $ @@;
 if complnc='N' then y=1;
 else y=0;

 %est_srs(sample=firm, npop=500, response=y,
 param=mean);
 run;
```

In the data step, the variable y is set equal to 1 if COMPLNC = 'N' and equal to 0, otherwise. The mean of the variable y is the sample proportion. The value of s^2 is equal to $n\hat{p}(1 - \hat{p})/(n - 1)$. The output is as follows:

```
--------------------------------------------------------------
              Estimate of the Population Mean
               Simple Random Sampling Design
                  Response Variable=comp

                 Standard                   s^2      Sample
    Estimate      Error        Bound      (comp)      Size

      0.3        0.10301     0.20601     0.22105       20
--------------------------------------------------------------
```

STRATIFIED RANDOM SAMPLES

In Example 5.2, an advertising firm interviews 40 households and records the TV-viewing time in hours per week. The data are in file TABLE5.1. The sample is stratified into three areas: A, B, and RURAL. The following program estimates the mean TV-viewing time for the entire population and for town B only:

```
/*--------------------------------------------------------

  EXAMPLE 5.2 (a) Estimating the mean from a stratified
  random sample and (b) estimating the mean from one stratum
----------------------------------------------------------*/
  %include 'est_strs';
  %include 'est_srs';

  data tv_time; infile 'table5.1';
  input area $ hours @@;

  data pop_info; input area $ popsize;
  cards;
   A       155
   B        62
   RURAL    93

  ;
  %est_strs(sample=tv_time, strata=area, setup=pop_info,
  param=mean, response=hours, npop=popsize);

  data part_b; set tv_time;
  if area='B';

  %est_srs(sample=part_b, response=hours, npop=62,
  param=mean);
```

In part (a), the population (strata) sizes are entered as values of a variable (namely, POPSIZE) in the data set POP_INFO. The data set is indicated by SETUP = POP_INFO, and the variable is defined by NPOP = POPSIZE. Note that the strata sizes are matched to their strata through the values of the class variable AREA.

In part (b), the subsetting IF statement creates a data set called PART_B with the information only from town B. The macro SRS can then be used on PART_B to estimate the mean for town B. The population size may be entered from the parameter list because it has a single value. The output is as follows:

```
----------------------------------------------------
           Actual Sample Sizes
      (Excludes Missing Data, if Any)
              AREA    n(i)
              A         20
              B          8
              RURAL     12
----------------------------------------------------
       Estimate of the Population Mean
       Stratified Random Sampling Design
          Response Variable=hours

                  Standard
          Estimate   Error     Bound

           27.675   1.40340   2.80679
----------------------------------------------------
       Estimate of the Population Mean
        Simple Random Sampling Design
          Response Variable=hours

            Standard                s^2     Sample
  Estimate    Error      Bound    (hours)    Size

   25.125   5.03019    10.0604   232.411      8
----------------------------------------------------
```

SAMPLE SIZE AND ALLOCATION

The macro ALLOC determines the overall sample size and allocation for various allocation schemes in a stratified sampling design. The following program is from Example 5.7, in which the sample sizes in each strata are determined under optimal allocation:

```
/*-----------------------------------------------------
 EXAMPLE. 5.7 Determining sample size and optimal
 allocation to obtain an error bound of 2 in estimation of
 the population mean
-----------------------------------------------------*/
 %include 'alloc';

 data summary;
 input area $ popsize s2 c;
 cards;

  A       155   25   9
  B        62  225   9
  RURAL    93  100  16

 ;
 %alloc(type=optimal, setup=summary, bound=2,
 param=mean, strata=area, npop=popsize, var=s2,
 cost=c);
```

There are three types of allocation, specified by TYPE=PROP, NEYMAN, or OPTIMAL. The strata information is stored in the data set SUMMARY. Note that both the desired bound and the parameter to be estimated must be specified:

```
----------------------------------------------------------
Stratum Weights and Sample Sizes for an Optimal Allocation
   to Obtain an Error Bound of 2 on the Population Mean
              Output Data Set=outalloc

          Sampling   Stratum   Stratum      Exact      Nearest
AREA       Costs      Sizes    Weights    Allocation   Integer

A             9        155     0.32258     18.7097       19
B             9         62     0.38710     22.4516       22
RURAL        16         93     0.29032     16.8387       17
                     =======   =======   ==========   =======
                       310     1.00000     58.000        58
----------------------------------------------------------
```

In Example 5.6, the sampling weights are required to be the same in each strata. Thus, the TYPE= option is omitted, and the OPT=FIXEDW is specified to set equal weights in each of the strata. The keyword parameter WI, which specifies the variable containing the fixed weights, must be used with this option. A variation of Neyman allocation, with a fixed weights restriction, is used to determine the allocation with the OPT=FIXEDW option:

```
/*-----------------------------------------------------
  EXAMPLE 5.6 Determining sample size and allocation to
  obtain a bound of 400 in estimation of the total, with
  equal sampling weights
------------------------------------------------------*/
  %include 'alloc';

  data one;
  input area $ npop s2 wts;
  cards;

   A        155   25   1
   B         62  225   1
   RURAL     93  100   1

  ;
  %alloc(opt=fixedw, setup=one, wi=wts, bound=400,
  param=total, strata=area, npop=npop, var=s2);

 ------------------------------------------------------
```

```
Stratum Weights and Sample Sizes for a Fixed Weights Allocation
  to Obtain an Error Bound of 400 on the Population Total
                 Output Data Set=weights

            Stratum      Stratum        Exact          Nearest
TOWN         Sizes       Weights      Allocation       Integer

A             155        0.33333        34.718           35
B              62        0.33333        34.718           35
RURAL          93        0.33333        34.718           35
            =======      =======      ==========       =======
              310        1.00000       104.153          105

 ------------------------------------------------------
```

In Example 5.11, optimal allocation is required with the total cost of sampling fixed at \$500. This is accomplished by specifying both TYPE=OPTIMAL and OPT=FIXEDC (for fixed cost), with MAXCOST=500.

```
/*-----------------------------------------------------
  EXAMPLE 5.11 Determining sample size and allocation to
  minimize variance with fixed sampling costs of $500.
------------------------------------------------------*/
  %include 'alloc';
```

```
data info;
input strata $ npop s2 c;
cards:

 A       155   25   9
 B        62  225   9
 RURAL    93  100  16


;
%alloc(type=optimal, opt=fixedc, maxcost=500, cost=c,
var=s2, setup=info);
```

```
------------------------------------------------------------------
Stratum Weights and Sample Sizes for an Optimal Allocation
       to Minimize Variance with Cost Fixed at $500
                 Output Data Set=weights

          Sampling   Stratum   Stratum      Exact      Nearest
TOWN        Costs     Sizes    Weights   Allocation    Integer

A             9        155     0.32258     14.5161        15
B             9         62     0.38710     17.4194        17
RURAL        16         93     0.29032     13.0645        13
                     =======   =======   ==========    =======
                       310     1.00000     45.0000        45
------------------------------------------------------------------
```

AUXILIARY VARIABLES IN ESTIMATION

The macro RATIO computes the ratio estimate of the population mean or total using an auxiliary variable x. The data for Example 6.4 are in the file EX6_4.DAT. The following program estimates the mean hours lost:

```
/*---------------------------------------------------------
  EXAMPLE 6.4 Finding the ratio estimate of a population
  mean
---------------------------------------------------------*/
  %include 'ratio';

  data hrs_lost; infile 'ex6_4.dat';
  input employee previous current @@;

  %ratio(sample=hrs_lost, x=previous, response=current,
  mu_x=16.3, param=ratio, npop=1000);
```

The auxiliary variable is specified by X = PREVIOUS. The population mean (μ_x) or total (τ_x), for the auxiliary variable X, is included in the parameter list:

```
---------------------------------------------------------------

              Ratio Estimate of Population R
               Simple Random Sample Design
                Response Variable=current
              (Auxiliary Variable=previous)

                 Standard                                Sample
      Estimate     Error        Bound       s(r)^2        Size

       1.05056   0.035995    0.071989     3.47707         10

---------------------------------------------------------------
```

In Example 6.7, the combined and separate ratio estimates are obtained from a stratified random sample of two companies, A and B. The data are in the file EX6_7.DAT. The following program computes both combined and separate ratio estimates. The output follows the program.

```
/*---------------------------------------------------------
 EXAMPLE 6.7 Finding the combined and separate ratio
 estimates of a population mean
---------------------------------------------------------*/
 %include 'c_ratio';
 %include 's_ratio';

 data hrs_lost; infile 'ex6_7.dat';
 input company $ employee previous current @@;

 data one; input company $ totalx str_size;
 cards;
  A  16300  1000
  B  12800  1500

 ;
 %s_ratio(sample=hrs_lost, setup=one, strata=company,
 param=mean, x=previous, response=current,
 tau_x=totalx, npop=str_size);

 %c_ratio(sample=hrs_lost, setup=one, strata=company,
 param=mean, x=previous, response=current,
 tau_x=totalx, npop=str_size);
```

```
------------------------------------------------------------------
              Ratio Estimates by Strata
             Response Variable=current
           (Auxiliary Variable=previous)

                    Sample
          Strata     Size      r(i)

             A        10     1.05056
             B        10     0.58974
------------------------------------------------------------------
           Separate Ratio Estimate of Mean
          Stratified Random Sampling Design
             Response Variable=current
           (Auxiliary Variable=previous)

                    Standard
          Estimate    Error      Bound

           9.86915   0.63475   1.26951
------------------------------------------------------------------
                Actual Sample Sizes
           (Excludes Missing Data, if Any)

                   COMPANY  n(i)

                      A      10
                      B      10
------------------------------------------------------------------
      Combined Ratio Estimate of Population Mean
          Stratified Random Sampling Design
             Response Variable=current
           (Auxiliary Variable=previous)

                  Standard
     Estimate       Error        Bound        r(c)

      10.1012      0.81459      1.62918     0.86780
------------------------------------------------------------------
```

SYSTEMATIC SAMPLES

The macro SYS selects a systematic 1-in-k random sample, and EST_SYS computes the successive difference estimator. The mean-per-unit estimate can be obtained from

EST_SRS. To illustrate SYS, the student responses to a class survey in file CLASSUR.DAT are treated as a population of $N=57$ elements. The following program uses the macros SYS and EST_SYS to select a systematic 1-in-5 sample from the file. The NOPRINT option is included in the call to SYS. The successive difference estimator is used to estimate the mean study hours in a typical week. The true mean for the population is $\mu=12.91$ hours:

```
/*--------------------------------------------------
  EXAMPLE Selecting a systematic 1-in-5 sample and
  estimation of the population mean
--------------------------------------------------*/

  %include 'sys';
  %include 'est_sys';

  data survey; infile 'classur.dat';
  input gender age gpa class ht wt stdhrs;

  %sys(noprint, frame=survey, k=5, npop=57,
  sample=outsys );

  %est_sys(sample=outsys, npop=57, param=mean,
  response=stdhrs);
---------------------------------------------------------
             Estimate of the Population Mean
               Systematic Sampling Design
           (Successive Difference Estimator)
               Response Variable=stdhrs

              Standard                          Actual
    Estimate    Error      Bound    s(d)^2  Differences

     12.5455   2.23767   4.47534   136.5        10
---------------------------------------------------------
```

3

CLUSTER DESIGNS

SINGLE-STAGE CLUSTER SAMPLES

There are three sets of single-stage cluster design macros. In this section, we include estimation examples for two types of cluster samples: (1) clusters selected with equal probability and (2) clusters selected with probability proportional to cluster size. The macro CL1STRS selects clusters within strata, and the options are described in the file named GLOSSARY.SAS on the data disk.

Clusters Selected with Equal Probability

The macro CL1 selects a single-stage cluster sample from raw data. Macro EST_CL1 computes either the ratio estimate of the population mean or total or the unbiased estimate of the population total. The choice of estimator is specified by PARAM = MEAN, TOTAL, or UNBIASED, and the data must be in the form of totals per cluster. The following program gives the solutions for Examples 8.2–8.4 with two calls to EST_CL1. In the first call, the mean cluster size is assumed to be unknown. In the second call, the total number of elements in the population is specified by the assignment $M = 2500$. The data are recorded in $1000s in file TABLE8.1. The SAS statement after the input line transforms income to dollars.

```
/*-----------------------------------------------------
  EXAMPLES 8.2-8.4 Selecting a single-stage cluster sample
  and estimating the population mean and total
-----------------------------------------------------*/

  %include 'est_cl1';

  data sample; infile 'table8.1';
  input cluster resids income @@;
  income=1000*income;

  %est_cl1( param=mean unbiased, mi=resids, npop=415,
  response=income );

  %est_cl1( param=total, mi=resids, npop=415,
  response=income, m=2500 );
 ----------------------------------------------------
                Single-Stage Cluster Design
                 Response Variable=income

                     Standard
           Estimate    Error     Bound      s(c)^2

            8801.32   808.570   1617.14  634501213.4
 ----------------------------------------------------
           Unbiased Estimate of the Population Total
                    (Does Not Depend on M)
                 Single-Stage Cluster Design
                  Response Variable=income
```

```
              Standard
   Estimate      Error        Bound        s(t)^2

   22061400   1752792.02   3505584.04   474556666.67
----------------------------------------------------------
          Ratio Estimate of Population Total
             Single-Stage Cluster Design
              Response Variable=income

                Standard
    Estimate      Error        Bound         s(c)^2

   22003311.26  2026761.17  4053522.35  634501213.4
----------------------------------------------------------
```

Probability Proportional to Size

The macro CL1PPS selects a single-stage cluster sample in which clusters are selected with probability proportional to cluster size. For CL1PPS, every element in each cluster is assumed to appear as an observation in the data set FRAME. The macro PPS selects a probability proportional to the sample size of the individual observations from a data set and can be used to select a single-stage cluster sample from a frame consisting of cluster totals.

The next program selects a single-stage cluster sample of states from the USPOP.DAT data set, with probability proportional to each state's 1990 population. The projected year 2000 population is estimated from this sample, using EST_CL1P. The response variable in this case is the projected 2000 population. Although this is not the typical application of cluster size for a single-stage cluster design, the 1990 population and the 2000 projected population are positively correlated. Note that the 2-standard-error bound easily encloses the total projected year 2000 U.S. population (namely, 282,055).

```
/*----------------------------------------------------
 EXAMPLE Selecting a single-stage cluster sample with a
 probability proportional to the size of the cluster
----------------------------------------------------*/
 %include 'est_cl1p';
 %include 'pps';
 data census; infile 'uspop.dat';
 input state $ 8-20 pop1990 34-38 pop2000 40-44;

 %pps(frame=census, n=5, mi=pop1990 );

 %est_cl1p(totals=pop2000, param=total, mi=pop1990,
 m=226503);
```

The output from PSS (and CL1PPS) contains a new variable REP, which is used to distinguish between multiple occurrences of a cluster in the sample. The macro EST_CL1P uses the variable REP for calculations. The variable REP can be renamed in the macros CL1PPS and EST_CL1P with the option REP=. The output is as follows:

```
---------------------------------------------------------------
         Probability Proportional to Size Sample
                 Output Data Set=sample
                  (Total Size=226503)

   OBS  STATE          POP1990  POP2000  MI       REP

    1   Connecticut     3108     3515     3108    1
    2   Ohio           10797    11183    10797    1
    3   Nebraska        1570     1563     1570    1
    4   California     23669    38112    23669    1
    5   California     23669    38112    23669    2
---------------------------------------------------------------
            Estimate of the Population Total
             Single-Stage PPS Cluster Design

          Response Variable (Totals)=pop2000

                    Standard
        Estimate     Error      Bound       s^2

       289138.30  31254.27  62508.54  0.095201
---------------------------------------------------------------
```

TWO-STAGE CLUSTER SAMPLES

There are two estimation macros for a two-stage cluster sample: EST_RAT computes ratio estimates of population parameters, and EST_UNB computes unbiased estimates. Sample data may be input as means, proportions, or counts. The cluster size and sample size for each cluster selected must be included in the data set. The data set containing this second-stage information is specified by the SETUP option. First-stage parameters are included as constants in the parameter list. The following program uses the data from Example 9.4 to compute the ratio estimator of the population proportion. In this example, the summary data are read 'instream' with the CARDS statement.

```
/*-----------------------------------------------------------
  EXAMPLE 9.4 Finding the ratio estimate of a proportion in a
  two-stage cluster sample
-----------------------------------------------------------*/
```

```
%include 'est_rat';

data samp; input plant capmi smallmi p @@;
cards;
 1  50  10  .40  2   65  13  .38
 3  45  9   .22  4   48  10  .30
 5  52  10  .50  6   58  12  .25
 7  42  8   .38  8   66  13  .31
 9  40  8   .25  10  56  11  .36

;
%est_rat(setup=samp, cluster=plant, mpopi=capmi,
mi=smallmi, npop=90, phat=p, param=prop);
```

```
-------------------------------------------------------------
          Ratio Estimate of Population Proportion
                 Two-Stage Cluster Design

                                 Standard
                     Estimate     Error            Bound

                     0.33732   0.028431         0.056861
-------------------------------------------------------------
```

4

SUMMARY STATISTICS

For EST_RAT and EST_UNB, the data must be in the form of summary statistics. For EST_SRS and EST_STRS, the data can also be entered as summary statistics, using the TYPE = SUMMARY option. If you are familiar with the MEANS procedure in SAS, raw data can be easily summarized in the main program. An output data set from PROC MEANS can then be used with the TYPE = SUMMARY option in these two macros. The data in Example 5.2 is reported both as raw data and as summary statistics. The next two programs use the TYPE = SUMMARY option of EST_STRS to compute the estimate of the mean. The first program uses the summarized data directly.

```
/*-----------------------------------------------------------
  EXAMPLE 5.2 Estimating from summary data
-----------------------------------------------------------*/
  %include 'est_strs';

  data one;
```

```
input region $ capn;
cards;
 A       155
 B        62
 RURAL    93

;
data two;
input region $ ymean yvar smalln;
cards:
 A       33.900   35.358  20
 B       25.125  232.411   8
 RURAL   19.000   87.636  93

%est_strs(setup=one, sample=two, ybar=ymean, var=yvar,
strata=region, n=smalln, npop=capn, param=mean,
type=summary);
```

The second program forms the summary statistics using PROC MEANS with an output statement. The output from both programs is the same and appears below the program.

```
/*------------------------------------------------------
    EXAMPLE 5.2 Creating summary data with Proc Means
------------------------------------------------------*/

%include 'est_strs';

data tv_time; infile 'table5.1';
input region $ hours @@;

data one;
input region $ capn smalln;
cards;
 A       155
 B        62
 RURAL    93

;
proc sort data=tv_time; by region;
proc sort data=one; by region;

proc means data=tv_time noprint;
by region; var hours;
```

```
output out=new mean=ymean var=yvar n=ni;

%est_strs(setup=one, sample=new, ybar=ymean,
var=yvar, strata=region, n=ni, npop=capn,
param=mean, type=summary);
```

```
-----------------------------------------------------------
          Estimate of the Population Mean
         Stratified Random Sampling Design
         Response Variable=ymean (Mean)

                      Standard
           Estimate    Error          Bound

            27.675    1.40340        2.80679
-----------------------------------------------------------
```

5

USING SEVERAL MACROS IN ONE PROGRAM

In the last section, one example combines several macros in one program. You should be careful to distinguish between the names of the different variables and data sets in the keyword parameter lists.

EXAMPLE

The objective in this example is to estimate the mean precipitation at 88 U.S. weather stations for the period January through May, with a bound of 0.25 on the error of estimation. The frame is in file TEMPS.DAT. A stratified random sample will be selected using the months as strata. In TEMPS.DAT, the data for the months are entered in five columns. A series of output statements is used to combine these into one variable, PRECIP, and create the indicator variable MONTH to distinguish between the strata.

First, a pilot sample of size 40 is selected with proportional allocation. The macro STATS is used with the pilot sample to determine the sample variances, which are in turn used as approximations to the strata variances for a Neyman allocation. ALLOC is then called to determine the final sample size and allocation. The option SETUP = OUTALLOC directs the macro STRS to use the output from ALLOC to select the final stratified sample. The last call to EST_STRS determines the final estimate and the error bound. The output follows the program.

```
/*---------------------------------------------------------
 EXAMPLE Using several macros in one program
---------------------------------------------------------*/

 %include 'strs';
 %include 'stats';
 %include 'est_strs';
 %include 'alloc';

 data weather; infile 'temps.dat';
 input station $ 2-22 jan 27-29 feb 36-38 mar 44-46
 apr 52-54 may 60-62;

 data weather(keep=station precip month); set weather;
 precip=jan; month='Jan'; output;
 precip=feb; month='Feb'; output;
 precip=mar; month='Mar'; output;
 precip=apr; month='Apr'; output;
 precip=may; month='May'; output;

 data info; input month $ capn smalln;
 cards;
 Jan  88  8
 Feb  88  8
 Mar  88  8
 Apr  88  8
 May  88  8

*SELECT PROPORTIONAL SAMPLE FOR PILOT STUDY;

 %strs(noprint, frame=weather, setup=info, npop=capn,
 n=smalln, strata=month, sample=pilot);

*OBTAIN PRELIMINARY ESTIMATES FROM PILOT SAMPLE;

 %est_strs(sample=pilot, strata=month, setup=info,
 response=precip, npop=capn, param=mean);

*COMPUTE BASIC STATISTICS FROM PILOT SAMPLE;

 %stats(sample=pilot, byvar=month, response=precip,
 setup=info);

*USE PILOT SAMPLE TO DETERMINE NEYMAN ALLOCATION;
```

```
 %alloc(type=neyman, setup=outstats, bound=.25,
 param=mean, strata=month, npop=capn, var=var);

*SELECT FINAL SAMPLE USING NEYMAN ALLOCATION;

 %strs(noprint, frame=weather, setup=outalloc,
 npop=capn, n=ni, strata=month);

*ESTIMATE MEAN PRECIPITATION FROM FINAL SAMPLE;

 %est_strs(strata=month, setup=info, response=precip,
 param=mean, npop=capn);
-------------------------------------------------------------
                  Actual Sample Sizes
           (Excludes Missing Data, if Any)

                     MONTH  n(i)

                      Apr     8
                      Feb     8
                      Jan     8
                      Mar     8
                      May     8

            Estimate of the Population Mean
           Stratified Random Sampling Design
              Response Variable=precip

                        Standard
              Estimate    Error     Bound

                2.74     0.19298   0.38596
-------------------------------------------------------------
             Statistics for precip by Month
                  Input Data Set=pilot
                Output Data Set=outstats

                                  Standard
MONTH  CAPN  SMALLN    Mean      Deviation   Variance   n

 Apr    88      8     3.2250    0.73436     0.53929   8
 Feb    88      8     1.9875    1.37054     1.87839   8
 Jan    88      8     2.7250    1.27588     1.62786   8
 Mar    88      8     3.1250    1.72440     2.97357   8
 May    88      8     2.6375    1.08356     1.17411   8
-------------------------------------------------------------
```

```
Stratum Weights and Sample Sizes for the Neyman Allocation
  to Obtain an Error Bound of .25 on the Population mean
                 Output Data Set=outalloc

                                                Nearest
              Stratum   Stratum      Exact      Integer
      MONTH    Sizes    Weights   Allocation     (ni)

       Apr       88     0.11866      9.4929         9
       Feb       88     0.22146     17.7166        18
       Jan       88     0.20616     16.4928        16
       Mar       88     0.27864     22.2908        22
       May       88     0.17509     14.0069        14
              =======   =======   ==========    =======
                440     1.00000     80.0000        79
----------------------------------------------------------
                   Actual Sample Sizes
           (Excludes Missing Data, if Any)
                      MONTH  n(i)

                       Apr     9
                       Feb    18
                       Jan    16
                       Mar    22
                       May    14

           Estimate of the Population Mean
          Stratified Random Sampling Design
              Response Variable=precip

                          Standard
            Estimate        Error     Bound

            2.73742       0.12845   0.25691
----------------------------------------------------------
```

The final recommended sample size is 79. The error bound in the example does not quite fall within the required 0.25 inches. Incidentally, the actual mean precipitation is $\mu = 2.67$ inches. A preliminary run was required to determine a reasonable value for the bound.

APPENDIX 4

DATA SETS

This section provides brief explanations of variables and shows some of the data for each of the eight large data sets used in the text. These data sets appear on the accompanying data disk.

CLASSSUR A survey of introductory statistics students at the University of Florida

CARS93 A summary of information on the popular cars for 1993

USPOP A summary of the U.S. population, current, past, and projected

WORLDPOP A summary of the population of most of the countries of the world

RECYCLE A recycling practices survey carried out in Florida

SCHOOLS A summary of data on the costs of education for the states of the United States

TEMPS A listing of monthly normal temperatures and precipitation for U.S. weather stations

RIVER A listing of daily flow rates for a Florida river

CLASSSUR: A SURVEY OF INTRODUCTORY STATISTICS STUDENTS

The following is the survey answered by introductory statistics students. The circled numbers indicate the numerical response for these categories.

1. Male ① Female ②
2. Age (in years) ______
3. Cumulative University of Florida GPA ______
   Check here if you have completed no classes at UF ______
4. Check your current class standing:
   Freshman ①
   Sophomore ②
   Junior ③
   Senior ④
   Other ⑤
5. Height (in inches) ______
   Weight (in pounds) ______
6. How many hours do you study during a typical week? ______
7. On the average, how many hours do you sleep per night during a typical week? ______
8. Do you have a job, outside of being a student? Yes ① No ②
9. How much did you pay for textbooks for the current term? ______
10. Is your permanent residence (check one):
    In Florida more than 150 miles from Gainesville? ①
    In Florida less than 150 miles from Gainesville? ②
    Outside Florida? ③

GENDER	AGE	GPA	CLASS	HEIGHT	WEIGHT	STUDY HRS	SLEEP HRS	JOB	TEXTPAY	RESIDE
2	19	2.50	2	70.50	147	12.0	7.0	2	200.00	2
2	20	2.30	3	71.00	158	11.8	7.0	2	170.00	1
2	17	*	1	65.00	140	6.0	6.4	1	200.00	2
1	23	2.80	4	72.00	160	9.0	7.0	2	111.00	2
2	33	3.45	4	65.00	155	10.0	7.0	1	150.00	2
2	20	2.50	3	60.00	138	10.0	8.0	1	250.00	1
2	32	3.67	4	66.00	125	14.0	6.0	2	150.00	2
2	21	2.50	3	65.00	116	20.0	8.0	2	260.00	1
2	20	*	3	68.60	138	13.0	6.0	1	90.00	1
2	17	3.50	1	64.00	110	15.0	7.0	2	130.00	2
2	20	3.10	3	67.00	120	7.0	7.0	1	180.00	3
2	21	*	3	68.00	127	*	8.0	2	200.00	1
2	19	*	2	61.00	115	11.0	6.8	1	51.47	1
2	21	2.56	3	64.00	133	7.0	7.5	1	100.00	2

continues

continued

GENDER	AGE	GPA	CLASS	HEIGHT	WEIGHT	STUDY HRS	SLEEP HRS	JOB	TEXTPAY	RESIDE
2	20	2.60	3	62.00	105	12.0	6.0	1	200.00	1
1	25	3.00	4	66.00	160	40.0	6.0	1	90.00	2
1	26	3.20	4	70.00	160	8.0	6.0	2	245.00	3
1	20	3.04	2	70.00	178	7.0	7.0	1	200.00	3
2	17	*	1	64.00	*	9.0	5.5	2	220.00	1
1	19	2.80	3	71.00	170	10.0	8.0	1	200.00	2
2	18	3.20	2	64.00	106	13.0	6.0	2	200.00	2
2	42	3.75	4	63.00	140	10.0	7.0	1	150.00	2
1	22	2.75	4	67.00	147	30.0	6.0	2	320.00	1
2	21	2.84	4	67.00	135	14.0	7.0	2	200.00	1
2	18	*	1	67.00	130	10.0	8.0	2	145.00	1
2	20	2.80	3	66.00	150	14.0	6.0	2	*	1
2	20	3.29	3	66.00	118	14.0	8.0	2	216.00	1
1	19	2.70	3	73.00	167	10.0	7.0	1	200.00	2
1	20	2.80	3	71.00	175	5.0	8.0	2	150.00	2
2	20	2.90	3	63.00	123	7.0	6.0	1	185.00	1
1	20	2.93	3	67.00	147	10.0	5.0	1	220.00	1
2	22	3.24	4	62.00	*	7.0	8.0	2	120.00	1
2	18	*	1	66.50	*	15.0	7.0	2	198.00	3
2	18	*	1	*	108	12.0	4.9	2	160.00	2
1	21	3.20	4	68.00	145	26.0	7.0	2	400.00	2
2	20	2.96	3	65.00	118	25.0	6.0	2	150.00	2
1	22	3.10	4	76.00	220	16.0	4.5	1	250.00	1
2	21	3.00	3	65.00	112	13.5	7.0	1	210.00	1
2	20	2.50	3	61.00	105	10.0	8.0	2	175.00	1
1	20	3.25	3	75.00	160	9.0	9.0	1	200.00	2
1	22	3.55	5	65.00	140	10.0	8.5	1	240.00	1
1	34	3.91	5	73.00	175	30.0	7.0	1	120.00	2
1	20	2.76	2	71.00	175	7.0	8.0	2	200.00	1
1	18	3.47	3	67.00	153	15.0	9.0	2	90.00	2
2	20	3.00	3	*	*	14.0	4.0	2	260.00	2
2	23	2.80	4	64.00	128	5.0	6.0	1	49.00	2
1	22	3.34	4	75.00	225	6.0	7.0	2	95.00	2
2	19	3.40	3	60.75	110	5.0	7.0	2	300.00	3
1	21	2.60	2	71.00	183	5.0	5.0	1	225.00	1
1	19	2.65	3	68.00	130	3.0	9.0	1	200.00	3
1	19	3.20	2	67.00	156	15.0	7.0	2	180.00	2
2	20	2.40	3	63.00	128	35.0	5.0	2	250.00	1
2	17	*	1	68.00	175	13.0	5.5	2	280.00	1
1	21	2.98	4	74.00	240	18.0	6.0	1	240.00	1
2	20	3.40	3	68.00	122	12.5	6.0	1	250.00	2
1	21	2.70	4	69.00	195	10.0	7.0	2	230.00	2
1	18	3.40	1	58.00	155	16.0	6.0	1	200.00	2

CARS93: A SURVEY OF POPULAR CARS FOR 1993

The following are the variables for which data are listed:

MANUFAC: manufacturer

MODEL: Model

TYPE: Small = 1

Compact = 2

Midsize = 3

Large = 4

Sporty = 5

Van = 6

MINPRICE: minimum price (in $1000s)—price for basic version of this model

MIDPRICE: midrange price (in $1000s)—average of minimum and maximum price

MAXPRICE: maximum price (in $1000s)—price for a premium version

MPGCITY: miles per gallon by EPA rating for city driving

MPGHIGH: miles per gallon for highway driving

AIRBAGS: standard air bags (0 = None, 1 = Driver only, 2 = Driver and passenger)

DRIVETR: drive-train type (0 = rear-wheel drive, 1 = front-wheel drive, 2 = all-wheel drive)

CYLINDR: number of cylinders

LITERS: engine size (in liters)

HPOWER: horsepower (maximum)

RPMMAX: revolutions per minute at maximum horsepower

US?: manufactured by a U.S. company

1 = Yes

0 = No

The printout that follows shows only the first 57 cars on the list.

ROW	MANUFAC	MODEL	TYPE	MINPRICE	MIDPRICE	MAXPRICE
1	Acura	Integra	Small	12.9	15.9	18.8
2	Acura	Legend	Midsize	29.2	33.9	38.7
3	Audi	90	Compact	25.9	29.1	32.3
4	Audi	100	Midsize	30.8	37.7	44.6
5	BMW	535i	Midsize	23.7	30.0	36.2
6	Buick	Century	Midsize	14.2	15.7	17.3
7	Buick	LeSabre	Large	19.9	20.8	21.7
8	Buick	Roadmaster	Large	22.6	23.7	24.9
9	Buick	Riviera	Midsize	26.3	26.3	26.3
10	Cadillac	DeVille	Large	33.0	34.7	36.3
11	Cadillac	Seville	Midsize	37.5	40.1	42.7
12	Chevrolet	Cavalier	Compact	8.5	13.4	18.3
13	Chevrolet	Corsica	Compact	11.4	11.4	11.4
14	Chevrolet	Camaro	Sporty	13.4	15.1	16.8
15	Chevrolet	Lumina	Midsize	13.4	15.9	18.4
16	Chevrolet	Lumina APV	Van	14.7	16.3	18.0
17	Chevrolet	Astro	Van	14.7	16.6	18.6
18	Chevrolet	Caprice	Large	18.0	18.8	19.6
19	Chevrolet	Corvette	Sporty	34.6	38.0	41.5
20	Chrysler	Concorde	Large	18.4	18.4	18.4
21	Chrysler	LeBaron	Compact	14.5	15.8	17.1
22	Chrysler	Imperial	Large	29.5	29.5	29.5
23	Dodge	Colt	Small	7.9	9.2	10.6
24	Dodge	Shadow	Small	8.4	11.3	14.2
25	Dodge	Spirit	Compact	11.9	13.3	14.7
26	Dodge	Caravan	Van	13.6	19.0	24.4
27	Dodge	Dynasty	Midsize	14.8	15.6	16.4
28	Dodge	Stealth	Sporty	18.5	25.8	33.1
29	Eagle	Vision	Large	17.5	19.3	21.2
30	Ford	Festiva	Small	6.9	7.4	7.9
31	Ford	Escort	Small	8.4	10.1	11.9
32	Ford	Tempo	Compact	10.4	11.3	12.2
33	Ford	Mustang	Sporty	10.8	15.9	21.0
34	Ford	Probe	Sporty	12.8	14.0	15.2
35	Ford	Aerostar	Van	14.5	19.9	25.3
36	Ford	Taurus	Midsize	15.6	20.2	24.8
37	Ford	Crown Victoria	Large	20.1	20.9	21.7
38	Geo	Metro	Small	6.7	8.4	10.0
39	Geo	Storm	Sporty	11.5	12.5	13.5
40	Honda	Prelude	Sporty	17.0	19.8	22.7

continues

continued

ROW	MANUFAC	MODEL	TYPE	MINPRICE	MIDPRICE	MAXPRICE
41	Honda	Civic	Small	8.4	12.1	15.8
42	Honda	Accord	Compact	13.8	17.5	21.2
43	Hyundai	Excel	Small	6.8	8.0	9.2
44	Hyundai	Elantra	Small	9.0	10.0	11.0
45	Hyundai	Scoupe	Sporty	9.1	10.0	11.0
46	Hyundai	Sonata	Midsize	12.4	13.9	15.3
47	Infiniti	Q45	Midsize	45.4	47.9	50.4
48	Lexus	ES300	Midsize	27.5	28.0	28.4
49	Lexus	SC300	Midsize	34.7	35.2	35.6
50	Lincoln	Continental	Midsize	33.3	34.3	35.3
51	Lincoln	Town Car	Large	34.4	36.1	37.8
52	Mazda	323	Small	7.4	8.3	9.1
53	Mazda	Protege	Small	10.9	11.6	12.3
54	Mazda	626	Compact	14.3	16.5	18.7
55	Mazda	MPV	Van	16.6	19.1	21.7
56	Mazda	RX-7	Sporty	32.5	32.5	32.5
57	Mercedes-Benz	190E	Compact	29.0	31.9	34.9

MPGCITY	MPGHIGH	AIRBAGS	DRIVETR	CYLINDR	LITERS	HPOWER	RPMMAX	US?	TYPECODE
25	31	0	1	4	1.8	140	6300	0	1
18	25	2	1	6	3.2	200	5500	0	3
20	26	1	1	6	2.8	172	5500	0	2
19	26	2	1	6	2.8	172	5500	0	3
22	30	1	0	4	3.5	208	5700	0	3
22	31	1	1	4	2.2	110	5200	1	3
19	28	1	1	6	3.8	170	4800	1	4
16	25	1	0	6	5.7	180	4000	1	4
19	27	1	1	6	3.8	170	4800	1	3
16	25	1	1	8	4.9	200	4100	1	4
16	25	2	1	8	4.6	295	6000	1	3
25	36	0	1	4	2.2	110	5200	1	2
25	34	1	1	4	2.2	110	5200	1	2
19	28	2	0	6	3.4	160	4600	1	5
21	29	0	1	4	2.2	110	5200	1	3
18	23	0	1	6	3.8	170	4800	1	6

continues

continued

MPGCITY	MPGHIGH	AIRBAGS	DRIVETR	CYLINDR	LITERS	HPOWER	RPMMAX	US?	TYPECODE
15	20	0	2	6	4.3	165	4000	1	6
17	26	1	0	8	5.0	170	4200	1	4
17	25	1	0	8	5.7	300	5000	1	5
20	28	2	1	6	3.3	153	5300	1	4
23	28	2	1	4	3.0	141	5000	1	2
20	26	1	1	6	3.3	147	4800	1	4
29	33	0	1	4	1.5	92	6000	1	1
23	29	1	1	4	2.2	93	4800	1	1
22	27	1	1	4	2.5	100	4800	1	2
17	21	1	2	6	3.0	142	5000	1	6
21	27	1	1	4	2.5	100	4800	1	3
18	24	1	2	6	3.0	300	6000	1	5
20	28	2	1	6	3.5	214	5800	1	4
31	33	0	1	4	1.3	63	5000	1	1
23	30	0	1	4	1.8	127	6500	1	1
22	27	0	1	4	2.3	96	4200	1	2
22	29	1	0	4	2.3	105	4600	1	5
24	30	1	1	4	2.0	115	5500	1	5
15	20	1	2	6	3.0	145	4800	1	6
21	30	1	1	6	3.0	140	4800	1	3
18	26	1	0	8	4.6	190	4200	1	4
46	50	0	1	3	1.0	55	5700	0	1
30	36	1	1	4	1.6	90	5400	0	5
24	31	2	1	4	2.3	160	5800	0	5
42	46	1	1	4	1.5	102	5900	0	1
24	31	2	1	4	2.2	140	5600	0	2
29	33	0	1	4	1.5	81	5500	0	1
22	29	0	1	4	1.8	124	6000	0	1
26	34	0	1	4	1.5	92	5550	0	5
20	27	0	1	4	2.0	128	6000	0	3
17	22	1	0	8	4.5	278	6000	0	3
18	24	1	1	6	3.0	185	5200	0	3
18	23	2	0	6	3.0	225	6000	0	3
17	26	2	1	6	3.8	160	4400	1	3
18	26	2	0	8	4.6	210	4600	1	4
29	37	0	1	4	1.6	82	5000	0	1
28	36	0	1	4	1.8	103	5500	0	1
26	34	1	1	4	2.5	164	5600	0	2
18	24	0	2	6	3.0	155	5000	0	6
17	25	1	0	*	1.3	255	6500	0	5
20	29	1	0	4	2.3	130	5100	0	2

USPOP: A SUMMARY OF THE U.S. POPULATION, CURRENT, PAST, AND PROJECTED*

The variables are the following:

1991: resident population in 1991 (in 1000s)
1990: resident population as of the 1990 census (in 1000s)
1980: resident population as of the 1980 census (in 1000s)
2010: projected population for 2010 (in 1000s)
Area: total area (1000s of square miles)
PopDens: population density (people per square mile)
MedAge: median age
< 18: percentage of the population under age 18 as of 1991
> 65: percentage of the population over age 65 as of 1991
Section: section of the country (coded)

* From Population Reference Bureau, Inc., Washington, D.C., 1994.

Row		1991	1990	1980	2010	Area	PopDens	MedAge	<18	>65	Section
1	UNITED STATES	252181	248710	226505	282056	3787.4	71.31	33.1	25.8	12.6	*
2	NORTHEAST	50977	50809	49137	51961	181.3	314.14	34.4	23.8	13.8	*
3	NEW ENGLAND	13198	13207	12349	14210	72.0	210.12	34.1	23.5	13.5	*
4	Maine	1235	1228	1125	1453	35.4	40.01	34.3	25.1	13.4	1
5	New Hampshire	1105	1109	921	1525	9.4	123.20	33.3	25.3	11.6	1
6	Vermont	567	563	511	654	9.6	61.30	33.4	25.6	11.8	1
7	Massachusetts	5996	6016	5737	5985	10.6	764.99	33.8	22.9	13.7	1
8	Rhode Island	1004	1003	947	1078	1.5	960.77	34.2	22.9	15.1	1
9	Connecticut	3291	3287	3108	3515	5.5	679.26	34.7	23.2	13.7	1
10	MIDDLE ATLANTIC	37779	37602	36788	37751	109.3	379.83	34.5	23.9	13.9	*
11	New York	18058	17990	17558	17259	54.5	382.39	34.0	24.2	13.1	1
12	New Jersey	7760	7730	7364	8455	8.7	1045.96	34.7	23.7	13.4	1
13	Pennsylvania	11961	11882	11867	12037	46.1	266.87	35.3	23.7	15.5	1
14	MIDWEST	60226	59669	58854	62744	821.9	80.14	33.3	26.3	13.0	*
15	EAST NORTH CENTRAL	42415	42009	41670	44142	301.5	174.16	33.2	26.1	12.7	*
16	Ohio	10939	10847	10797	11183	44.8	267.11	33.7	25.8	13.1	2
17	Indiana	5610	5544	5490	5973	36.4	156.40	33.2	26.1	12.6	2
18	Illinois	11543	11431	11418	12000	57.9	207.63	33.1	26.0	12.5	2
19	Michigan	9368	9295	9258	9941	96.8	164.90	32.9	26.5	12.1	2
20	Wisconsin	4955	4892	4705	5045	65.5	91.23	33.2	26.5	13.3	2
21	WEST NORTH CENTRAL	17811	17660	17184	18602	520.4	35.06	33.4	26.6	13.9	*
22	Minnesota	4432	4375	4077	4773	86.9	55.67	32.8	26.8	12.5	2
23	Iowa	2795	2777	2913	2555	56.3	50.02	34.3	25.9	15.4	2
24	Missouri	5158	5117	4917	5708	69.7	74.86	33.8	26.0	14.1	2

continues

continued

Row	1991	1990	1980	2010	Area	PopDens	MedAge	<18	>65	Section
25 North Dakota	635	639	653	588	70.7	9.20	32.8	27.2	14.5	2
26 South Dakota	703	696	690	732	77.1	9.26	32.9	28.4	14.8	2
27 Nebraska	1593	1578	1570	1563	77.4	20.72	33.3	27.3	14.1	2
28 Kansas	2495	2478	2363	2683	82.3	30.49	33.1	26.9	13.9	2
29 SOUTH	86917	85446	75349	101007	919.9	99.78	33.0	25.9	12.6	*
30 SOUTH ATLANTIC	44422	43566	36943	54725	293.0	166.86	33.9	24.4	13.5	*
31 Delaware	680	666	595	826	2.5	347.83	33.2	24.7	12.2	3
32 Maryland	4860	4781	4216	5836	12.4	497.19	33.3	24.7	10.9	3
33 District of Columbia	598	607	638	601	0.1	9803.28	33.4	20.2	12.9	3
34 Virginia	6286	6187	5346	7700	42.8	158.75	32.9	24.5	10.8	3
35 West Virginia	1801	1793	1950	1622	24.2	74.77	35.7	24.3	15.0	3
36 North Carolina	6737	6629	5874	8193	53.8	138.29	33.4	24.4	12.3	3
37 South Carolina	3560	3487	3119	4292	32.0	118.23	32.3	26.3	11.4	3
38 Georgia	6623	6478	5464	8671	59.4	114.35	31.8	26.8	10.1	3
39 Florida	13277	12938	9740	16984	65.8	245.88	36.5	22.6	18.3	3
40 EAST SOUTH CENTRAL	15347	15176	14663	17504	183.4	85.92	33.2	26.1	12.7	*
41 Kentucky	3713	3685	3661	3861	40.4	93.45	33.3	25.8	12.7	3
42 Tennessee	4953	4877	4591	5961	42.1	120.16	33.9	24.8	12.7	3
43 Alabama	4089	4041	3890	4777	52.4	80.57	33.3	26.2	12.9	3
44 Mississippi	2592	2573	2521	2905	48.4	55.25	31.4	29.0	12.5	3
45 WEST SOUTH CENTRAL	27148	26704	23743	28778	443.5	63.69	31.5	28.3	11.1	*
46 Arkansas	2372	2351	2286	2670	53.2	45.55	34.1	26.4	14.9	3
47 Louisiana	4252	4220	4204	4183	51.8	97.62	31.3	29.0	11.1	3
48 Oklahoma	3175	3146	3025	2980	69.9	46.23	33.4	26.6	13.5	3
49 Texas	17349	16987	14228	18946	268.6	66.24	31.0	28.6	10.1	3
50 WEST	54061	52786	43165	66344	1864.4	30.87	32.0	27.2	10.9	*
51 MOUNTAIN	14036	13659	11368	16650	863.6	16.39	31.9	28.4	11.3	*
52 Montana	808	799	787	752	147.0	5.55	34.3	27.7	13.4	4
53 Idaho	1039	1007	944	1097	83.6	12.56	31.8	30.6	11.9	4
54 Wyoming	460	454	471	413	97.8	4.74	32.4	29.6	10.7	4
55 Colorado	3377	3294	2889	3694	104.1	32.56	32.9	26.1	10.1	4
56 New Mexico	1548	1515	1300	1881	121.6	12.75	31.5	29.6	10.9	4
57 Arizona	3750	3665	2718	5086	114.0	33.00	32.5	26.9	13.3	4
58 Utah	1770	1723	1461	2085	84.9	21.54	26.3	36.3	8.8	4
59 Nevada	1284	1202	799	1641	110.6	11.69	33.4	25.0	10.7	4
60 PACIFIC	40025	39127	31797	49694	1000.8	44.70	32.0	26.7	10.8	*
61 Washington	5018	4867	4130	6060	71.3	75.37	33.4	26.2	11.8	4
62 Oregon	2922	2842	2633	3335	98.4	30.44	34.8	25.6	13.7	4
63 California	30380	29760	23669	38112	163.7	194.78	31.5	26.9	10.5	4
64 Alaska	570	550	400	578	656.4	1.00	29.8	31.6	4.2	4
65 Hawaii	1135	1108	965	1608	10.9	176.71	32.9	25.4	11.4	4

WORLDPOP: A SUMMARY OF THE WORLD POPULATION BY COUNTRY*

The variables are the following:

1994: resident population in 1994 (1000s)
BirRate: birth rate per 1000 of population
DeaRate: death rate per 1000 of population
2010: projected population for 2010
LifeExpt: life expectancy at birth, total
LifeExpM: life expectancy at birth, males
LifeExpF: life expectancy at birth, females
GNP: per-capita gross national product for 1992 (in U.S. dollars)

The printout that follows shows only the first page of data.

* From Population Reference Bureau, Inc., Washington, D.C., 1994.

Row	Region	1994	BirRate	DeaRate	2010	LifeExpT	LifeExpM	LifeExpF	GNP
1	WORLD	5606989	25.11	9.15	7021701	65.32	63.36	67.26	4340
2	MORE DEVELOPED	1164377	12.42	9.80	1227568	74.56	70.66	78.21	16610
3	LESS DEVELOPED	4442612	28.43	8.98	5794133	62.90	61.45	64.39	950
4	LESS DEVELOPED (excl. China)	3250636	32.22	9.84	4418041	60.24	58.82	61.67	1180
5	AFRICA	699602	42.13	13.17	1078280	54.58	53.10	56.04	650
6	SUB-SAHARAN AFRICA	570310	44.75	14.49	900784	52.36	50.83	53.86	540
7	NORTHERN AFRICA	157467	33.04	8.36	220541	62.28	61.10	63.48	1070
8	Algeria	27895	31.60	6.80	38186	66.71	65.70	67.70	1830
9	Egypt	58873	30.30	7.80	80688	61.58	60.40	62.80	630
10	Libya	5057	41.90	8.10	8913	63.22	61.60	65.00	*
11	Morocco	28559	29.80	6.90	38112	66.85	65.10	68.60	1040
12	Sudan	28175	44.20	13.20	43045	53.00	52.20	53.80	*
13	Tunisia	8696	25.20	6.20	11296	67.79	66.90	68.70	1740
14	Western Sahara	212	47.90	20.10	301	44.30	43.30	45.30	*
15	WESTERN AFRICA	193356	45.76	14.95	313294	51.88	50.58	53.17	380
16	Benin	5265	48.90	17.80	8357	46.42	44.80	48.00	410
17	Burkina Faso	10069	49.60	18.20	15474	48.41	47.70	49.10	290
18	Cape Verde	407	35.50	6.80	600	67.72	66.60	68.70	850
19	Cote d'Ivoire	13896	49.90	14.70	23657	51.52	50.20	52.90	670
20	Gambia	1061	48.00	21.00	1545	45.02	43.40	46.60	390
21	Ghana	16944	41.70	11.70	26594	56.01	54.20	57.80	450
22	Guinea	6392	46.00	21.00	9303	42.79	40.70	44.90	510
23	Guinea Bissau	1050	42.70	21.30	1473	43.52	41.90	45.10	210
24	Liberia	2941	47.30	14.20	4829	55.40	54.00	57.00	*
25	Mali	9113	51.70	21.30	14966	45.08	43.50	46.60	300

continues

continued

Row	Region	1994	BirRate	DeaRate	2010	LifeExpT	LifeExpM	LifeExpF	GNP
26	Mauritania	2270	46.00	17.50	3491	48.07	46.40	49.70	530
27	Niger	8813	53.00	19.00	14326	46.52	44.90	48.10	300
28	Nigeria	98091	44.10	13.30	161969	54.11	53.00	55.20	320
29	Senegal	8165	43.00	16.00	12352	49.30	48.30	50.30	780
30	Sierra Leone	4616	48.20	21.60	6944	43.02	41.40	44.60	170
31	Togo	4255	48.50	12.20	7401	56.02	54.00	58.00	400
32	St. Helena	8	13.60	8.10	13	74.00	72.00	76.00	*
33	EASTERN AFRICA	221223	45.82	15.26	350728	50.71	49.24	52.17	210
34	Burundi	6039	46.00	16.90	9103	48.09	46.40	49.70	210
35	Comoros	530	47.30	12.10	919	56.48	54.40	58.60	510
36	Djibouti	569	46.50	16.50	905	48.98	47.40	50.60	*
37	Eritrea	3482	42.40	16.40	5449	*	*	*	*
38	Ethiopia	55228	45.70	14.60	89047	51.80	50.20	53.40	110
39	Kenya	26975	43.70	10.30	44387	58.95	57.10	60.80	330
40	Madagascar	13702	45.50	12.70	22431	55.52	54.00	57.00	230
41	Malawi	9532	47.40	20.10	14696	43.80	43.50	44.10	210
42	Mauritius	1100	21.10	6.50	1284	69.30	65.00	73.00	2700
43	Mozambique	15823	45.10	18.20	25406	46.82	45.20	48.40	60
44	Reunion	644	23.10	5.60	783	73.84	69.40	78.10	*
45	Rwanda	7664	40.10	17.10	10400	46.26	44.80	47.70	250
46	Seychelles	72	22.60	7.40	81	68.96	65.10	72.70	5480
47	Somalia	9845	50.20	18.50	15915	47.02	45.40	48.60	*
48	Tanzania	29755	48.10	14.60	48371	50.86	49.40	52.30	110
49	Uganda	19823	51.00	21.00	30690	41.86	40.80	42.90	170
50	Zambia	9132	46.40	18.00	13885	44.16	43.50	44.80	290
51	Zimbabwe	11215	40.60	11.00	16808	55.86	54.40	57.30	570
52	Mayotte	93	50.00	12.00	168	56.50	54.40	58.60	*
53	MIDDLE AFRICA	80562	45.61	15.48	126434	50.94	49.20	52.64	*
54	Angola	11206	47.20	20.20	17561	46.01	44.00	48.00	*
55	Cameroon	13132	40.80	11.90	21165	56.32	54.30	58.30	820
56	Central African Republic	3142	43.70	20.10	3898	43.85	42.30	45.30	410
57	Chad	6494	43.70	18.00	9788	47.52	45.90	49.10	220

RECYCLE: A SURVEY ON RECYCLING PRACTICES

The survey is discussed in Chapter 5, and the questionnaire for this survey is printed at the end of that chapter.* The following printout shows the recorded data through Question 2(d) for the first 40 respondents.

* From Florida Survey Research Center, University of Florida.

ROW	ID Code	Intervie	County	City	Q1	Q2 (a)	Q2 (b)	Q2 (c)	Q2 (d)
1	1	4	17	27	3	3	3	3	3
2	2	32	17	27	1	2	2	1	1
3	3	7	17	28	2	1	2	1	1
4	4	22	17	28	2	3	3	1	1
5	5	36	17	27	3	3	3	1	1
6	6	10	17	28	1	2	2	2	2
7	7	4	17	27	2	1	1	3	3
8	8	35	17	28	3	3	2	1	1
9	9	36	13	21	1	1	2	1	1
10	10	32	17	28	2	1	3	1	1
11	11	22	17	28	5	3	3	3	3
12	12	10	13	21	1	1	1	1	1
13	13	36	13	21	4	1	2	1	1
14	14	7	13	21	2	1	1	1	1
15	15	7	13	21	1	1	2	1	1
16	16	7	13	21	1	3	3	1	1
17	17	4	13	21	5	1	1	1	1
18	18	27	13	21	5	1	3	3	3
19	19	37	15	25	1	2	2	2	2
20	20	36	15	25	2	1	3	1	1
21	21	32	15	25	2	1	3	1	1
22	22	22	15	25	3	1	2	1	1
23	23	36	15	25	2	1	3	1	1
24	24	10	15	25	1	1	1	1	2
25	25	37	15	25	1	1	1	1	1
26	26	35	15	25	2	1	3	1	1
27	27	37	15	25	2	1	1	1	1
28	28	32	17	27	2	3	2	1	1
29	29	22	17	28	3	2	2	3	3
30	30	35	15	25	1	1	3	1	1
31	31	7	13	21	1	1	2	1	1
32	32	37	15	25	2	1	1	1	1
33	33	10	13	21	1	1	1	1	1
34	34	22	13	25	2	1	2	1	1
35	35	32	15	25	4	1	1	1	1
36	36	35	15	25	4	1	2	1	2
37	37	32	15	25	3	1	2	1	1
38	38	22	12	20	3	3	2	1	3
39	39	37	12	20	1	1	2	1	1
40	40	35	15	25	2	1	2	1	1

SCHOOLS: A SUMMARY OF THE COSTS OF EDUCATION FOR THE UNITED STATES

The data set contains information on schools for all 50 states and the District of Columbia.* The variables are the following:

ExpPP90: expenditures per pupil for 1990 (in dollars)

ExpPC90: expenditures per capita for 1990 (in dollars)

TeaSal90: average teacher salary for 1990 (in $1000s)

%Compl: percentage of residents over the age of 25 who completed high school as of the 1990 census

ExpPP92: expenditures per pupil for 1992 (in dollars)

ExpPC92: expenditures per capita for 1992 (in dollars)

TeaSal92: average teacher salary for 1992 (in $1000s)

%Dropout: percentage of 16- to 19-year-olds who dropped out of school, as measured in 1990

Region: 1 = Northeast, 2 = Midwest, 3 = South, 4 = West

Popultn: approximate population in 1992 (in millions)

Enroll: number of students enrolled in 1990 (in thousands)

Teachers: number of teachers as of 1990 (in thousands)

* From U.S. Bureau of the Census, *Statistical Abstract of the United States, 1993–94*, Washington, D.C., 1995.

Row	State	ExpPP90	ExpPC90	TeaSal90	%Compl	ExpPP92	ExpPC92
1	Alabama	3314	624	25.5	63.2	3675	684
2	Alaska	7252	1566	43.2	86.9	9248	1798
3	Arizona	3853	867	29.4	80.6	4750	961
4	Arkansas	3272	621	22.0	67.6	3770	758
5	California	4598	811	36.4	78.6	4686	930
6	Colorado	4580	787	30.8	83.2	5259	961
7	Connecticut	7934	1129	40.5	80.6	8299	1227
8	Delaware	5848	835	33.4	80.7	6080	899
9	DC	7407	945	38.0	72.9	8116	1059
10	Florida	5051	787	28.8	77.9	5639	934
11	Georgia	4456	796	27.9	71.1	4720	884
12	Hawaii	4504	710	32.0	82.3	5453	898
13	Idaho	3037	668	23.9	77.3	3528	803
14	Illinois	4853	728	32.8	77.2	5248	842
15	Indiana	4126	728	30.5	78.0	5429	1003

continues

continued

Row	State	ExpPP90	ExpPC90	TeaSal90	%Compl	ExpPP92	ExpPC92
16	Iowa	4590	777	26.7	83.4	4949	882
17	Kansas	4706	819	28.7	82.2	5131	941
18	Kentucky	3824	629	26.3	64.7	4616	773
19	Louisiana	3457	620	24.3	70.9	4378	806
20	Maine	5577	974	26.9	76.9	5969	1087
21	Maryland	5887	891	36.6	80.7	6273	967
22	Massachusetts	6170	826	34.2	80.7	6323	863
23	Michigan	5073	888	36.0	77.0	5630	987
24	Minnesota	5114	933	32.2	85.5	5510	1054
25	Mississippi	3151	626	24.4	67.7	3344	670
26	Missouri	4226	676	27.2	75.9	4534	780
27	Montana	4147	801	25.1	83.6	5127	943
28	Nebraska	3874	652	25.5	82.2	4676	866
29	Nevada	4387	878	30.6	84.0	4910	1006
30	New Hampshire	5149	827	29.0	82.2	5500	888
31	New Jersey	8439	1114	35.7	79.4	10219	1337
32	New Mexico	4180	857	25.1	74.6	4692	971
33	New York	8094	1091	38.9	76.7	8658	1247
34	N. Carolina	4386	737	27.8	71.3	4857	835
35	N. Dakota	3581	672	23.0	81.1	4119	794
36	Ohio	4394	720	31.2	77.6	5451	921
37	Oklahoma	3484	683	23.1	75.4	3939	814
38	Oregon	5085	873	30.8	83.9	5927	1018
39	Pennsylvania	5670	789	33.3	76.8	6980	954
40	Rhode Island	6523	827	36.1	72.7	6834	903
41	S. Carolina	3731	678	27.2	69.8	4537	861
42	S. Dakota	3312	618	21.3	78.3	4255	835
43	Tennessee	3503	575	27.1	64.7	3736	611
44	Texas	4056	849	27.5	74.3	4651	988
45	Utah	2733	738	23.7	88.2	3092	854
46	Vermont	5418	921	28.8	81.8	6992	1205
47	Virginia	5000	852	30.9	74.3	5487	950
48	Washington	4638	966	30.5	88.2	5331	1160
49	West Virginia	4146	725	22.8	68.0	5415	962
50	Wisconsin	5703	887	31.9	81.1	5972	976
51	Wyoming	5281	1149	28.2	85.6	5333	1189

Row	TeaSa192	%Dropout	Region	Popultn	Enroll	Teachers
1	27.0	12.6	3	4.1	722	41.1
2	44.7	9.6	4	0.6	114	7.2
3	31.2	14.3	4	3.8	640	33.5
4	26.6	10.9	3	2.4	433	25.9
5	40.2	14.3	4	30.9	4951	221.0
6	33.1	9.6	4	3.5	574	33.1
7	47.0	9.2	1	3.3	469	34.8
8	34.5	11.2	3	0.7	106	6.1
9	41.3	19.1	3	0.6	80	6.2
10	31.1	14.2	3	13.5	1862	109.9
11	29.5	14.1	3	6.8	1152	70.3
12	34.5	7.0	4	1.2	172	10.0
13	26.3	9.6	4	1.1	221	11.6
14	36.5	10.4	2	11.6	1822	109.9
15	34.8	11.4	2	5.7	955	55.1
16	29.2	6.5	2	2.8	484	31.5
17	30.7	8.4	2	2.5	437	29.3
18	30.9	13.0	3	3.8	636	37.4
19	27.0	11.9	3	4.3	785	44.0
20	30.1	8.4	1	1.2	215	14.8
21	39.5	11.0	3	4.9	715	43.3
22	37.3	9.5	1	6.0	834	56.0
23	41.1	9.9	2	9.4	1582	80.2
24	33.7	6.1	2	4.5	754	44.9
25	24.4	11.7	3	2.6	503	27.9
26	28.9	11.2	2	5.2	812	52.3
27	27.6	7.1	4	0.8	153	9.9
28	27.2	6.6	2	1.6	274	18.9
29	33.9	14.9	4	1.3	201	11.4
30	33.2	9.9	1	1.1	172	11.5
31	41.0	9.3	1	7.8	1090	80.5
32	26.7	10.8	4	1.6	302	17.1
33	43.3	10.1	1	18.1	2598	180.3
34	29.2	13.2	3	6.8	1087	64.8
35	24.5	4.3	2	0.6	118	7.8
36	33.3	8.8	2	11.0	1772	103.2
37	25.3	9.9	3	3.2	579	37.7
38	34.1	11.0	4	3.0	485	26.7
39	38.7	9.4	1	12.0	1668	100.5
40	36.0	12.9	1	1.0	139	9.8
41	28.3	11.9	3	3.6	622	36.3

continues

continued

Row	TeaSal92	%Dropout	Region	Popultn	Enroll	Teachers
42	23.3	7.1	2	0.7	129	8.6
43	28.6	13.6	3	5.0	824	43.1
44	29.0	12.5	3	17.6	3383	212.6
45	26.5	7.9	4	1.8	448	18.3
46	33.6	8.7	1	0.6	96	7.0
47	31.9	10.4	3	6.4	998	64.7
48	34.8	10.2	4	5.1	840	42.9
49	27.4	10.6	3	1.8	322	21.0
50	35.2	6.9	2	5.0	798	52.0
51	30.4	6.3	4	0.5	98	6.4

TEMPS: MONTHLY NORMAL TEMPERATURES (T) AND PRECIPITATION (P) FOR U.S. WEATHER STATIONS

	Jan.		*Feb.*		*Mar.*		*Apr.*		*May*	
Station	**T**	**P***	**T**	**P**	**T**	**P**	**T**	**P**	**T**	**P**
Albany, N.Y.	22	2.2	24	2.1	33	2.6	47	2.7	58	3.3
Albuquerque, N.M.	35	0.3	40	0.4	46	0.5	56	0.5	65	0.5
Anchorage, Alaska	12	0.8	18	0.8	24	0.6	35	0.6	46	0.6
Asheville, N.C.	38	3.4	39	3.6	46	4.7	56	3.5	64	3.3
Atlanta, Ga.	42	4.3	45	4.4	51	5.8	61	4.6	69	3.7
Baltimore, Md.	33	2.9	35	2.8	43	3.7	54	3.1	64	3.6
Barrow, Alaska	–15	0.2	–19	0.2	–15	0.2	–1	0.2	19	0.2
Birmingham, Ala.	44	4.8	47	5.3	53	6.2	63	4.6	71	3.6
Bismarck, N.D.	8	0.5	14	0.4	25	0.7	43	1.4	54	2.2
Boise, Idaho	29	1.5	36	1.2	41	1.0	49	1.1	57	1.3
Boston, Mass.	29	3.7	30	3.5	38	4.0	49	3.5	59	3.5
Buffalo, N.Y.	24	2.9	24	2.6	32	2.9	45	3.2	55	3.0
Burlington, Vt.	17	1.7	19	1.7	29	1.9	43	2.6	55	3.0
Caribou, Maine	11	2.0	13	2.1	24	2.2	37	2.4	50	3.0
Charleston, S.C.	49	2.9	51	3.3	56	4.8	65	3.0	72	3.8
Chicago, Ill.	24	1.9	27	1.6	37	2.7	50	3.8	60	3.4
Cincinnati, Ohio	32	3.4	34	3.0	43	4.1	55	3.9	64	4.0
Cleveland, Ohio	27	2.6	28	2.2	36	3.1	48	3.5	58	3.5
Columbus, Ohio	28	2.9	30	2.3	39	3.4	51	3.7	61	4.1
Dallas–Ft. Worth, Texas	45	1.8	49	2.4	55	2.5	65	4.3	73	4.5
Denver, Col.	30	0.6	33	0.7	37	1.2	48	1.9	57	2.6
Des Moines, Iowa	19	1.1	24	1.1	34	2.3	50	2.9	61	4.2
Detroit, Mich.	26	1.9	27	1.8	35	2.3	48	3.1	58	3.4
Dodge City, Kans.	31	0.5	35	0.6	41	1.1	54	1.7	64	3.1

continues

continued

Station	*Jan.* T	*Jan.* P*	*Feb.* T	*Feb.* P	*Mar.* T	*Mar.* P	*Apr.* T	*Apr.* P	*May* T	*May* P
Duluth, Minn.	9	1.2	12	0.9	24	1.8	39	2.6	49	3.4
Eureka, Calif.	47	7.4	48	5.2	48	4.8	50	3.0	53	2.1
Fairbanks, Alaska	−12	0.6	−3	0.5	10	0.5	29	0.3	47	0.7
Fresno, Calif.	45	1.8	50	1.7	54	1.6	60	1.2	67	0.3
Galveston, Tex.	54	3.0	56	2.7	61	2.6	69	2.6	76	3.2
Grand Junction, Colo.	27	0.6	34	0.6	41	0.8	52	0.8	62	0.6
Grand Rapids, Mich.	23	1.9	25	1.5	33	2.5	47	3.4	57	3.2
Hartford, Conn.	25	3.3	27	3.2	36	3.8	48	3.8	58	3.5
Helena, Mont.	18	0.6	25	0.4	31	0.7	43	0.9	52	1.8
Honolulu, Hawaii	72	4.4	72	2.5	73	3.2	75	1.4	77	1.0
Houston, Tex.	52	3.6	55	3.5	61	2.7	69	3.5	76	5.1
Huron, S.D.	13	0.4	18	0.8	29	1.1	46	2.0	57	2.8
Indianapolis, Ind.	28	2.9	31	2.4	40	3.8	52	3.9	62	4.1
Jackson, Miss.	47	4.5	50	4.6	56	5.6	66	4.7	73	4.4
Jacksonville, Fla.	55	2.8	56	3.6	61	3.6	68	3.1	74	3.2
Juneau, Alaska	24	3.9	28	3.4	32	3.6	39	3.0	47	3.3
Kansas City, Mo.	27	1.3	32	1.3	41	2.6	54	3.5	64	4.3
Knoxville, Tenn.	41	4.7	43	4.7	50	4.9	60	3.6	68	3.3
Lander, Wyo.	20	0.5	26	0.7	31	1.2	43	2.4	53	2.6
Little Rock, Ark.	40	4.2	43	4.4	50	4.9	62	5.3	70	5.3
Los Angeles, Calif.	57	3.0	58	2.8	59	2.2	62	1.3	65	0.1
Louisville, Ky.	33	3.5	36	3.5	44	5.1	56	4.1	65	4.2
Marquette, Mich.	18	1.5	20	1.5	27	1.9	40	2.6	50	2.9
Memphis, Tenn.	41	4.9	44	4.7	51	5.1	63	5.4	71	4.4
Miami, Fla.	67	2.2	68	2.0	71	2.1	75	3.6	78	6.1
Milwaukee, Wis.	19	1.6	23	1.1	31	2.2	45	2.8	54	2.9
Minneapolis, Minn.	12	0.7	17	0.8	28	1.7	45	2.0	57	3.4
Mobile, Ala.	51	4.7	54	4.8	59	7.1	68	5.6	75	4.5
Moline, Ill.	22	1.7	26	1.3	36	2.6	51	3.8	61	3.9
Nashville, Tenn.	38	4.8	41	4.4	49	5.0	60	4.1	69	4.1
Newark, N.J.	31	2.9	33	3.0	41	3.9	52	3.4	62	3.6
New Orleans, La.	53	4.5	56	4.8	61	5.5	69	4.2	75	4.2
New York, N.Y.	32	2.7	33	2.9	41	3.7	52	3.3	62	3.5
Nome, Alaska	6	0.9	5	0.8	7	0.8	19	0.7	35	0.7
Norfolk, Va.	41	3.4	41	3.3	48	3.4	58	2.7	67	3.3
Okla. City, Okla.	37	1.1	41	1.3	48	2.1	60	3.5	68	5.2
Omaha, Nebr.	23	0.8	28	1.0	37	1.6	52	3.0	63	4.1
Parkersburg, W. Va.	33	3.1	35	2.8	43	3.8	55	3.5	64	3.6
Philadelphia, Pa.	32	2.8	34	2.6	42	3.7	53	3.3	63	3.4
Phoenix, Ariz.	51	0.7	55	0.6	60	0.8	68	0.3	76	0.1
Pittsburgh, Pa.	28	2.8	29	2.4	38	3.6	50	3.4	60	3.6
Portland, Maine.	22	3.4	23	3.5	32	3.6	43	3.3	53	3.3
Portland, Oreg.	38	5.9	43	4.1	46	3.6	51	2.2	57	2.1
Providence, R.I.	28	3.5	29	3.5	37	4.0	47	3.7	57	3.5
Raleigh, N.C.	41	3.2	42	3.3	49	3.4	60	3.1	67	3.3
Rapid City, S.D.	22	0.5	26	0.6	31	1.0	45	2.1	55	2.8
Reno, Nev.	32	1.2	37	0.9	40	0.7	47	0.5	55	0.7

continues

continued

	Jan.		Feb.		Mar.		Apr.		May	
Station	**T**	**P***	**T**	**P**	**T**	**P**	**T**	**P**	**T**	**P**
Richmond, Va.	38	2.9	39	3.0	47	3.4	58	2.8	67	3.4
St. Louis, Mo.	31	1.9	35	2.1	43	3.0	57	3.9	66	3.9
Salt Lake City, Utah	28	1.3	33	1.2	40	1.6	49	2.1	58	1.5
San Antonio, Tex.	51	1.7	55	2.1	61	1.5	70	2.5	76	3.1
San Diego, Calif.	55	1.9	57	1.5	58	1.6	61	0.8	63	0.2
San Francisco, Calif.	48	4.4	51	3.0	53	2.5	55	1.6	58	0.4
San Juan, P.R.	75	3.7	75	2.5	76	2.0	78	3.4	79	6.5
Sault Ste. Marie, Mich.	14	1.9	15	1.5	24	1.7	38	2.2	49	3.0
Savannah, Ga.	50	2.9	52	2.9	58	4.4	66	2.9	73	4.2
Seattle, Wash.	38	5.8	42	4.2	44	3.6	49	2.5	55	1.7
Spokane, Wash.	25	2.5	32	1.7	38	1.5	46	1.1	55	1.5
Springfield, Mo.	33	1.7	37	2.2	44	3.0	57	4.3	65	4.9
Syracuse, N.Y.	24	2.7	25	2.8	33	3.0	47	3.1	57	3.0
Tampa, Fla.	60	2.3	62	2.9	66	3.9	72	2.1	77	2.4
Trenton, N.J.	32	2.8	33	2.7	41	3.8	52	3.2	62	3.4
Washington, D.C.	36	2.6	37	2.5	45	3.3	56	2.9	66	3.7
Wilmington, Del.	32	2.9	34	2.8	42	3.7	52	3.2	62	3.4

Source: The World Almanac & Book of Facts, 1984 edition. Copyright © 1983 Newspaper Enterprise Association, Inc., New York.

* T is in degrees Fahrenheit, and P is in inches.

RIVER: FLOW RATES FOR CRYSTAL RIVER, FLORIDA

Discharge water (in cubic feet per second), October 1977–September 1979 (mean values)

	1977–1978											
Day	Oct.	Nov.	Dec.	Jan.	Feb.	Mar.	Apr.	May	June	July	Aug.	Sept.
1	51	11	3.6	6.1	13	37	9.8	16	3.6	9.3	26	127
2	45	11	3.6	6.1	14	26	8.6	18	4.1	3.4	47	122
3	42	11	4.0	5.7	17	26	8.4	18	3.6	2.5	48	126
4	38	11	3.3	4.8	16	54	8.0	25	9.1	1.6	40	132
5	35	9.5	2.9	4.4	14	55	8.7	27	12	1.3	34	136
6	32	7.4	3.5	5.3	13	45	8.4	19	12	1.9	33	139
7	29	5.4	2.6	5.9	12	37	9.2	50	12	2.6	36	143
8	28	5.5	2.6	6.6	12	34	9.5	49	9.9	2.2	38	133
9	28	5.5	2.8	9.0	31	111	9.9	28	9.1	4.1	46	126
10	27	6.3	3.1	8.8	47	216	10	22	8.4	5.3	63	123
11	26	6.0	3.1	7.4	44	134	11	20	6.8	6.6	78	120
12	27	4.8	2.8	6.1	35	91	11	18	6.8	11	85	116
13	26	4.7	2.6	11	28	77	12	18	6.8	5.8	88	114
14	24	5.0	2.6	15	25	68	13	17	6.1	6.8	89	114
15	23	5.5	2.8	12	24	64	13	15	6.1	17	91	98
16	23	5.5	5.4	9.3	35	62	12	14	6.8	3.8	96	86
17	23	6.0	6.2	11	112	50	13	13	6.8	7.7	98	81
18	22	6.0	5.6	17	234	42	13	14	5.4	19	103	77
19	21	5.7	4.7	18	243	36	14	20	4.7	48	108	78
20	21	4.8	4.0	29	138	30	17	14	4.7	65	127	75
21	20	5.4	4.1	29	125	27	17	13	4.1	35	119	60
22	19	5.6	4.2	29	113	21	16	12	3.3	20	104	49
23	16	7.5	4.9	26	98	16	14	9.6	4.1	17	96	40
24	17	4.8	5.1	24	82	14	14	7.7	4.1	14	93	31
25	17	4.8	5.0	23	69	13	14	7.1	3.7	13	91	24
26	16	4.4	6.1	23	58	12	16	6.9	3.0	22	90	24
27	15	4.3	5.9	23	49	12	17	11	2.3	29	91	19
28	14	3.6	5.0	19	41	11	17	12	2.7	24	98	13
29	13	3.8	5.2	18	—	11	17	9.5	2.8	22	95	9.6
30	13	3.8	5.4	16	—	9.5	16	4.6	7.9	19	100	8.5
31	12	—	6.8	14	—	9.6	—	2.6	—	20	114	—
Total	763	185.6	129.5	442.3	1742	1451.1	377.5	531.2	182.8	459.9	2465	2544.1
Mean	24.6	6.19	4.18	14.3	62.2	46.8	12.6	17.1	6.09	14.8	79.5	84.8
Max.	51	11	6.8	29	243	216	17	50	12	65	127	143
Min.	12	3.6	2.6	4.4	12	9.5	8.0	2.6	2.3	1.3	26	8.5

continues

continued

	1978–1979											
Day	**Oct.**	**Nov.**	**Dec.**	**Jan.**	**Feb.**	**Mar.**	**Apr.**	**May**	**June**	**July**	**Aug.**	**Sept.**
1	10	1.2	2.6	7.3	23	29	11	3.5	38	6.1	13	521
2	8.4	.77	1.8	9.5	21	23	11	2.3	36	4.3	13	508
3	9.1	.63	2.6	11	19	21	9.2	1.5	34	3.3	14	401
4	8.1	.60	2.5	8.3	18	19	6.7	1.4	39	4.0	16	345
5	7.1	.50	4.1	6.5	19	18	6.8	1.4	40	8.2	26	284
6	9.4	.65	4.1	4.7	19	40	8.0	2.1	32	11	27	237
7	8.7	.56	3.7	6.0	19	101	7.6	2.6	26	6.8	58	205
8	6.1	.64	3.3	7.6	20	85	7.9	504	22	17	84	183
9	6.1	.57	3.0	8.5	19	65	8.7	1300	19	14	133	165
10	5.4	.56	3.0	8.0	17	50	9.0	489	16	8.5	114	155
11	4.6	.70	3.0	7.5	16	44	9.4	275	14	6.2	150	150
12	5.6	.76	3.4	27	16	37	9.4	220	12	6.8	490	145
13	6.8	.74	3.1	59	15	33	9.9	185	11	11	379	226
14	9.3	.80	4.3	49	15	31	8.5	185	9.6	14	296	314
15	5.7	.87	3.5	30	15	27	8.5	160	8.6	18	296	287
16	5.0	.96	3.1	22	15	25	7.8	140	7.8	12	264	333
17	4.7	1.0	3.0	19	15	21	6.4	126	7.3	14	281	324
18	4.5	1.1	2.6	17	14	21	6.1	116	6.4	11	261	290
19	4.4	1.1	3.1	14	14	19	6.3	95	6.4	10	228	276
20	5.8	1.1	4.7	13	13	16	6.1	78	5.9	10	337	254
21	6.6	1.1	5.6	17	12	16	6.8	76	5.8	16	331	235
22	7.9	1.3	7.7	19	12	15	5.9	74	5.6	14	281	472
23	9.6	1.6	8.6	17	12	17	4.8	62	5.2	14	288	472
24	6.6	1.8	10	33	13	20	5.0	74	5.1	19	321	897
25	5.6	2.2	11	40	56	18	5.9	82	4.7	25	305	764
26	5.2	2.0	12	33	106	16	4.8	70	3.7	17	296	878
27	5.8	2.0	15	26	62	13	4.7	60	3.9	12	328	786
28	7.6	2.1	28	24	36	12	4.5	54	7.2	11	305	680
29	3.8	2.6	19	23	—	12	3.7	49	14	10	287	580
30	2.7	.93	12	23	—	11	3.6	45	9.1	11	465	500
31	1.4	—	8.8	24	—	12	—	41	—	12	492	—
Total	197.6	33.44	202.2	613.9	651	887	214.0	4574.8	455.3	357.2	7189	11867
Mean	6.37	1.11	6.52	19.8	23.3	28.6	7.13	148	15.2	11.5	232	396
Max.	10	2.6	28	59	106	101	11	1300	40	25	492	897
Min.	1.4	.50	1.8	4.7	12	11	3.6	1.4	3.7	3.3	13	145

Source: U.S. Department of Interior, Geological Survey.

SELECTED ANSWERS

CHAPTER 4 EXERCISES

4.1 $\sigma^2 = 2$, $V(\bar{y}) = .75$

4.5 $\hat{p} = .83$, $B = .131$

4.6 $n = 128$

4.7 $\hat{\mu} = 12.5$, $B = 7.04$

4.8 $\hat{\tau} = 125{,}000$, $B = 70{,}412.50$

4.9 $\hat{\mu}_1 = 2.30$, $B_1 = .070$
$\hat{\mu}_2 = 4.52$, $B = .086$

4.10 $\hat{p} = .625$, $B = .1535$

4.11 $\hat{\mu} = 2.0$, $B = .938$

4.12 $\hat{p} = .430$, $B = .0312$

4.13 $n = 2392$

4.14 $N\hat{p} = 100$, $B = 31.30$

4.15 $\hat{\mu} = 2.1$, $B = .17$

4.16 $n = 4$

4.17 $\hat{\mu} = 5.01$, $B = .871$

4.18 $\hat{p} = .183$, $B = .096$

4.19 $n = 87$

4.20 $\hat{\tau} = 37{,}800$, $B = 3379.94$

4.21 $n = 400$

4.22 $\hat{\tau} = \$3{,}898{,}000$, $B = \$263{,}918.17$

4.29 $\hat{\tau} = \$17{,}333.33$, $B = \$4{,}479.23$

4.30 $\hat{\tau} = \$98{,}550.00$, $B = \$19{,}905.83$

4.31 $\hat{p} = .3$, $B = .206$

4.34 $(\hat{p}_1 - \hat{p}_2) = -.05$, $B = .06$, no

4.39 **a.** $\hat{p} = .22$, $B = .0893$

b. $\hat{p} = .63$, $B = .1041$

c. $\hat{p} = .10$, $B = .0880$

d. $\hat{p} = .90$, $B = .0880$

4.43 **a.** $\hat{\mu} = 19.9$, $B = .773$

b. $\hat{\mu} = 18.0$, $B = .717$

c. Large firms
$\hat{\mu}_1 = 5.3$, $B_1 = .161$
$\hat{\mu}_2 = 12.7$, $B_2 = .693$
Small firms
$\hat{\mu}_1 = 4.9$, $B_1 = .173$
$\hat{\mu}_2 = 11.2$, $B_2 = .633$

d. $\hat{p} = .585$, $B = .016$

4.44 **a.** $\hat{\mu}_1 = 68.04$, $B_1 = 4.52$
$\hat{\mu}_2 = 56.74$, $B_2 = 4.52$

b. $\hat{\mu}_1 = 48.79$, $B_1 = 3.88$
$\hat{\mu}_2 = 66.39$, $B_2 = 3.91$

CHAPTER 5 EXERCISES

5.1 $\hat{p}_{st} = .30$, $B = .117$

5.2 $n_1 = 18$, $n_2 = 10$, $n_3 = 2$

5.3 $\hat{\tau} = 1903.9$, $B = 676.8$

5.5 $n = 26$, $n_1 = 16$, $n_2 = 7$, $n_3 = 3$

5.6 **a.** $\hat{\mu} = 59.99$, $B = 3.03$;

c. $\hat{\mu}_1 - \hat{\mu}_2 = 14.96$, $B = 6.79$

5.7 $n_1 = 12$, $n_2 = 20$, $n_3 = 18$

5.8 $n = 33$

5.9 $n = 32$

5.10 $\hat{\tau} = 50{,}505.60$, $B = 8663.12$

5.11 $n = 60$

5.12 $n = 29$

5.13 $n = 158$, $n_1 = 39$, $n_2 = 17$, $n_3 = 69$, $n_4 = 33$

5.14 $\hat{p}_{st} = .701$, $B = .0503$

5.15 $n = 61$, $n_1 = 15$, $n_2 = 7$, $n_3 = 27$, $n_4 = 12$

5.16 **a.** $\hat{\mu} = 251.07$, $\hat{V}(\bar{y}_{st}) = 141.88$

b. $\hat{\mu} = 250.05$, $\hat{V}(\bar{y}_{st}) = 181.22$

5.17 Dividing points = 40, 70, 90

5.18 Stratum 1 = \$0–\$200
Stratum 2 = \$201–\$350

5.19 $\hat{\mu} = 63.88$, $B = .628$

5.21 **a.** $\hat{p} = .160$, $B = .074$

b. $\hat{p}_{st} = .159$, $B = .081$

5.27 **a.** $n_1 = .29n$, $n_2 = .71n$

b. $n = 41$, $n_1 = 12$, $n_2 = 29$

5.28 $\hat{\mu} = 189.93$, $\hat{V}(\bar{y}_{st}) = 1378.82$

5.29 $\hat{p}'_{st} = .68$, $\hat{V}(\hat{p}) = .00204$

CHAPTER 6 EXERCISES

6.1 $\hat{\tau}_y = 1589.55$, $B = 186.32$

6.2 $\hat{\tau}_y = 2958.33$, $B = 730.12$

6.3 $r = .147$, $B = .0102$

6.4 $\hat{\tau}_y = 145{,}943.78$, $B = 7353.67$

6.5 $\hat{\mu}_y = 1186.53$, $B = 59.79$

6.6 $\hat{\mu}_y = 17.59$, $B = .271$

6.7 $\hat{\mu}_y = 4.16$, $B = .085$

6.8 $r = .283$, $B = .0615$

6.9 $\hat{\tau}_y = 5492.31$, $B = 428.44$

6.10 $r = 1.072$, $B = .0064$

6.11 $\hat{\mu}_y = 997.92$, $B = 53.65$

6.12 $\hat{\tau}_y = 231{,}611.86$, $B = 3073.83$

6.13 $n = 14$

6.14 $\hat{\mu}_{yL} = 1186.55$, $B = 61.33$

6.16 $\hat{\tau}_{yL} = 5515.50$, $B = 448.61$

6.17 $r = .835$, $B = .012$

6.21 **a.** $r = 1.043$, $B = .0733$

b. $r = .870$, $B = .176$

6.25 $\hat{\tau}_{yRc} = 48{,}209.84$, $\hat{V}(\hat{\tau}_{yRc}) = 557{,}095.07$

6.26 **a.** $\text{RE}(\hat{\tau}_y/\hat{\tau}) = 15.36$
b. $\text{RE}(\hat{\tau}_{yL}/\hat{\tau}) = 16.03$
c. $\text{RE}(\hat{\tau}_{yL}/\hat{\tau}_y) = 1.04$

6.27 **a.** $\text{RE}(\hat{\mu}_{yL}/\hat{\mu}_y) = 1.195$
b. $\text{RE}(\hat{\mu}_{yL}/\hat{\mu}_{yD}) = 1.086$
c. $\text{RE}(\hat{\mu}_y/\hat{\mu}_{yD}) = .908$

6.28 For Exercise 6.9
a. $\text{RE}(\hat{\tau}_{yL}/\hat{\tau}_y) = .912$
b. $\text{RE}(\hat{\tau}_{yL}/\hat{\tau}_{YD}) = 1.56$
c. $\text{RE}(\hat{\tau}_y/\hat{\tau}_{YD}) = 1.708$
For Exercise 6.11
a. $\text{RE}(\hat{\mu}_{yL}/\hat{\mu}_y) = 1.95$
b. $\text{RE}(\hat{\mu}_{yL}/\hat{\mu}_{YD}) = 1.20$
c. $\text{RE}(\hat{\mu}_y/\hat{\mu}_{yD}) = .61$

6.30 $r = .0056,\quad B = .022$

6.32 **a.** $r = 3.595,\quad B = 3.221$
b. $r = 1.042,\quad B = .305$

6.33 $\hat{\mu}_{yL} = 196.97$

6.34 $\hat{\mu}_{yL} = .396,\quad B = .023$

CHAPTER 7 EXERCISES

7.4 $\hat{p}_{sy} = .66,\quad B = .0637$

7.5 $n = 1636$

7.6 $\hat{\mu}_{sy} = 11.94,\quad B = .026$

7.7 $n = 28$

7.8 $\hat{\mu}_{sy} = 2007.11,\quad B = 74.505$

7.9 $\hat{p}_{sy} = .81,\quad B = .036$

7.10 $n = 1432$

7.11 $\hat{\tau}_{sy} = 127{,}500,\quad B = 30{,}137.06$

7.12 $n = 259$

7.14 $\hat{\mu}_{sy} = 225.47,\quad B = 6.75$

7.15 $\hat{\tau}_{sy} = 48{,}680,\quad B = 1370.34$

7.16 $\hat{\mu}_{sy} = 7038.10,\quad B = 108.74$

7.17 $\hat{p}_{sy} = .738,\quad B = .104$

7.18 $\hat{\tau}_{sy} = 4400,\quad B = 784.08$

7.20 **a.** $\hat{\tau}_{sy} = 155{,}586,\quad B = 8223$
b. $\hat{\mu}_{sy} = 19.29,\quad B = 2.36$

CHAPTER 8 EXERCISES

8.2 $\hat{\mu} = 19.73,\quad B = 1.78$

8.3 $\hat{\tau} = 12{,}312,\quad B = 3175.07$

8.4 $\hat{\tau} = 14{,}008.85,\quad B = 1110.78$

8.5 $n = 14$

8.6 $\hat{\mu} = 51.56,\quad B = 1.344$

8.7 $n = 13$

8.8 $\hat{p} = .709,\quad B = .048$

8.9 $n = 7$

8.10 $\hat{\mu} = 40.17,\quad B = .640$

8.11 $\hat{\tau} = 157{,}020,\quad B = 6927.88$

8.12 $n = 30$

8.13 $\hat{\mu} = 16.005,\quad B = .0215$

8.14 $\hat{p} = .5701,\quad B = .0307$

8.15 $n = 21$

8.16 $\hat{\mu} = 5.91,\quad B = .322$

8.17 $\hat{p} = .40,\quad B = .116$

8.18 $\hat{\tau} = 3532.8,\quad B = 539.50$

8.19 $\hat{\mu} = 2.685,\quad \hat{V}(\bar{y}^*) = .056$

8.20 **a.** $\hat{\mu} = 3.153,\quad B = .460$
b. $\hat{\mu} = 1.90,\quad B = .232$

8.21 $\hat{p} = .133,\quad B = .075$

8.22 $\hat{\tau} = 80,\quad B = 40.44$

8.23 $\hat{p} = .0918,\quad B = .0390$

8.31 $\hat{\tau} = 600,\quad B = 308.22$

8.32 $\hat{\mu} = .25,\quad B = .128$

8.36 **a.** $\hat{p} = .391$
b. $\hat{V}(\hat{p}) = .0527$

CHAPTER 9 EXERCISES

9.2 $\hat{\mu}_r = 9.379$, $B = 1.455$

9.3 $\hat{\mu} = 9.559$, $B = 1.367$

9.4 $\hat{p} = .287$, $B = .112$

9.5 $\hat{p} = .351$, $B = .177$

9.6 $\hat{\tau} = 3980.7$, $B = 274.73$

9.7 $\hat{p} = .120$, $B = .067$

9.8 $\hat{\tau} = 1276.25$, $B = 333.44$

9.9 $\hat{\mu} = 7.933$, $B = .0923$

9.11 $\hat{\mu}_r = 97.97$, $B = 10.996$

9.12 $\hat{\tau} = 57{,}608$, $B = 6465.37$

9.14 $\hat{\tau} = 3900$, $B = 896.10$

9.15 $\hat{\mu}_r = .9811$, $B = .225$

9.16 $\hat{\mu} = 13.0$, $B = 2.987$

CHAPTER 10 EXERCISES

10.4 $\hat{N} = 445$, $B = 150.60$

10.5 $\hat{N} = 1811$, $B = 344.51$

10.6 $\hat{N} = 10{,}868$, $B = 715.82$

10.7 $\hat{N} = 3349$, $B = 455.11$

10.8 $\hat{N} = 200$, $B = 78.88$

10.9 $n = 625$

10.10 $\hat{N} = 1067$, $B = 507.72$

10.11 $\hat{N} = 750$, $B = 441.59$

10.13 $\hat{N} = 250$, $B = 52.04$

10.14 $\hat{\lambda} = 2.1$, $B = .0748$

10.15 $\hat{\lambda} = .0171$, $B = .00191$

10.16 $\hat{M} = 1920$, $B = 135.76$

10.19 $\hat{\lambda} = 2.792$, $B = .216$

10.21 ***a.*** $\hat{\lambda} = 6.0$

b. $\hat{V}(\hat{\lambda}) = 2.4$

CHAPTER 11 EXERCISES

11.1 $\bar{y} = 407.125$, $B = 93.70$

11.2 $\bar{y} = 5.26$, $B = .789$

11.3 $\bar{y} = 23.61$, $B = 9.097$

11.4 $\hat{\tau} = 1794.455$, $B = 778.15$

11.5 $\hat{\tau} = 1959.73$, $B = 763.51$

11.6 $\bar{y} = 9.804$, $B = 2.376$

11.7 $\hat{\tau} = 3866.763$, $B = 1163.892$

11.8 $\hat{\tau} = 4117.75$, $B = 999.81$

11.9 $\hat{p} = .875$, $B = .105$

11.10 $\hat{p} = .125$, $B = .1377$

INDEX